Microbial Production and Consumption of Greenhouse Gases: Methane, Nitrogen Oxides, and Halomethanes

Microbial Production and Consumption of
Greenhouse Gases:
Methane, Nitrogen Oxides, and Halomethanes

Editors:

John E. Rogers

Environmental Research Laboratory
U.S. Environmental Protection Agency
Athens, Georgia

William B. Whitman

Department of Microbiology
University of Georgia, Athens

American Society for Microbiology
Washington, D.C.

1325 Massachusetts Ave., N.W.
Washington, D.C. 20005

Library of Congress Cataloging-in-Publication Data

Microbial production and consumption of greenhouse gases: methane, nitrogen oxides, and halomethanes / [edited by] John E. Rogers, William B. Whitman.
p. cm.
Includes index.
ISBN 1-55581-035-7
1. Atmospheric chemistry. 2. Greenhouse gases—Environmental aspects. 3. Methane—Environmental aspects. 4. Nitrogen oxides—Environmental aspects. 5. Microbiology. I. Rogers, J. E. (John E.) II. Whitman, William Barnaby.
QC879.6.M53 1991
574.5'223—dc20 91-15621
CIP

Printed in the United States of America

Contents

Contributors

Peter S. Bakwin • Department of Earth and Planetary Sciences and Division of Applied Sciences, Harvard University, 29 Oxford Street, Cambridge, Massachusetts 02138

Karen B. Bartlett • Complex Systems Research Center, Institute for the Study of Earth, Oceans and Space, University of New Hampshire, Durham, New Hampshire 03824

David R. Boone • Environmental Science and Engineering, Oregon Graduate Institute of Science and Technology, 19600 N.W. von Neumann Drive, Beaverton, Oregon 97006-1999

Douglas G. Capone • Center for Environmental and Estuarine Studies, Chesapeake Biological Laboratory, University of Maryland, Solomons, Maryland 20688-0038

Patrick M. Crill • Complex Systems Research Center, Institute for the Study of Earth, Oceans and Space, University of New Hampshire, Durham, New Hampshire 03824

Eric A. Davidson • Woods Hole Research Center, P.O. Box 296, Woods Hole, Massachusetts 02543

Peter M. Groffman • Department of Natural Resources Science, University of Rhode Island, Kingston, Rhode Island 02881

R. S. Hanson • Gray Freshwater Biological Institute, University of Minnesota, Navarre, Minnesota 55392

Robert C. Harriss • Complex Systems Research Center, Institute for the Study of Earth, Oceans and Space, University of New Hampshire, Durham, New Hampshire 03824

Daniel J. Jacob • Department of Earth and Planetary Sciences and Division of Applied Sciences, Harvard University, 29 Oxford Street, Cambridge, Massachusetts 02138

W. Jack Jones • Environmental Research Laboratory, Athens, U.S. Environmental Protection Agency, College Station Road, Athens, Georgia 30613-7799

Ronald P. Kiene • University of Georgia Marine Institute, Sapelo Island, Georgia 31327

J. Gijs Kuenen • Kluyver Laboratory for Biotechnology, Delft University of Technology, Julianalaan 67, 2628 BC Delft, The Netherlands

Terry L. Miller • Wadsworth Center for Laboratories and Research, New York State Department of Health, Albany, New York 12201-0509

Lesley A. Robertson • Kluyver Laboratory for Biotechnology, Delft University of Technology, Julianalaan 67, 2628 BC Delft, The Netherlands

John E. Rogers • Environmental Research Laboratory, Athens, U.S. Environmental Protection Agency, College Station Road, Athens, Georgia 30613-7799

John McNeill Sieburth • Graduate School of Oceanography, University of Rhode Island Bay Campus, Narragansett, Rhode Island 02882-1197

E. Topp • Land Resources Research Center, Research Branch, Agriculture Canada, C.E.F. Ottawa, Ontario K1A 0C6, Canada

Stanley C. Tyler • National Center for Atmospheric Research, 1850 Table Mesa Drive, Boulder, Colorado 80303

R. Wever • E. C. Slater Institute for Biochemical Research and Biotechnological Center, University of Amsterdam, Plantage Muidergracht 12, 1018 TV Amsterdam, The Netherlands

William B. Whitman • Department of Microbiology, University of Georgia, Athens, Georgia 30602

Acknowledgments

During the period 14–16 November 1989, the University of Georgia's Office of Academic Affairs, Department of Microbiology, and Center for Continuing Education and the U.S. Environmental Protection Agency's Environmental Research Laboratory, Athens, Georgia, conducted a Workshop on the Microbial Production and Consumption of Radiatively Important Trace Gases as part of EPA's Global Climate Change Research Program. This book expands on the portions of the workshop dealing with methane, nitrogen oxides, and halomethanes.

Several people assisted in the development of the workshop program. We thank Ray Lassiter, David Lewis, William Steen, and Richard Zepp for their able assistance. We also thank Peter Simpson for invaluable assistance in conducting the Workshop and in facilitating the timely completion of the book.

Chapter 1

Introduction

John E. Rogers and William B. Whitman

The buildup of carbon dioxide in the troposphere, coupled with the pronounced depletion of stratospheric ozone observed in the Antarctic in recent years, has heightened public awareness of the global impacts of greenhouse gases on the atmosphere. Although ozone and carbon dioxide have received a great deal of attention in the press, there are a number of trace gases, including methane, nitrous oxide, nitric oxide, carbon monoxide, carbon-sulfur gases, and halocarbons, that also have an important impact on global climate. The impact of these trace gases may equal that of carbon dioxide.

Methane, nitrous oxide, nitric oxide, carbon monoxide, carbon-sulfur gases, and halocarbons affect global climate by a number of mechanisms. Next to carbon dioxide, methane has the greatest influence on the global radiative balance (Andreae and Crutzen, 1985). The infrared radiative heating effect of a methane increase from 0.7 ppm (pre-Industrial Revolution) to 1.7 ppm (1988 concentration) is about half as large as the comparable effect of the simultaneous increase in CO_2 from 275 ppm to 345 ppm. The influence of nitrous oxide on the atmosphere is through its interactions with oxygen. Nitrous oxide can react with molecular oxygen to produce nitric oxide, which can then react with stratospheric ozone by a sequence of catalytic reactions leading to a reduction in the ozone concentration (Bolle et al., 1986). Assuming present atmospheric conditions, a doubling of the nitrous oxide mixing ratio could lead to a 3 to 5% reduction in ozone abundance. Halocarbons participate in reactions that cause depletion of stratospheric ozone (Watson, 1986).

Carbon monoxide and the carbon-sulfur gases, although not addressed in this book, also have important atmospheric impacts. Carbon monoxide (CO), although not itself radiatively important, has important effects on the atmospheric concentrations of methane and tropospheric ozone. The affects of CO and methane are interactive. The major atmospheric sink for both is reactions with OH; thus, the increase in one will lead to an increase in the other because of a decrease in OH (Logan et al., 1981). Carbon monoxide is also a product of atmospheric methane oxidation. The carbon-sulfur trace gases dimethyl sulfide and carbonyl sulfide have

John E. Rogers • Environmental Research Laboratory, Athens, U.S. Environmental Protection Agency, College Station Road, Athens, Georgia 30613-7799. *William B. Whitman* • Department of Microbiology, University of Georgia, Athens, Georgia 30602.

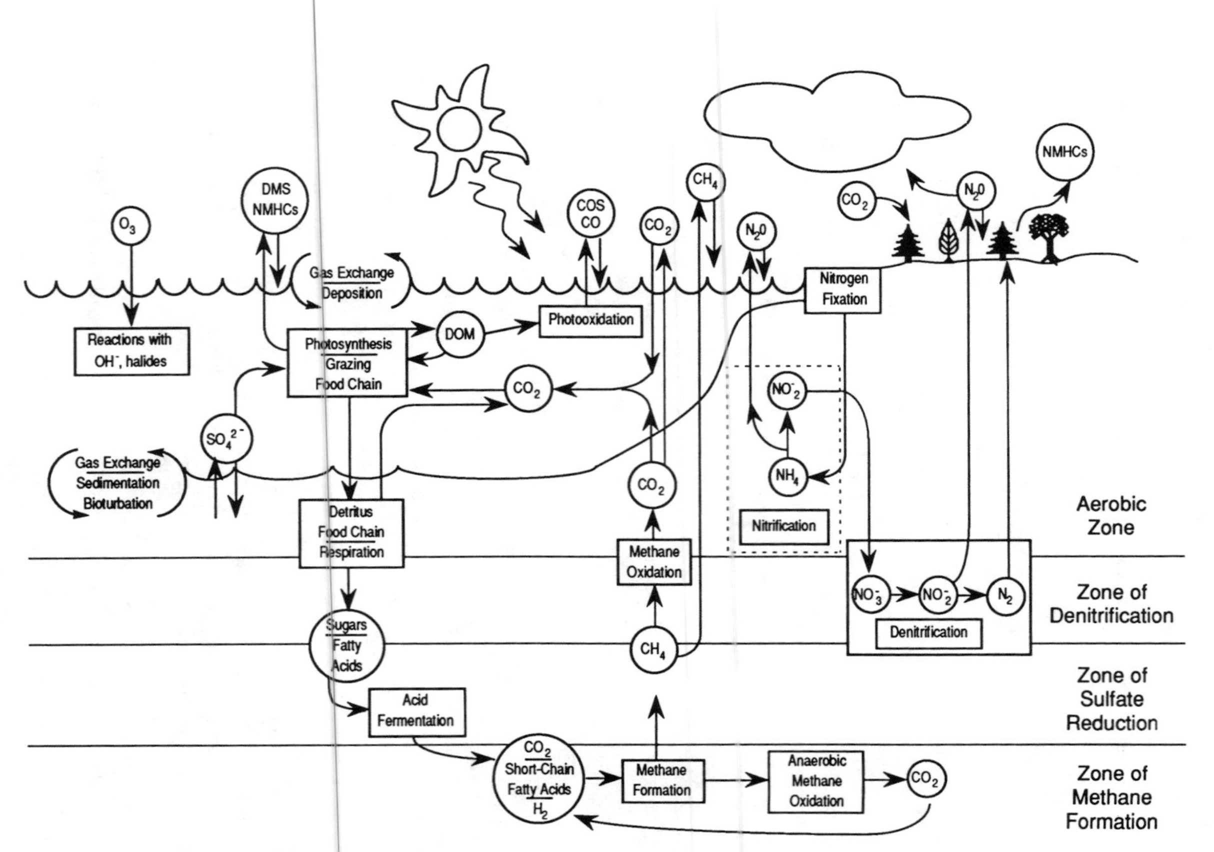

Figure 1. Biogeochemical cycles affecting atmospheric concentrations of selected gases and processes important in global climate change.

important influences on global albedo and mean temperature (Andreae and Crutzen, 1985; Watson, 1986).

Biogeochemical processes that affect the atmospheric concentrations of the gases important in global change are depicted within the boxes in Fig. 1. As indicated, the biosphere helps regulate the composition of the atmosphere, acting as both a source and a sink for the gases. Preliminary estimates of emission factors based on field observations indicate that all of these gases have significant sources and some have significant sinks in aquatic and terrestrial ecosystems (Penner et al., 1988). Competition between their production, biological uptake, and volatilization determines the net flux of the gases to the atmosphere.

Methane is produced microbiologically in anaerobic environments where oxygen and sulfate are scarce: swamps, peat bogs, wetlands, paddies, and the intestinal tracts of cattle and termites. Microorganisms can also remove methane from the environment through aerobic (Rudd and Taylor, 1980) and anaerobic (Alperin and Reeburgh, 1984) oxidation. Many studies have shown that in lakes aerobic methane oxidation occurs primarily in the water column whereas anaerobic oxidation occurs in the sediments. Soil is also thought to be an aerobic sink for methane due to the activities of oxidizing microorganisms.

In anaerobic habitats, organic carbon is converted to CH_4 and CO_2 by an anaerobic microbial food chain that includes fermentative, acetogenic, and methanogenic bacteria. The methanogens are the terminal bacteria in this food chain. The metabolic activity of these organisms is coupled through a process termed "interspecies hydrogen transfer." Fermentative organisms can convert large organic molecules to low-molecular-weight acids, CO_2, and molecular hydrogen, which acts as a feedback inhibitor of the process. Methanogens enhance the activity of the fermentation process by removing hydrogen and reducing the feedback inhibition. Interspecies hydrogen transfer results in increased carbon turnover, production of more oxidized end products, greater energy conservation for the fermentative organisms, increased growth of all organisms, and the displacement of unfavorable reaction equilibria. A consequence of this process is the nearly complete transformation of organic carbon in anaerobic environments to CH_4.

The same conditions (low porosity, water saturation, and high organic content) which result in anaerobicity also slow the diffusion of methane to the atmosphere. Because of the slow diffusion, a significant fraction may be lost through aerobic and anaerobic oxidation before it leaves the sediment. Although the anaerobic oxidation of methane has been well documented in sulfate-containing sediments and anoxic waters (Cicerone and Oremland, 1988), little is known about the organisms that carry out this process. In contrast, much is known about the organisms that oxidize methane in aerobic environments. A substantial part of the gas that diffuses into the aerobic zone is metabolized by organisms, such as the methanotrophic bacteria, that are typically present in large numbers in or at the periphery of anaerobic zones. Methanotrophs (family *Methylococcaceae*) can obtain all of their carbon and energy from CH_4 under aerobic conditions (Whittenbury and Kreig, 1984). For example, 85% of the methane produced in deep sediments of freshwater lakes may be consumed by methanotrophs in the overlying water column before it reaches the surface.

Diffusion is not the only mechanism for release of trace gases from anaerobic environments to the atmosphere. Methane, for example, is somewhat water insoluble, and when its concentration exceeds its solubility in anaerobic environments, it forms bubbles. Perturbations of the bubbles, such as bioturbation by sediment-dwelling organisms or some physical disturbance, result in their escape into the overlying water and into the atmosphere (Kuivila et al., 1988). Aquatic plants also can provide an important pathway for the transfer of gases between anaerobic environments and the atmosphere (Dacey and Klug, 1979). Gases can move from the root zone up through the stems into the atmosphere or from the atmosphere into the root zone.

Unlike the formation of methane, which is the direct product of the anaerobic degradation of organic matter, the formation of nitrous oxide results from the inefficient conversion of ammonium ion to nitrate or nitrate to molecular nitrogen. Denitrification has been considered the principal source of nitrous oxide to the atmosphere (Scharrong et al., 1984; Yoshida et al., 1989). Nitrification, however, also contributes a significant amount of nitrous oxide to the atmosphere (Yoshida et al., 1989). Several laboratories (Rasmussen and Khalil, 1981; Khalil and Rasmussen, 1983; Bolle et al., 1986) have reported and confirmed that nitrous oxide in the atmosphere is increasing by 0.3% per year. The reasons for these increases are only speculative.

Denitrification has classically been defined as the reduction of nitrate to molecular nitrogen (Payne, 1973); however, a more appropriate definition of denitrification is the reduction of oxidized nitrogen compounds to gaseous products. Nitrate is converted to N_2 through a general pathway that includes the intermediates nitrite, nitric oxide, and nitrous oxide. Certain bacteria contain the entire pathway, whereas other bacteria may only use a few steps of the pathway (Ingraham, 1981). Depending on the type and efficiency of the organisms present, N_2O can be released to the environment. Under field conditions, for example, the yield of N_2O relative to N_2 can range from negligible to 20% (Stefanson, 1972, 1973; Rolston et al., 1976). Manipulation of various factors such as pH, reactant concentration, and temperature is known to affect the N_2O yield. Similar changes may result from biofeedbacks resulting from climate change.

For many years denitrification was thought to be the only source of N_2O. However, it is now well recognized that N_2O can be produced during nitrification (Focht, 1974). Production of NO_2^- and NO_3^- from NH_4^+ via nitrification can result from a number of different pathways (Firestone and Davidson, 1989). Chemoautotrophic nitrifying bacteria can obtain energy from the oxidation of NH_4^+ to NO_2^- and NO_3^-. The most thoroughly investigated of these bacteria are the genera *Nitrosomonas* and *Nitrobacter* (Kuenen and Robertson, 1987). A second important group are the heterotrophic nitrifying bacteria that oxidize ammonium ion at the expense of a carbon substrate (Robertson and Kuenen, 1988). Chemoautotrophic nitrifying bacteria have specific activities 10^2 to 10^3 greater than heterotrophic nitrifying bacteria; however, heterotrophs vastly outnumber chemoautotrophs in the environment. Therefore their contributions to nitrogen turnover in the environment may be comparable, with the relative difference susceptible to changes in global climate.

The effects of halogenated gases on the atmosphere are now well recognized (Barrie et al., 1988). Concern raised about these compounds has centered on man-made halocarbons; however, there is also a large natural source. In terms of global production, methyl iodide (Lovelock et al., 1973), methyl bromide (Singh et al., 1983), and methyl chloride (Cicerone, 1981) are the most important. Both fungi (Harper, 1985) and algae (Lovelock, 1975) are known to produce methyl halides.

Our objective in this book is to provide an overview of the biological processes that contribute to the increase in trace gases (methane, nitrous oxide, nitric oxide, and halocarbons) in the atmosphere. Physical and chemical processes are discussed as they relate to biological processes. We do not attempt here to provide a complete discussion of this research area but intend to provide an introduction to biological processes that contribute to changes in global climate and processes that can be influenced by biofeedback mechanisms as climate changes occur. Several excellent reviews, such as the papers by Andreae and Crutzen (1985), Bolle et al. (1986), and Cicerone and Oremland (1988) are available which address other aspects of global climate change.

REFERENCES

Alperin, M. J., and W. S. Reeburgh. 1984. Geochemical observations supporting anaerobic methane oxidation, p. 282–289. *In* R. L. Crawford and R. S. Hanson (ed.), *Microbial Growth of C_1 Compounds*. American Society for Microbiology, Washington, D.C.

Andreae, M. O., and P. J. Crutzen. 1985. Atmospheric chemistry, p. 75–113. *In* T. F. Malone and J. G. Roederer (ed.), *Global Change*. ICSU Press/Cambridge University Press, Cambridge.

Barrie, L. A., J. W. Botenheim, R. C. Schell, P. J. Crutzen, and R. A. Rasmussen. 1988. Ozone destruction and photochemical reactions at polar sunrise in the lower Arctic atmosphere. *Nature* (London) **334:**138–141.

Bolle, H.-J., W. Seiler, and B. Bolin. 1986. Other greenhouse gases and aerosols, p. 157–205. *In* B. Bolin, B. R. Doos, J. Jager, and R. A. Warrick (ed.), *The Greenhouse Effect, Climate Change, and Ecosystems*. John Wiley & Sons, New York.

Cicerone, R. J. 1981. Halogens in the atmosphere. *Rev. Geophys. Space Physics* **19:**123–139.

Cicerone, R. J., and R. S. Oremland. 1988. Biochemical aspects of atmospheric methane. *Global Biogeochem. Cycles* **2:**299–327.

Dacey, J. W. H., and M. J. Klug. 1979. Methane efflux from lake sediments through water lilies. *Science* **203:**1253–1255.

Dacey, J. W. H., and S. G. Wakeham. 1986. Oceanic dimethylsulfide: production during zooplankton grazing on phytoplankton. *Science* **233:**1314–1316.

Firestone, M. K., and E. A. Davidson. 1989. Microbiological basis of NO and N_2O production and consumption in soil, p. 7–21. *In* M. O. Andreae and D. S. Schimel (ed.), *Exchange of Trace Gases between Terrestrial Ecosystems and the Atmosphere*. John Wiley & Sons, New York.

Focht, D. D. 1974. The effect of temperature, pH, and aeration on the production of nitrous oxide and gaseous nitrogen, a zero-order kinetic model. *Soil Sci.* **118:**173–179.

Harper, D. B. 1985. Halomethane from halide ion—a highly efficient fungal conversation of environmental significance. *Nature* (London) **315:**55–57.

Ingraham, J. L. 1981. Microbiology and biochemistry of denitrifiers, p. 45–65. *In* C. C. Delwiche (ed.), *Denitrification, Nitrification, and Atmospheric Nitrous Oxide*. Wiley-Interscience, New York.

Khalil, M. A. K., and R. A. Rasmussen. 1983. Increase and seasonal cycles of nitrous oxide in the earth's atmosphere. *Tellus* **35B:**161–169.

Kuenen, J. G., and L. A. Robertson. 1987. Ecology of nitrification and denitrification, p. 162–218. *In* J. A. Cole and S. Furgeson (ed.), *The Nitrogen and Sulfur Cycles*. Cambridge University Press, Cambridge.

Kuivila, K. M., J. W. Murray, A. H. Devol, M. E. Lidstrom, and C. E. Reimers. 1988. Methane cycling in the sediments of Lake Washington. *Limnol. Oceanogr.* **33:**571–581.

Logan, J. A., M. J. Prather, S. C. Wofsy, and M. B. McElroy. 1981. Tropospheric chemistry: a global perspective. *J. Geophys. Res.* **86:**7210–7254.

Lovelock, J. E. 1975. Natural halocarbons in the air and in the sea. *Nature* (London) **256:**193–194.

Lovelock, J. E., R. J. Maggs, and R. J. Wade. 1973. Halogenated hydrocarbons in and over the Atlantic. *Nature* (London) **241:**194–196.

Payne, W. J. 1973. Reduction of nitrogenous oxide by microorganisms. *Bacteriol. Rev.* **37:**409–452.

Penner, J. E., P. S. Connell, D. J. Wuebbles, and C. C. Covey. 1988. Climate change and its interactions with air chemistry: perspectives and research needs. Background Report for EPA Report to Congress on effects of global climate change. IAG no. DW89932676-01-1 with Lawrence Livermore Laboratory.

Rasmussen, R. A., and M. A. K. Khalil. 1981. Atmospheric methane: trends and seasonal cycles. *J. Geophys. Res.* **86:**9826–9832.

Robertson, L. A., and J. G. Kuenen. 1988. Heterotrophic nitrification in *Thiosphaera pantotropha*—oxygen uptake and enzyme studies. *J. Gen. Microbiol.* **134:**351–354.

Rolston, D. E., M. Fried, and D. A. Goldhamer. 1976. Denitrification measured directly from nitrogen and nitrous oxide gas fluxes. *Soil Sci. Soc. Am. J.* **40:**259–266.

Rudd, J. W. M., and C. D. Taylor. 1980. Methane cycling in aquatic environments. *Adv. Aquat. Microbiol.* **2:**77–150.

Scharrong, J., H. B. Jensen, and J. Sorensen. 1984. Nitrous oxide production, nitrification and denitrification in various sediments at low oxygen concentrations. *Can. J. Microbiol.* **30:**895–897.

Singh, H. B., L. J. Salas, and R. E. Stiles. 1983. Methyl iodides in and over the eastern Pacific. *J. Geophys. Res.* **88:**3684–3690.

Stefanson, R. C. 1972. Soil denitrification in sealed soil-plant systems. III. Effect on distributed and undistributed soil samples. *Plant Soil* **37:**141–149.

Stefanson, R. C. 1973. Evolution patterns of nitrous oxide and nitrogen in sealed soil-plant systems. *Soil Biol. Biochem.* **5:**167–169.

Watson, R. 1986. Atmospheric ozone, p. 69–82. *In* J. G. Titus (ed.), *Effects of Changes in Stratospheric Ozone and Global Climate*, vol. 1. U.S. Environmental Protection Agency, Washington, D.C., and United Nations Environment Programme, New York.

Whittenbury, R., and N. R. Kreig. 1984. *Methanococcaceae*, p. 256–261. *In* N. R. Krieg (ed.), *Bergey's Manual of Systematic Bacteriology*, vol. 1. The Williams and Wilkins Co., Baltimore.

Yoshida, N., H. Morimoto, M. Hirano, I. Koike, S. Matsuo, E. Wada, T. Saino, and A. Hattori. 1989. Nitrification rates and ^{15}N abundances of N_2O and NO_3^- in the western North Pacific. *Nature* (London) **342:**895–897.

Chapter 2

The Global Methane Budget

Stanley C. Tyler

Regardless of whether they believe that increasing greenhouse gas concentrations are now affecting climate, most climate modelers agree that if present emission trends for greenhouse gases continue, global temperatures will rise (Schneider, 1989; Kerr, 1989). One of the gases that scientists are very concerned about regarding global warming is methane. It is both naturally occurring and a product of human activities. Its current concentration in the atmosphere is about 1.7 ppmv (parts per million by volume). At this concentration methane exerts a strong influence over the earth's climate and the chemistry of the troposphere and stratosphere. Because of its interaction with planetary infrared radiation, methane plays a direct role in climate as an effective greenhouse gas. Key atmospheric chemistry involving methane includes its oxidation in the troposphere to affect ozone, hydroxyl radicals, and carbon monoxide. In the stratosphere, it is a major source of CH_2O and H_2 and of upper stratospheric H_2O. The most abundant hydrocarbon in the atmosphere, methane is a major regulator of OH radicals and is the primary sink for Cl atoms which take part in ozone-destroying chain mechanisms.

These reasons alone are sufficient to make methane the object of study by atmospheric scientists, but there is one more. Atmospheric methane has been increasing at a rate of about 1% per year for at least the last decade (Rasmussen and Khalil, 1981; Steele et al., 1987; Blake and Rowland, 1988). This increase began earlier in historic time but has accelerated in recent years. Ice core samples and other evidence show that methane has more than doubled overall in the last few hundred years after being relatively stable at a value of about 0.7 ppmv during the time since the last ice age up through about 300 years before the present (Craig and Chou, 1982; Rasmussen and Khalil, 1984; Pearman et al., 1986). Recent analyses of air bubbles from even older ice cores extend the record back to 100,000 years before present (Stauffer et al., 1988) and 160,000 years before present (Raynaud et al., 1988). During glacial maxima, the CH_4 concentration fell to about 0.35 ppmv while interglacial values were about 0.65 ppmv, very near the preindustrial value.

Exactly why atmospheric methane is increasing cannot be completely explained with existing information. However, it seems clear that the rise in methane

Stanley C. Tyler • National Center for Atmospheric Research, 1850 Table Mesa Drive, Boulder, Colorado 80303.

during the last 300 years or so is unprecedented for the time span in which we have data and is related to a rise in human population and its accompanying activities. Explanations of the recent increase include increases in some methane sources, decreases in methane sinks, or a combination of the two (Ehhalt and Schmidt, 1978; Khalil and Rasmussen, 1985; Cicerone and Oremland, 1988).

Because of its rapid increase and its important chemistry, scientists from numerous interdisciplinary fields are studying methane. Topics of investigation include methods of consumption and production, measurements of emission fluxes, isotopic signatures of carbon and hydrogen in various methane sources, and atmospheric models which attempt to provide realistic output that correlates with the experimental data. The ultimate goal is to provide an accurate methane budget. Such a budget could be monitored as changing ecosystems and anthropogenic activity are noted. Appropriate models could then be used to predict future trends in atmospheric methane and related chemistry. In the text that follows, some of the important information that has been determined regarding methane and some of the methods used to obtain it are discussed.

METHANE ORIGINS

Methane is formed in the earth's interior and at the surface of the earth. The two major types of formation are by thermogenic processes in buried carbon and by biological processes performed by bacteria at or near the earth's surface. Methane processes are sometimes labeled as abiogenic or biogenic, although the distinction between the two is not always clear. By definition, abiogenic methane is not produced by the actions of living organisms, while biogenic methane is.

One process which produces methane that is clearly of abiogenic origin is that of thermogenic reactions of simple gases trapped in the earth's crust or upper mantle. The methane produced is from a high-temperature process where CO_2 is reduced in the presence of H_2 (Schoell, 1980). This kind of production occurs in volcanoes and geothermal vents and is thought to be small compared to other sources of methane (e.g., Welhan and Craig, 1979).

Carbon buried in the earth's crust can be another source of methane. This carbon is in the form of fossil fuels such as coal, petroleum, and natural gas and is often called abiogenic when it is formed by thermocatalytic reactions. However, it could well be considered as biogenic because its origin initiates from the product of once-living organisms (Ourisson et al., 1984). Coal originated in vast primeval swamps, where partial decay of dead trees and plants formed thick beds of peat. Petroleum began for the most part as dispersed organic material, including plankton and plant material, in the sediments of inland seas or coastal marine basins. Coal is mostly carbon but contains small amounts of methane which are released when it is mined. Petroleum and natural gas can have methane and a wide variety of other hydrocarbons in their mixture depending upon the conditions of formation.

While methane from coal forms from geochemical reactions, the natural gas formed with petroleum deposits can be of bacterial or thermogenic origin. Shallow petroleum deposits can undergo partial bacterial decomposition to release what is known as biogenic natural gas (Schoell, 1980; Coleman et al., 1981). Unlike thermogenic natural gas, which includes significant amounts of higher alkanes in addition to methane, biogenically formed natural gas is about 99% methane with only trace amounts of higher alkanes. About 20% of the natural gas deposits are of microbial origin (Rice and Claypool, 1981).

In addition to the methane which escapes during the mining and transportation processes that bring petroleum and natural gas hydrocarbons (including methane) up to the surface, some methane is also emitted to the atmosphere as the materials are refined into methane and/or burned as fuels (where there is incomplete combustion). In summation, fossil fuel methane sources include coal and lignite mining, pipeline and transmission losses of natural gas, venting and flaring of gas wells and deposits, and burning of petroleum products for fuel.

Biogenic methane formation results from complex biochemical reactions involving bacteria during the decomposition of organic matter (Mah et al., 1977; Cicerone and Oremland, 1988). These bacteria are strict anaerobes and therefore require reducing environments for growth. For this reason, ecosystems such as water-covered soils and sediments (e.g., Koyama, 1963), the intestinal tracts of animals (e.g., Miller and Wolin, 1986), and landfills (e.g., Games and Hayes, 1975) are prime habitats for methane production.

The principal substrates used by methanogenic bacteria to promote their growth include H_2, CO_2, acetate, and formate. These substrates are called competitive substrates because the methanogenic bacteria have to compete with sulfate-reducing bacteria for them (e.g., Oremland and Polcin, 1982; Whiticar et al., 1986). As a result, methanogenesis occurs in sediment zones with no sulfate or when the available dissolved sulfate is exhausted. Other substrates which are sometimes used by methanogenic bacteria to make methane include methanol, carbon monoxide, methylated amines, and dimethyl sulfide (e.g., Oremland et al., 1982; King et al., 1983; King, 1984). Sulfate-reducing bacteria do not have as strong an affinity for these substrates as they do for the competitive substrates. The full importance of these noncompetitive substrates toward methane formation is not yet determined.

One last source of methane is burning biomass. Methane is released from this source due to the incomplete combustion of organic material (Crutzen et al., 1979; Crutzen et al., 1985). This source has been described as abiogenic in the literature (Seiler, 1984). The combustion process itself is a high-temperature thermogenic reaction but, since the starting carbon material in all cases is plant matter, it could be classified as a biogenic source. Placed in either category, biomass burning is both man made (land clearing) and naturally occurring (lightning strikes, natural fires), and it produces varying amounts of methane depending on the conditions of the fire (Greenberg et al., 1984).

It is clear that methane is formed and released to the atmosphere both by natural and anthropogenic means. However, nearly every source can be affected by anthropogenic activities. This is because of the unprecedented growth in human population in the last few hundred years. This growth has been accompanied by an increased need for fossil fuels for energy use as well as an increased need to manage land and animals for agricultural and other uses.

METHANE CHEMISTRY

The most abundant hydrocarbon in the atmosphere, methane is centrally involved in atmospheric chemistry. Its chemistry is initiated by reaction with OH radicals to form methyl radicals and water as in reaction 1:

$$CH_4 + OH \Rightarrow CH_3 + H_2O \tag{1}$$

Even though reaction 1 is comparatively slow (CH_4 atmospheric lifetime is 8 to 10 years), the relatively high methane concentration makes this reaction important in regulating OH radical concentration (Levy, 1981; Logan et al., 1981). Increasing methane concentrations can deplete the oxidizing power of the atmosphere and lead to longer lifetimes for other atmospheric pollutants.

The complete oxidation of methane is a series of reactions leading to increased water vapor and carbon dioxide, the two most important greenhouse gases. The intermediate products formed during methane oxidation depend upon the nitrogen oxide (NO_x) concentration (Levine et al., 1985; Thompson and Cicerone, 1986). In high-NO_x environments, as in polluted tropospheric air and all of the stratosphere, methane oxidation produces ozone (O_3) and hydrogen oxides (HO_x). After reaction 1, the process continues with reactions 2 and 3 below (where *M* represents any third molecule which acts as an energy-absorbing collision partner to help stabilize the reaction product).

$$CH_3 + O_2 + M \Rightarrow CH_3O_2 + M \tag{2}$$

$$CH_3O_2 + NO \Rightarrow CH_3O + NO_2 \tag{3}$$

NO_2 is then photolysed, and a subsequent reaction produces O_3. An important intermediate product is formaldehyde (CH_2O), which is formed by reaction 4.

$$CH_3O + O_2 \Rightarrow CH_2O + HO_2 \tag{4}$$

Reaction 4 returns HO_2 to the atmosphere to balance the HO_x lost in the form of OH in reaction 1. Formaldehyde is then oxidized to CO and, depending on its pathway of oxidation, may lead to increased OH radicals. In low NO_x environments, methane oxidation consumes O_3 and HO_x in producing CO_2, H_2O, and H_2 (Crutzen, 1987). That sequence of reactions follows methane oxidation, as in reactions 1 and 2, with reaction 5 below:

$$CH_3O_2 + HO_2 \Rightarrow CH_3OOH + O_2 \tag{5}$$

From there additional reactions consume various amounts of O_3 and HO_x depending upon atmospheric OH concentrations, pathways of CH_3OOH reactions, and removal rates for intermediate species like CH_3OOH.

The methane that reaches the stratosphere is almost entirely consumed there by OH, Cl atoms, and other oxidants. In the stratosphere, methane oxidation may increase H_2O vapor concentrations by as much as 50% over conditions which would exist if the stratosphere had no methane (Ramanathan et al., 1987). The resulting water vapor is known to be involved in heterogeneous chemistry such as that responsible for the Antarctic ozone hole. In contrast, methane reaction with Cl atoms to form HCl is one of the main pathways which take Cl out of the Cl-catalytic chain which destroys ozone. Reaction 6 sequesters ozone-destroying Cl atoms into a temporary reservoir, HCl, which can then be rained out of the stratosphere (Rowland and Molina, 1975).

$$CH_4 + Cl \Rightarrow CH_3 + HCl \quad (6)$$

THE GREENHOUSE EFFECT AND METHANE

Our sun and earth have maintained a distinct and critical energy balance that has allowed life on the planet to develop and evolve to its present state. The air, clouds, soils, plants, rocks, and oceans all reflect, absorb, and re-emit radiant energy that arrives from the sun. This energy arrives in electromagnetic waves that include energies from the entire solar spectrum. Thirty percent of the sun's incoming energy is reflected directly back to space (Schneider, 1989), while the remaining energy heats the solids, liquids, and gases on the planet which absorb it.

The term greenhouse effect means that the following additional effects occur (e.g., Hileman, 1989; Schneider, 1989). Some of the absorbed solar energy is reradiated to space by the absorbing material. This reradiated energy is in the infrared portion of the electromagnetic spectrum. (The absorption of the portion of the sun's radiant energy in the infrared range is small since the sun emits little infrared radiation. But our earth's infrared-absorbing materials become much more important when we consider radiation re-emitted by the earth.) Gases in our atmosphere such as H_2O, CO_2, CH_4, and N_2O are very effective at both absorbing this energy and re-emitting some of it up to space and down toward the earth's surface again, thus trapping and keeping a greater portion of the sun's energy with the earth's atmosphere and surface. This delicate balance is just right for maintaining an average temperature of about 15°C over the surface of the earth. Without these infrared-absorbing gases, the earth's average surface temperature would be about −18°C (Schneider, 1989).

It is now known that many greenhouse gases which are naturally occurring in the atmosphere can have their concentrations increased by human activities. It is this enhancement of greenhouse gases and their effect on global warming that has scientists and many other people concerned. One of the first sets of data which show an increase in a greenhouse gas over a long time period was taken by Keeling (Keeling et al., 1976a; Keeling et al., 1976b). By measuring CO_2 concentrations continuously since 1958, Keeling showed that CO_2 has increased from about 315 ppmv to 350 ppmv between 1958 and 1989 (Keeling et al., 1989). Furthermore, by analyzing older air trapped in polar ice cores, we now know that about 200 years ago the concentration was 280 ppmv (Barnola et al., 1987). The growth in CO_2 began at the same time as the Industrial Revolution. This increase in CO_2 coincides with an increase in the burning of fossil fuels and deforestation. It is almost certainly attributable to anthropogenic activities.

Methane's role in global warming is both direct and indirect. Methane is a strong absorber in the infrared portion of the electromagnetic spectrum. More importantly, it has a strong absorption centered at about 7.7 μm, a wavelength at which no other atmospheric gases are strongly absorbing. For this reason methane is becoming a proportionately larger contributor to the total greenhouse effect. For example, the infrared radiative heating effect of a methane increase from 0.7 ppmv (pre-Industrial Revolution concentration) to 1.7 ppmv (1989 concentration) is about half as large as the comparable effect of simultaneously increasing CO_2 from 275 ppmv to 345 ppmv (Ramanathan et al., 1985; Dickinson and Cicerone, 1986). Because of this, methane's contribution to greenhouse warming is comparable (although smaller) to that of CO_2 despite its considerably lower concentration. Based solely on the ratio of the infrared absorption and mean residence time terms, Kiehl and Dickinson (1987) derived formulas that show methane to be about 25 to 30 times more effective as a greenhouse gas than carbon dioxide on a per-molecular basis. Donner and Ramanathan (1980) calculated that the presence of 1.3 ppmv of CH_4 in the atmosphere causes the globally averaged surface temperature to be about 1.3°C higher than it would be with zero methane.

There are also several potentially important indirect ways that increasing methane can affect global warming. One results from the aforementioned oxidation of CH_4, which produces CO which is converted further to CO_2. Cicerone and Oremland (1988) calculate that in this way the atmospheric production of CO_2 from atmospheric CH_4 is about 6% as much as the direct annual release of CO_2 from anthropogenic sources. Another indirect effect includes the chemical production of tropospheric O_3, another greenhouse gas, in the presence of NO_x. This becomes important especially if O_3 concentrations should increase in the upper troposphere where ozone is a particularly effective greenhouse gas (Cicerone and Oremland, 1988).

Overall, the relative greenhouse warming from methane must take into account its residence time in the atmosphere and its subsequent production to produce other greenhouse gases such as CO_2. When a comparison of the relative greenhouse warming effect of methane versus that of CO_2 takes into account the longer residence time of CO_2, the methane is about 2.1 to 4.9 more effective as a greenhouse gas than carbon dioxide (Lassey et al., 1990).

METHANE EMISSIONS

Determining methane fluxes from chamber experiments in source regions is one of the main methods used to estimate source strengths. Individual flux values from specific regions are extrapolated to global values by determining the world-wide distribution and area of like regions. There are still large uncertainties associated with these sources because of the difficulty in extrapolating to a global sum. For instance, within each source region flux may vary with time of day, temperature, season, water cover, or organic input. In addition, world distributions of the many source types are not known precisely.

Table 1 contains data from four different studies of rice paddy methane

Table 1. Methane flux measurements for rice paddies (compiled from the recent literature)

Source area	Avg flux (g of $CH_4/m^2/day$)	Reference
California	0.075	Cicerone and Shetter, 1981
California	0.25	Cicerone et al., 1983
Spain	0.096	Seiler et al., 1984b
Italy	0.38	Holzapfel-Pschorn and Seiler, 1986

emissions during the course of a growing season. The average flux emitted for the entire season differs by as much as a factor of 5 between the studies. More importantly, observed variations in flux within one study were quite different from other studies. In a paddy field in Spain (Seiler et al., 1984b), the flux peaked around the time of ripening, while in the study of Italian paddies (Holzapfel-Pschorn and Seiler, 1986), two relative maxima were observed for methane flux, one at the time of tillering and one at the time of flowering. In contrast, the California field studied in 1982 (Cicerone et al., 1983) had a tremendous spike in methane flux only a few days before harvest.

Dacey and Klug (1979) showed that the principal means of methane escape to the atmosphere for the water lily *Nuphar luteum* is through the plant vascular system. Cicerone and Shetter (1981) investigated this possiblity for methane emissions from rice plants. Since then, the rice paddy studies cited above as well as many studies of other plants have focused on the role of emergent aquatic plants in wetland ecosystems with regard to methane emissions (e.g., Sebacher et al., 1985; Raskin and Kende, 1985; Dacey, 1987; and Chanton et al., in press). These studies demonstrate that methane is emitted to the atmosphere from wetlands via three primary modes: (i) diffusion of dissolved methane across the water-air interface, (ii) bubble ebullition, and (iii) air circulation between the atmosphere and buried tissues of aquatic plants, with the stems and leaves serving as conduits (see Fig. 1).

Studies similar to the rice paddy investigations have taken place in Alaskan

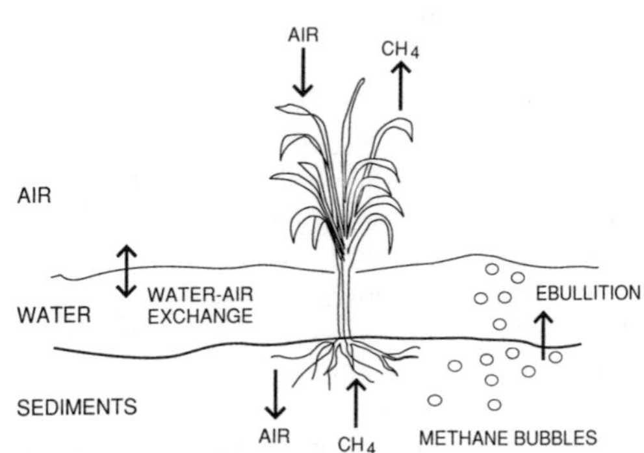

Figure 1. Primary modes of methane transfer to the atmosphere from aquatic environments.

tundra and water-covered wetlands such as tropical floodplains and southeastern United States swamps (e.g., Harriss and Sebacher, 1981; Sebacher et al., 1986; Bartlett et al., 1988; Devol et al., 1988; and Crill et al., 1988). Recognizing that the emission rates might be quite different depending on the mechanism of release of methane to the surface, the researchers in many cases determined the relative contribution of each mode of methane transport to the total emissions. They differentiated between ecosystem categories such as open lake, flooded forest, and grass mats for floodplain sites, and several different categories of tundra and bog in Alaska.

Harriss and Sebacher (1981) reported a range of from 0.0046 to 0.068 g of CH_4 per m^2 per day as the flux from cypress swamp habitat in the southeastern United States. Differences in nutrient input and organic accumulation seemed to be probable causes for the wide range. Working in the Amazon floodplain, Devol et al. (1988) determined that ebullition was the dominant mechanism of emission for CH_4, accounting for 85% of the total. Bartlett et al. (1988) found that all three transport mechanisms for methane emissions to the atmosphere were present in varying amounts in the Amazonian floodplain. The significance of each type of mechanism depended on the type of terrain (flooded forest, floating grass mats, open water).

For sources not readily studied by flux chambers, methane emissions are estimated using a combination of release estimates, population densities, and land use data. For instance, Crutzen et al. (1986) combined published information on animal populations and animal dietary intake and its relation to methane production to deduce per-animal production rates and aggregate methane production. They determined that the current yearly input of methane to the atmosphere from domestic animals and humans is about 74 Tg (1 Tg $= 10^{12}$ g) with an uncertainty of about 15%. Lerner et al. (1988) used a high-resolution global data base of animal population with associated methane emission studies to arrive at another estimate of emissions from enteric fermentation. A global emission of 75.8 Tg/year was estimated for 1984 for domestic animals. Bingemer and Crutzen (1987) provided a detailed estimate of landfill methane emissions. They considered separately municipal solid wastes and industrial and commercial wastes and differentiated between developed and undeveloped countries to arrive at their estimate of 30 to 70 Tg/year.

Natural gas exploration and distribution and coal mining contribute uncertain amounts of methane. Koyama (1963) estimated methane emissions from coal mining using figures for the annual production of coal and the production rate of coal field gas. His emission estimate was 20 Tg/year for this source. Since then, his figures have been updated by extrapolations based on higher coal production (Seiler, 1984; Cicerone and Oremland, 1988) to arrive at a value of 35 Tg/year. Losses due to natural gas exploration and transmission are primarily estimated from industry data and are estimated to be about 45 Tg/year by Cicerone and Oremland (1988). Previous estimates from Ehhalt (1974), Seiler (1984), and Crutzen (1987) using figures from industrial representatives for annual production and assumed loss rates during transportation resulted in slightly lower emissions estimates.

METHANE BUDGET

A budget for an atmospheric gas must take into account all the production and destruction processes for that gas in such a way that the amount calculated to be present in the atmosphere agrees with the amount found by measurement data. Budgets for methane must reconcile a range of possible lifetimes for methane in the atmosphere and a known globally averaged methane concentration of about 1.7 ppmv. In 1989, the average methane concentration in the northern hemisphere was about 1.75 ppmv, while in the southern hemisphere it was 1.67 ppmv. The tropospheric lifetime is between 8 and 10 years, based on reaction 1, the principal sink for methane. According to Cicerone and Oremland (1988), this accounts for about 85% of the atmospheric sink, with most of the rest being reactions with OH, Cl, and $O(^1D)$ in the stratosphere (where OH reaction still dominates). An additional sink for methane is from microbial uptake of atmospheric CH_4 in soils at the earth's surface. Presently, most budgets consider this sink to be very small ($<0.5\%$ to a few percent of the total sink). However, this sink is not well characterized, and recent work indicates it may play a larger role than previously thought (Steudler et al., 1989; Born et al., 1989).

A likely budget for the methane sources is shown in Table 2. This source budget is a consensus of several others which have been constructed in recent years (i.e., Ehhalt, 1974; Ehhalt and Schmidt, 1978; Seiler et al., 1984a; Cicerone and Oremland, 1988; and Tyler, 1989a). Methane sources are often divided so that certain aspects can be isolated and studied. In this case, the sources have been separated according to whether they contain radioactive carbon-14 (all living or recently dead things do) or are radiocarbon "dead" (where all the carbon-14 has had time to decay and no new carbon can replenish it, e.g., fossil fuel sources of methane).

As Table 2 shows, there are still large uncertainties in the methane budget because of the difficulty in extrapolating each of several individual source measurements to a global sum. For example, termite data are very uncertain because researchers do not agree on emissions data or on the number of termites (Zimmerman et al., 1982; Rasmussen and Khalil, 1983; Zimmerman and Greenberg, 1983; Seiler et al., 1984a; Fraser et al., 1986). In the studies cited, termite emissions of methane range from a high of about 150 Tg/year to a low of 2 Tg/year.

In addition to absolute source strength estimates, source distribution estimates are needed. One estimate of this type which needs further study is the distribution of natural wetlands. These wetlands include bog, fen, marsh, swamp, floodplain, and shallow lake, according to Aselmann and Crutzen (1989). Their data are compiled from published information and maps and are spatially resolved on a 2.5° latitude by 5° longitude grid. Their results indicate that major source regions are located in the subtropics between 20 and 30°N, the tropics between 0 and 10°S, and the temperate-boreal region between 50 and 70°N. When the areas are weighted for

Table 2. Potential annual methane release for identified sources

Source of CH_4	Range of strength (Tg of CH_4/year)
Radiocarbon live	
Natural wetlands	120–200
Rice paddies	70–170
Livestock	80–100
Termites	25–150
Solid wastes	5–70
Biomass burning	10–40
Oceans	1–20
Tundra	1–5
Subtotal	312–755
Radiocarbon dead	
Coal mining	10–35
Venting and flaring	15–30
Industrial losses	5–25
Pipeline losses	10–20
Methane hydrates	2–4
Volcanoes	0.5
Automobiles	0.5
Subtotal	43–115
Total	355–870

CH_4 emissions, the tropical and temperate regions are equally important contributors with a total emission range of 60 to 140 Tg of methane per year. In contrast, Matthews and Fung (1987) published a compilation of wetlands from three independent data sets on vegetation, soil, and inundation. Their global estimate for wetland areas classifies 50% of all wetlands as forested or nonforested bogs mostly in latitudes north of 50°N. When combined with emissions data, this estimate indicates that the total emission is about 110 Tg, with 60% of all wetland methane emissions poleward of 50°N.

ISOTOPIC WORK

One of the investigative tools used to study methane involves the isotopes of carbon and hydrogen. The different isotopic forms of methane exhibit virtually identical chemical behavior but have different masses. Measurements of the ratios of carbon-13 to carbon-12, carbon-14 to carbon-12, and deuterium to hydrogen in the individual atoms of methane can be used to reveal clues as to the origins of atmospheric methane and its sources. In the case of stable isotopic discrimination, ratios vary because (i) kinetic processes, such as photosynthesis and bacterial reactions, preferentially use the lighter isotope of an element because of a lower activation energy for bond breaking, and (ii) isotopic exchange occurs between different chemical substances, different phases, or individual molecules as chemical processes move toward isotopic equilibrium. In the case of carbon-14, which is unstable, the ratios vary for the reasons noted above, but the ratio of carbon-14 to carbon-12 varies most noticeably as the carbon-14 disappears following the principles of radioactive decay.

The amount of ^{13}C (carbon-13) in methane from clean background air, which varies only slightly in different locations globally, can be compared with the amount in individual methane sources. The results of such comparisons are used to make estimates of relative source strengths of methane. In addition, $^{13}C/^{12}C$ ratios can give us clues regarding mechanisms of methane release to the atmosphere and formation and consumption prior to its release to the atmosphere. The measurement of $^{13}C/^{12}C$ is expressed relative to PDB (Pee Dee Belemnite) carbonate, a conventional standard, using the equation described by Craig (1957) where:

$$\delta^{13}C = [(^{13}C/^{12}C)_{sam}/(^{13}C/^{12}C)_{std} - 1] \times 1{,}000$$

In this notation, sam and std refer to isotopic ratios of carbon in the sample and standard, respectively. Isotopic values calculated from this formula which are negative indicate a sample which is relatively depleted in ^{13}C with respect to the standard.

Stevens and Rust (1982) originally proposed that the mass-weighted average composition of all sources of methane should equal the mean $\delta^{13}C$ of atmospheric methane corrected for any fractionation effects in the methane sink reactions. Figure 2 shows many of the measurements of $\delta^{13}C$ in methane which are available. With the $\delta^{13}C$ value of atmospheric methane around $-47‰$ in 1989 (see discussion below), one can see that most sources of methane are relatively depleted in ^{13}C with respect to atmospheric methane.

The $\delta^{13}CH_4$ value for clean, well-mixed background air has been measured by several researchers. Table 3 shows the results of an ongoing study to compare clean background air from each hemisphere. The sites chosen, Niwot Ridge, Colorado, and Baring Head, New Zealand, differed by about 0.5‰ after about 1 full year of sample collections (Lowe et al., 1989). Care has been taken to maintain a strict intercalibration between the two laboratories near the clean air sites. Methods of sampling, processing samples, and analysis are as nearly identical as possible. The southern hemispheric value at Baring Head, New Zealand ($-47.16 \pm 0.15‰$) agrees quite closely with clean air samples taken at Scott Base, Antarctica, by the same group (within 0.1‰ of Antarctic air). However, while the northern hemispheric value is clearly more enriched ($-46.69 \pm 0.16‰$) than the southern, it may not be representative of all northern hemispheric air. The northern hemisphere is expected to be more varied than the southern since it contains more methane sources and has a more intricate distribution pattern for them. Too, the sample site at Niwot Ridge is at about 3.1 km altitude. This means the air being sampled may include descending air from above the boundary layer which has been in contact with hydroxyl radicals for a longer time than well-mixed tropospheric air. This could affect the isotopic ratio.

The above study must be contrasted with another study by Stevens (1988)

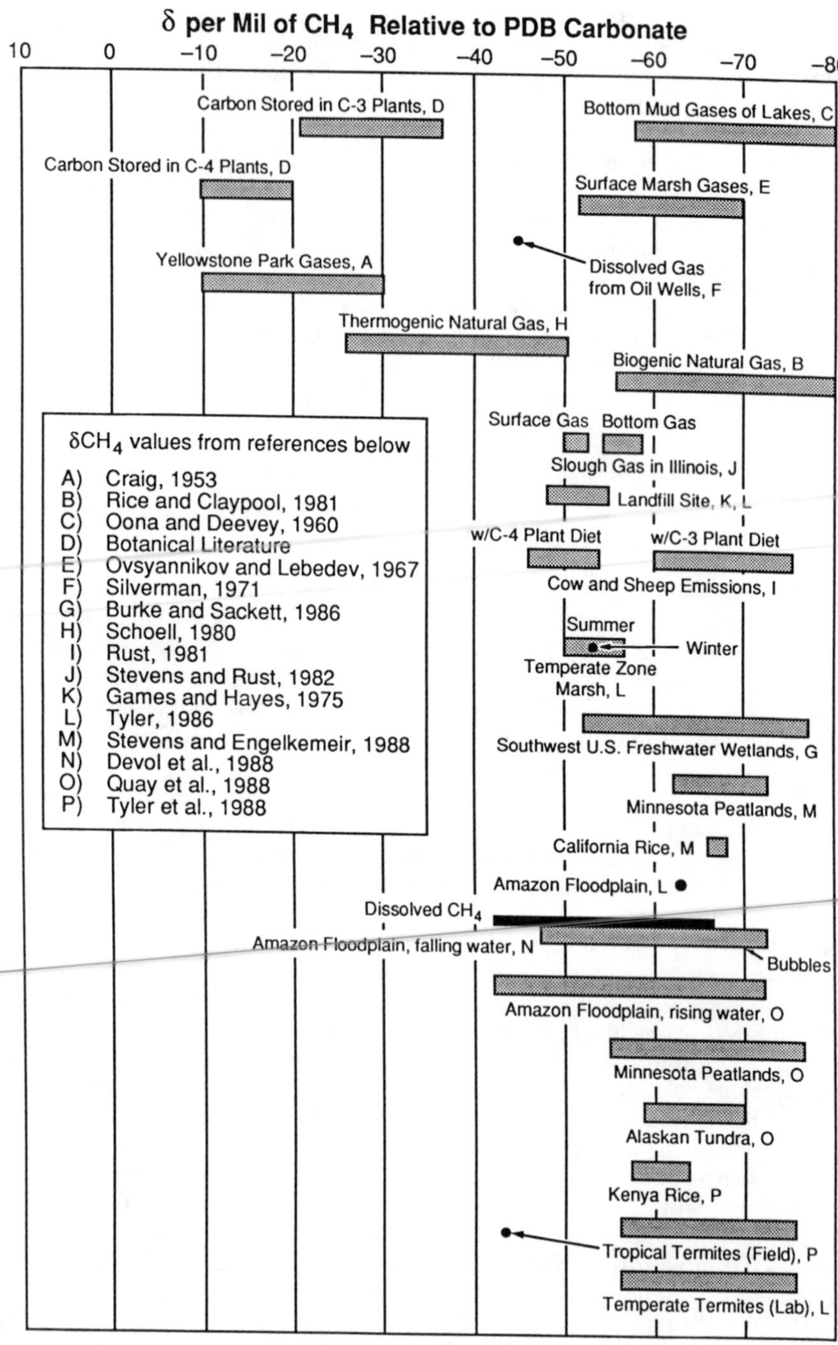
δ per Mil of CH_4 Relative to PDB Carbonate
10
0
−10
−20
−30
−40
−50
−60
−70
−80
Carbon Stored in C-3 Plants, D
Bottom Mud Gases of Lakes, C
Carbon Stored in C-4 Plants, D
Surface Marsh Gases, E
Dissolved Gas from Oil Wells, F
Yellowstone Park Gases, A
Thermogenic Natural Gas, H
Biogenic Natural Gas, B
δCH_4 values from references below
A) Craig, 1953
B) Rice and Claypool, 1981
C) Oona and Deevey, 1960
D) Botanical Literature
E) Ovsyannikov and Lebedev, 1967
F) Silverman, 1971
G) Burke and Sackett, 1986
H) Schoell, 1980
I) Rust, 1981
J) Stevens and Rust, 1982
K) Games and Hayes, 1975
L) Tyler, 1986
M) Stevens and Engelkemeir, 1988
N) Devol et al., 1988
O) Quay et al., 1988
P) Tyler et al., 1988
Surface Gas
Bottom Gas
Slough Gas in Illinois, J
Landfill Site, K, L
w/C-4 Plant Diet
w/C-3 Plant Diet
Cow and Sheep Emissions, I
Summer
Winter
Temperate Zone Marsh, L
Southwest U.S. Freshwater Wetlands, G
Minnesota Peatlands, M
California Rice, M
Amazon Floodplain, L
Dissolved CH_4
Amazon Floodplain, falling water, N
Bubbles
Amazon Floodplain, rising water, O
Minnesota Peatlands, O
Alaskan Tundra, O
Kenya Rice, P
Tropical Termites (Field), P
Temperate Termites (Lab), L

Table 3. Remote sampling of clean background air for methane interhemispheric comparisons[a]

Site	$\delta^{13}CH_4$ (‰ ± 1σ)	^{14}C-pMC
Niwot Ridge, Colorado (40°N) (6 January–15 September 1989)	− 46.69 ± 0.16 (32)	120.6 ± 2.9 (13)
Baring Head, New Zealand (41°S) (21 October 1988–26 September 1989)	− 47.16 ± 0.15 (22)	117.0 ± 3.1 (20)

[a] 1989 average values of $\delta^{13}CH_4$ and ^{14}C for methane in clean background air at two clean air sites. Number of samples in parentheses.

which indicates the opposite trend between the two hemispheres. In late 1987, Stevens' values were about − 47.3‰ in samples mostly from rural Illinois and − 46.7‰ in samples from American Samoa (14°S) and Canberra, Australia (35°S). Some of this difference may be from Stevens' choice of Samoa as the southern hemispheric site. This western Pacific site may be strongly influenced by influxes of northern hemispheric air at certain times of the year. Too, Stevens' data are part of a long-term time trend study that began in 1978 but has more samples and better precision in progressing years. The indication is that the northern hemispheric values of Stevens and co-workers (around − 47.0 ± 0.2‰ for 1989) and Lowe and co-workers (at − 46.69 ± 0.16‰ between January and December, 1989) may be in reasonable agreement for the year 1989 (C. Stevens, personal communication, 1990).

Clearly, discrepancies between Lowe's work and similar work by Stevens must be resolved in order to fully utilize isotopic data in methane studies. Strict intercalibration among different laboratories will be needed to determine real differences in $\delta^{13}CH_4$ in different geographic locations. Comparisons will also be needed to correct any errors in older data sets arising from different methods of sample collection, processing, and measurement.

To balance the average δ^{13}C value of methane sources with the background atmospheric value, it is necessary to correct the atmospheric value for fractionation by the methane sink processes. Fractionation due to the principal sink, reaction 1, has been measured several times in recent years with somewhat varying results. Rust and Stevens (1980) reported a value of $\alpha = 1.0028$ for the kinetic isotope effect in the reaction of OH with CH_4 at 297 K (room temperature), where $\alpha = k_{12}/k_{13}$ (the ratio of the rate constants for OH reaction with methane containing a particular carbon isotope). Work by Davidson et al. (1987) was generally accepted as having a superior experimental method and resulted in the value $\alpha = 1.010 \pm 0.007$ (2σ). However, a redetermination of this work in the same laboratory (Cantrell et al., 1990) now indicates the value as $\alpha = 1.0054 \pm 0.0009$ (2σ) with a negligible temperature dependence. These studies were all made in a laboratory environ-

Figure 2. δ^{13}C values of CH_4 relative to PDB carbonate. Values for sources of methane appear as bars, showing range of measured values, or single points depicting individual measurements.

ment. Studies of this type require a suitable OH source and allow for the opportunity to isolate the reactions of interest from myriad other interfering reactions which take place in the atmosphere.

In yet another study by Wahlen et al. (1989a), α was determined using upper tropospheric and lower stratospheric air samples. Measurements of CH_4 concentration and $\delta^{13}CH_4$ were made in these samples along with values for temperature and ^{85}Kr, a vertical transport tracer. The values measured in the vertical profile were then used along with transport and chemical kinetic data to arrive at a carbon kinetic fractionation factor for CH_4 reacting with OH. This work indicates that α = 1.0143 $\pm$ 0.0008 at stratospheric temperatures of about 220 to 230 K. Wahlen et al. also made calculations which indicated a temperature dependence would make the value of α about 1.0109 at 297 K.

The study of Wahlen and co-workers is an interesting cross-check to the laboratory results. There are some disadvantages to studying chemical reactions directly from atmospheric samples. In studying reaction 1, for example, the CH_4 vertical profile is subject to seasonal uncertainty, so a measurement over a short time span may not be appropriate. One must also average over all stratospheric variables (dynamics and winds) in making the calculations leading to α. It is also necessary to have fairly precise knowledge of rates of reactions for OH, $O(^1D)$, and Cl with CH_4 and for vertical transport coefficients. In spite of these problems, this kind of approach to studying an atmospheric chemical reaction can provide clues which point the way toward processes overlooked or misinterpreted in a laboratory systems study. Because the value for α is critical to understanding methane source balancing using carbon isotope ratios, as yet no value is universally accepted.

Because many sources of methane have overlapping ranges of $\delta^{13}CH_4$, many recent studies of $\delta^{13}CH_4$ have focused on pathways of methane release to the atmosphere, methods of consumption and production, seasonal effects of emissions from natural and man-made wetlands, and effects of methane-oxidizing bacteria. This type of information will help to further characterize factors affecting $\delta^{13}CH_4$ in individual sources. The resulting isotopic data can then be better applied when extrapolating regional source characteristics to global values.

For example, data from Tyler (1989b) obtained from field collections in Panama showed no appreciable seasonal effects (Table 4). The sites included both wet and dry season samplings of rain-forest marshes, lake and river coves, a seasonal swamp, and flooded forest floor. The water temperature change with season was only a few degrees. No consistent trend relating $\delta^{13}CH_4$ to methane flux or depth of water cover was found. The results indicated that nearly all the methane release was from ebullition rather than diffusion of dissolved methane through the water column or transport through the plants. Additional observations indicated that the water-covered sediments received the same type of leaf litter and decaying matter from floating and rooted aquatic plants all year round. The combination of factors mentioned above therefore leads to little variation in $\delta^{13}CH_4$ from anaerobic bacteria producing the methane.

The above seasonal study in a rain forest is in contrast to results from a seasonal study of North Carolina coastal wetlands by Martens et al. (1986) (see

Table 4. $\delta^{13}CH_4$ values from a seasonal study of emissions from Panamanian tropical wetlands

Date	Season	$\sigma^{13}CH_4$[a] (‰ ± 1σ)
June, 1986	Wet	− 61.4 ± 2.1 (6)
November, 1986	Wet	− 62.9 ± 3.0 (13)
February, 1987	Dry	− 61.5 ± 2.9 (13)
May–June, 1987	Wet	− 63.2 ± 3.4 (14)
September, 1987	Wet	− 60.8 ± 3.3 (14)
February–March, 1988	Dry	− 61.4 ± 4.2 (10)

[a] Number of samples in parentheses.

Table 5). Martens found as much as a 10‰ difference between methane emissions in summer months as compared with winter months, with the summer being more enriched in $\delta^{13}CH_4$. The summer months had temperatures around 30°C while the winter average was around 10°C. Martens and co-workers concluded that seasonal changes in $\delta^{13}CH_4$ at the coastal site resulted from changes in the pathways of microbial methane production and cycling of key substrates such as acetate and hydrogen.

Additional studies of mechanisms of release and/or consumption and production of methane which use $\delta^{13}CH_4$ ratios include work by Tyler et al. (1988), Chanton et al. (1988), and King et al. (1989). Working in Kenya, Tyler et al. (1988) measured $\delta^{13}CH_4$ for tropical termites and rice paddies. The majority of all termite $\delta^{13}CH_4$ measured was in the range − 55.7 to − 63.7‰, with another smaller group of measurements centered around − 73‰. While $\delta^{13}CO_2$ of gaseous emissions, $\delta^{13}C$ of carbon stored in the bodies, and $\delta^{13}C$ of the food source are closely correlated, $\delta^{13}C$ of methane emitted by termites is not correlated with the $\delta^{13}C$ of the food source. The rice data show little variation in $\delta^{13}CH_4$ with variety of rice planted and no significant changes in $\delta^{13}CH_4$ throughout the growth cycle in spite of significant changes in CH_4 flux throughout the growth cycle. This indicates that, at least in these Kenyan fields, the substrate carbon being made into methane and potentially oxidized by bacteria is not changing isotopically at different stages in the growth cycle. In addition, if methane-oxidizing bacteria are present and oxidation

Table 5. $\delta^{13}CH_4$ values from a seasonal study of emissions from North Carolina temperate coastal wetlands[a]

Season-year	Avg temp (°C)	$\delta^{13}CH_4$ (‰)
Summer, 1983 and 1984	30	− 57.3
Winter, 1983	10	− 68.5
Full year weighted avg		− 60.0

[a] Martens et al., 1986.

takes place in the rhizosphere, their effect appears to be unchanging throughout the growth cycle.

Chanton et al. (1988) studied $\delta^{13}CH_4$ in the Florida Everglades. Methane collected in bubbles from the submerged soils varied from − 70 to − 63‰ while organic carbon in the soils and dominant plants varied only from − 28 to − 25‰. Emergent aquatic plants were found to transport methane of $\delta^{13}C$ content up to 12‰ more enriched in ^{13}C than methane from the soil bubble reservoir. Because of the relatively large difference between $\delta^{13}C$ of the plant stem methane and that of the sedimentary bubbles, Chanton and co-workers hypothesized that the fractionation was due to mass-dependent gas transport mechanisms. However, they did not rule out fractionation by methane oxidation in the rhizosphere.

Coleman et al. (1981) have shown in laboratory culture studies that one might expect a large fractionation effect by methanotropic bacteria consuming methane. The residual methane was enriched in both ^{13}C and D. Coleman observed a fractionation factor of about 1.025 at 26°C and 1.013 at 11.5°C for α in the carbon kinetic isotope effect. This effect could be important in helping to balance a mass-weighted isotopic budget of methane sources and sinks if soil uptake of methane proves to be more important than previously thought. King et al. (1989) made two measurements of soil uptake of methane from Alaskan tundra soils. Methane collected from chambers after exposure to soil was measured and the kinetic isotope effect was calculated using the Rayleigh distillation equation. At an air temperature of 14°C the fractionation factor was 1.027, while at 4°C it was 1.016. The King study indicates that the fractionation effect on methane by bacteria in soils is comparable to the laboratory studies.

Similar to the delta notation for carbon isotopes, deuterium/hydrogen ratios are expressed in delta notation with respect to a standard (Hoefs, 1987), usually Standard Mean Ocean Water (SMOW). At present, there is less information available about the δ(D/H) of atmospheric methane and its sources than there is for $\delta^{13}C$. Some measurements of D/H in atmospheric methane have been made (e.g., Ehhalt, 1973; Wahlen et al., 1987). Based on these few measurements, δ(D/H) in clean background air is between − 70 and − 94‰ with respect to SMOW. Some information also exists on δ(D/H) of individual sources of methane. Thermogenic gases have D/H ratios from − 260 to − 150‰, while microbially produced methanes from a variety of gas deposits including glacial drift deposits, marsh gas, and shallow dry gas deposits have a range of − 280 to − 180‰ (Schoell, 1980). Biogenic methane from recent organic material has a wide range of δ(D/H) values. Burke and Sackett (1986) reported δ(D/H) of − 300 ± 40‰ for shallow aquatic environments. Whiticar et al. (1986) found δ(D/H) values from − 250 to − 170‰ in marine sediments where the CO_2 reduction pathway for methane production predominated, while δ(D/H) values were − 400 to − 250‰ in freshwater sediments where the acetate fermentation pathway predominated. No published values exist for δ(D/H) of methane from intestinal fermentation in ruminants or biomass burning to date.

With additional information on δ(D/H) in sources, similar to data using ^{13}C, the ratios can be used to differentiate between many of the sources of methane and to provide clues as to its formation. A mass-weighted hydrogen isotope balance between the sources of methane and the atmospheric value could be very useful in constraining various theories as to why atmospheric methane is increasing. One reason is that some CH_4 sources, for which $\delta^{13}CH_4$ signals overlap, have markedly different D/H isotope signals. To make such data useful in balancing the methane budget, the kinetic isotope effect in reaction 7 must be determined.

$$CH_3D + OH \Rightarrow \text{products (including } CH_2D + H_2O) \tag{7}$$

The single existing measurement by Gordon and Mulac (1975) was made at only one temperature, 416 K. Using that value for the D/H kinetic isotope effect in the reaction of methane with hydroxyl radicals yields an estimate of about − 320‰ as the present-day average δ(D/H) of all methane sources, according to Senum and Gaffney (1985). This is somewhat more depleted in D than what is expected from the known measurements to date. However, until reaction 7 is measured at tropospheric temperatures, it may be premature to attempt a balance of the methane budget using D/H ratios.

There is an additional advantage in having both $\delta^{13}C$ and δ(D/H) data for the same sources of methane. The two primary pathways for methane formation are CO_2 reduction as in reaction 8 and acetate fermentation as in the general reaction 9, where acetate fermentation refers collectively to methanogenesis involving transfer of a methyl group from any substrate.

$$CO_2 + 4H_2 \Rightarrow CH_4 + 2H_2O \tag{8}$$

$$^*CH_3COOH \Rightarrow {}^*CH_4 + CO_2 \tag{9}$$

Using both $^{13}C/^{12}C$ and D/H ratios to characterize sources, relative contributions to methane production by CO_2 reduction and acetate dissimilation may be discerned (e.g., Schoell, 1980, and Whiticar et al., 1986). In these studies, data on the $\delta^{13}C$ of CH_4 and CO_2 and on the δ(D/H) of CH_4 and formation H_2O from marine and freshwater ecosystems were used to arrive at empirically derived formulas which relate $\delta^{13}C$ and δ(D/H) to the methane formation pathway. In using the measurement data, Schoell and Whiticar et al. assumed that most of the marine production of methane was from CO_2 reduction and that most of the freshwater production was from fermentation. This is not always the case. Therefore, the principles derived from the study are somewhat complicated by the existence of mixed methanogenic pathways as well as by the presence of sulfate-reducing bacteria in some systems in the study.

Nevertheless, the following holds true. In marine sediments the methane formed by CO_2 reduction is often more depleted in ^{13}C than is the methane product of acetate fermentation in freshwater sediments. Methane formed by the transfer of a methyl group is more depleted in D than that formed by reduction of CO_2. In terms of δ values, the $\delta^{13}CH_4$ range is about − 60 to − 110‰ for CO_2 reduction and − 50 to − 70‰ for fermentation. The δ(D/H) range is about − 150 to − 250‰ for CO_2 reduction and − 300 to − 350‰ for fermentation. Researchers have successfully used the above results along with other methods of interpretation to study several shallow aquatic environments. Burke and Sackett (1986) evaluated

$\delta^{13}CH_4$ and δ(D/H) along with model studies to determine that acetate dissimilation accounted for between 50 and 80% of the total methane production in wetland areas in southeastern United States. Burke et al. (1988) showed that seasonal changes in $\delta^{13}CH_4$ of coastal marine sediments in North Carolina were related to acetate concentrations in the sediment and were not related to microbial consumption of methane.

Carbon-14 measurements in methane provide a different kind of information from that obtained by carbon-13 measurements. Carbon-14 is radioactive and therefore decays with a half-life of 5,730 years. The measurement unit is known as percent modern carbon (pMC) in which comparisons are made against contempory carbon. Carbon-14 is made in the upper atmosphere following a series of reactions which begin with solar cosmic rays producing neutrons. In turn, these neutrons are captured by ^{14}N, which then emits a proton to become ^{14}C. After the ^{14}C is oxidized by O_2, it eventually ends up as $^{14}CO_2$ in the lower stratosphere and upper troposphere. This means that radioactive ^{14}C as $^{14}CO_2$ is incorporated into the biosphere including oceans and other water cover and terrestrial plants and animals. At that point other carbon compounds involved in the CO_2 cycle can incorporate the ^{14}C. The amount present in a methane source at any given time gives us clues as to the history of the carbon. If the carbon is of recent origin it is said to be 100 pMC. (At present, modern carbon is about 120 pMC because of additional ^{14}C in the atmosphere from nuclear bomb tests and pressurized-water nuclear reactors). Only two possibilities allow us to find radiocarbon-dead methane. Either the carbon has been buried, as in the form of fossil fuels, and the ^{14}C has been allowed to decay away, or the methane is from primordial methane which is from deeper in the earth's interior. While the amount of primordial methane released may prove to be important (e.g., Gold, 1979), most scientists think this source is insignificant (e.g., Cicerone and Oremland, 1988), at least at present.

When Ehhalt (1974) reviewed the literature and made additional calculations to derive one of the first methane budgets, he kept track of the portion of methane released which contained no ^{14}C. Ehhalt made estimates of the contribution of fossil methane sources to atmospheric CH_4 using data obtained before the atmosphere was contaminated with ^{14}C from nuclear explosions (before about 1950). He determined that the fossil fuel contribution to atmospheric methane was at most 20% (Ehhalt, 1974). Later he revised his calculations to arrive at a most probable value of 10% (Ehhalt, 1979). Recent calculations take into account increased fossil fuel use, the decay of bomb test radiation cycling through the atmosphere, and more contamination from pressurized-water nuclear power plants. The indication is that current fossil fuel sources may be more important than previously thought. Lowe et al. (1988), Wahlen et al. (1989b), and Manning et al. (1990) have used ^{14}C measurements of atmospheric methane to demonstrate that 20 to 30% of it is radiocarbon dead (or the equivalent of this in ^{14}C-depleted methane).

Because of the revised estimates of the importance of fossil fuel methane, several new or overlooked potentially important methane sources have been investigated in recent years. These include methane hydrates, emissions from asphalt pavement, and the possibility that methane released in natural gas production and transmission has been greatly underestimated.

Methane hydrates are solids which form as icelike particles when gaseous methane is trapped in a lattice of water molecules. According to Kvenvolden (1988), conditions for the occurrence of gas hydrates are met in three distinct regions: (i) offshore in sediments of outer continental margins, (ii) onshore in areas of continuous permafrost, and (iii) on nearshore continental shelfs (sometimes called offshore permafrost). The source for the methane can be biogenic or thermogenic natural gas. Because of this, methane hydrates are sometimes called gas hydrates although the gas is primarily methane.

The amount of methane estimated to be stored in these regions is enormous. Current estimates by Kvenvolden (1990) put the quantity at 10,000 GT (a GT is a gigaton, where 1 GT $= 10^{15}$ g), with the vast majority being stored in sediments of outer continental margins. He estimates that under the present climatic conditions only 2 to 4 MT (MT $=$ Tg $= 10^{12}$ g; see Table 2), a small contribution from offshore permafrost, is released to the atmosphere each year. Although the amount of methane currently thought to be released from these regions is small, the amount of methane that could be released during global warming from gas hydrate destabilization is potentially large enough to dwarf all other methane sources. It is probable that only the permafrost-associated gas hydrates are vulnerable to continued global warming on a foreseeable basis. These regions account for about 400 GT of the total estimate given above, potentially a formidable portion. Although it is unlikely that gas hydrates will contribute much to atmospheric methane in the near future, continued studies of gas hydrates, including their characterization by stable isotope measurements, will be important as indicators of the extent of global warming.

Sackett and Barber (1988) proposed that an overlooked radiocarbon-dead source of atmospheric methane might be derivatives of fossil fuel carbon such as asphalts and road tars. However, findings by Tyler et al. (1990) indicate that methane emissions from asphalt pavement cannot be a significant source of atmospheric methane as compared to other identified methane sources. Their study, which included fresh pavement (a few days old) and older pavement (up to 3 years), concluded that less than 0.1 Tg of methane per year was emitted globally from this source. Other petroleum-derived products with uses such as roofing and other construction applications must be less important still because their total surface area exposed to the atmosphere is much less than that of asphalt pavement.

Emissions of methane from natural gas production and transmission have already been mentioned as a potential source (estimates for its range of source strength appear in Table 2). However, there is a growing suspicion that difficult-to-estimate loss processes such as venting and flaring at wells, leakage from abandoned wells, and transmission losses, especially from low-pressure distribution lines, may be underestimated. In part this suspicion results by default as a likely way to explain the radiocarbon methane budget. However, there is growing evidence that practices which lead to methane release in these loss processes vary greatly from one geographic location to another. Even recent estimates of losses may have relied too heavily on available data from industry in North America and Western Europe, which are much easier to come by than data from most other regions. For example, Rowland et al. (1990) have made collections of well-mixed city air in parts of eastern Europe and Asia as well as in several U.S. cities. Their data suggest that leakage during natural gas transmission is significantly higher in the non-U.S. cities.

In differentiating between ^{14}C-depleted methane as opposed to radiocarbon-dead methane, the possibility is introduced that there is ^{14}C-depleted carbon in biogenic sources where the carbon has been in long-term storage. Such regions include temperate and arctic boreal forests, where vast amounts of peat are known to be stored. Basal dates for peat accumulated in Canada and western Siberia can approach 12,000 years or so, roughly two carbon-14 half-lives (Sue Short, personal communication, Institute of Arctic and Alpine Research, University of Colorado, Boulder, 1988). Isotopic measurements of ^{14}C and ^{13}C will be a key to detecting changes in this type of ecosystem. If the older (deeper) carbon is being used as a substrate to make methane, the isotopic signal in such a biogenic source would be distinct in that it would be both ^{14}C depleted (from the older carbon) and ^{13}C depleted (from its biogenic origin). This would help explain recent estimates which indicate that radiocarbon-dead methane is a larger part of the total emitted than previously thought. The dead fraction would not all have to be from fossil fuel sources.

If this older carbon is not a significant source of atmospheric methane now, it could become more important as global warming takes effect (Harriss, 1989a). The variability in prehistoric CH_4 concentrations between glacial and interglacial times is hypothesized to be due to the expansion of arctic and boreal peatlands following glacial retreat (Harriss et al., 1985). Data that are available on temperature sensitivity of CH_4 sources from organic soils and sediments indicate that these peatlands are particularly sensitive to an enhanced greenhouse effect. The initial response to warming might be an increase in CH_4 flux in the atmosphere. However, complicated negative feedback mechanisms, such as the drying of wetland soils due to increased temperature, must be resolved.

MODELING METHANE

Mathematical models are one way to test experimental and observational data to see what parts of a complex system are understood and where more information is needed. Over the years the complexity of models used to study methane has grown from simple one-dimensional (1-D) models solving only equations involving methane to complex three-dimensional (3-D) models solving interactively the equations of all important compounds.

An example of a 1-D tropospheric mixing model (latitudinal direction) is that used at the University of California at Irvine (Blake, 1983). A set of equations relating methane concentration to production, loss, and mixing in each of several boxes representing latitudinal bands was solved numerically by iterating the calculated results through a sequence of time steps. Latitudinal mixing coefficients

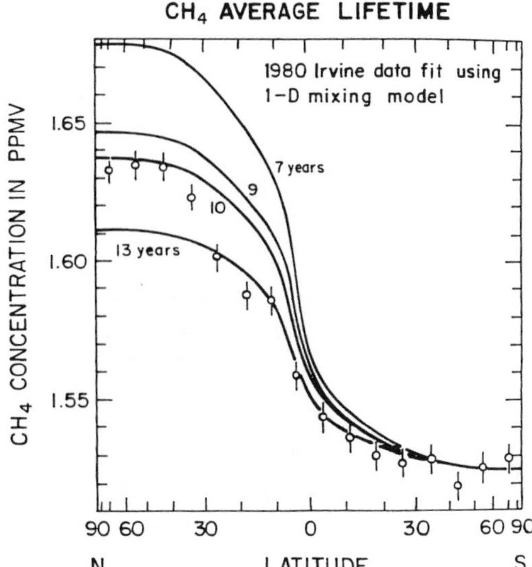

Figure 3. Comparison of concentration gradients of methane, generated by 1-D mixing model with different methane lifetimes, versus actual measurement values for methane gradient. (From Blake [1983], p. 187.)

between boxes were determined empirically from previous models relating calculated concentrations with experimental observations for fluorochlorocarbons (Tyler, 1983). By running the model with different loss rates for methane in the atmosphere, several curves of methane concentration as a function of latitude were generated (see Fig. 3). These curves were compared with concentration measurements made by the Irvine group. In this case, a 9- or 10-year lifetime for methane is the best fit to the measurements, although none was a perfect fit.

In recent years more sophisticated 2-D and 3-D models of atmospheric methane and its sources and sinks have been constructed. These models depend partly on increasingly accurate information in the form of measured, observed, and calculated parameters. One such model is a 3-D model developed by John Taylor and co-workers (Taylor et al., 1991). This is a global Lagrangian transport model of tropospheric methane which moves air parcels by wind fields. The OH sink is specified at a level consistent with methylchloroform data. Data and other model parameters were set to 1980 values, and the model was run until 1985.

Inputs to the model include two types of source functions. One source function is based on global net primary productivity (NPP). The NPP source function assumes that methane release is proportional to NPP. Calculatons have been made using temperature and precipitation to determine the NPP. Figure 4 shows the results as a contour plot of the predicted methane concentrations over the model surface layer for the months January and July. Above each contour plot is a 3-D representation of the same data. The other source function uses data from individual source regions from high-resolution land use data (e.g., Matthews and Fung, 1987). Figure 5 shows a contour plot of the predicted methane concentrations in a manner similar to Fig. 4.

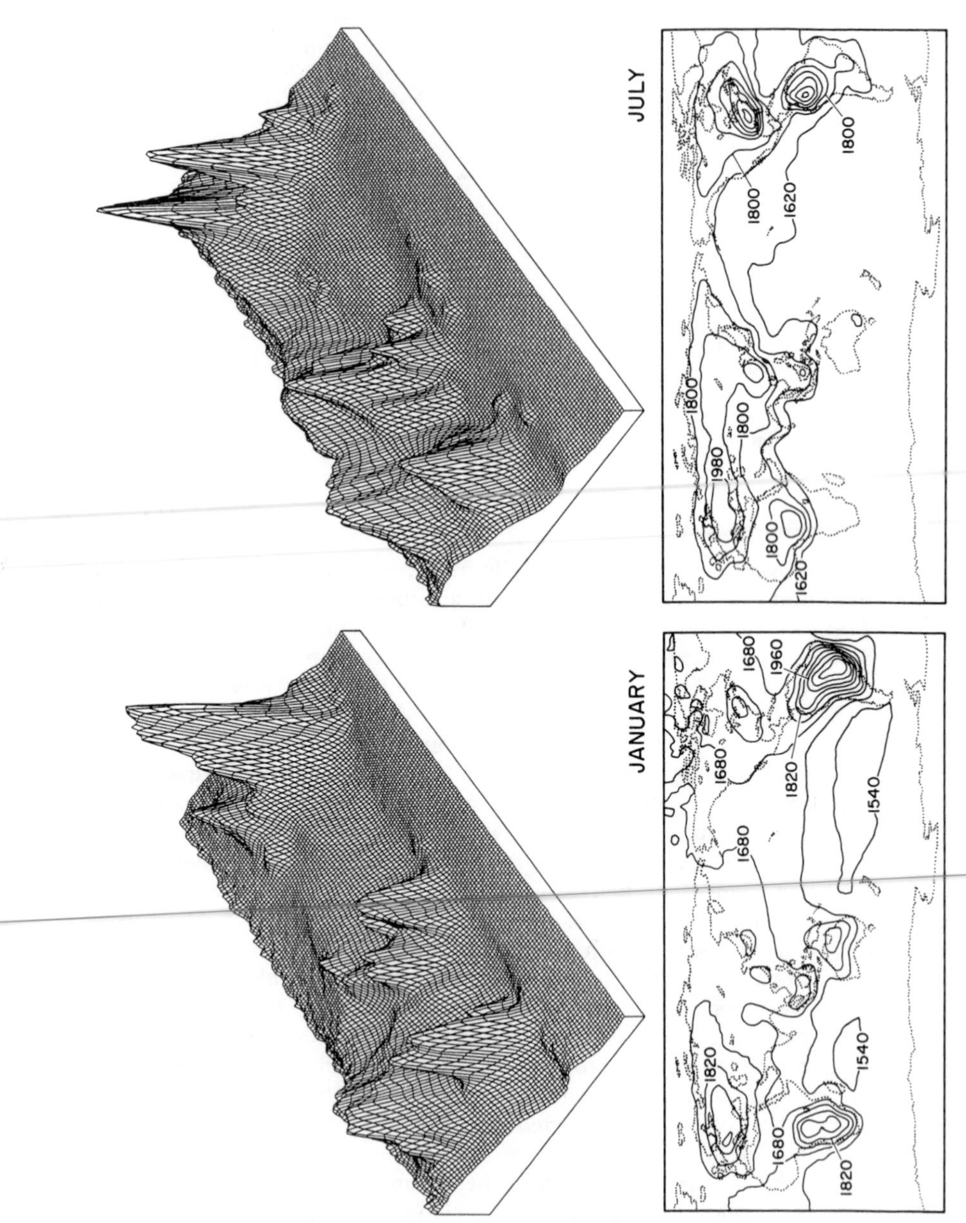
JULY
1800
1620
1800
1800
1800
1980
1800
1620
JANUARY
1680
1960
1680
1820
1540
1680
1820
1540
1680
1820

As one can see from the figures, the two inputs to the model give similar results but have some major discrepancies to resolve. For example, the region around China, India, and parts of Indonesia has much higher methane concentrations indicated in the contour plot of Fig. 5 than in Fig. 4. This kind of result means that rice paddy regions are not properly represented. One possible explanation for the discrepancy in the rice results is that land use data are overestimating the contribution from paddy fields. Alternatively, the NPP function, which does not include the effects of soil type or fertilizer for agricultural areas, is probably underestimating rice production.

Figures 4 and 5 also show that the NPP source function leads to higher methane concentrations in South America and central Africa (tropical wetlands) than does the land use data source function. Speculation along similar lines to that for the paddy field results can be made.

The methane concentration values generated by the model of Taylor et al. have also been checked against experimental measurements for CH_4 concentration at field measurement stations in both January and July, 1985. The results are promising, although the agreement is not always good. There is quite a lot of variation in the degree of agreement between measured and calculated values at many of the measurement stations at certain times of the year.

Until recently, complex models like the one described above have used isotopic methane data only in a supporting role; i.e., the data are used to help determine distribution of sources and relative source strengths in a qualitative way. However, several recent models have used measured isotopic values of CH_4 as key data in computational models. These include the studies by Lowe et al. (1988), Wahlen et al. (1989b), and Manning et al. (1990), which determined that fossil fuel sources of methane (or the equivalent of this in ^{14}C-depleted methane) may be between 20 and 30% of the total methane source. Their results were obtained by mixing isotopically distinct methane sources in a two-box latitudinal mixing model. One source is taken to be the average of all fossil methane sources, and the other is the average of all biogenic methane where both ^{14}C and ^{13}C are considered together. The models weighted the relative importance of each source by comparing the computer-generated $\delta^{13}C$ and ^{14}C-pMC CH_4 values to known values for clean background air in each hemisphere. Similar to other atmospheric models, the accuracy of the results depend on parameters such as north/south mixing times and methane lifetime. In addition, the accuracy is dependent on knowing the kinetic isotope effect of OH reacting with CH_4, and a proper correction for the production of $^{14}CH_4$ by pressurized-water nuclear reactors.

Two-box models of atmospheric mixing have also been used to study relative source strengths and hemispheric release distributions of methane sources (Stevens, 1988; Lowe et al., 1990). These models use $\delta^{13}CH_4$ mass balance equations

---

Figure 4. Contour plot of the predicted methane concentrations (parts per billion by volume [ppbv]) over the model surface layer for the months January and July based on net primary productivity source function (from Taylor et al. [1991]). Above each contour plot is a 3-D representation of the same data (courtesy J. A. Taylor).

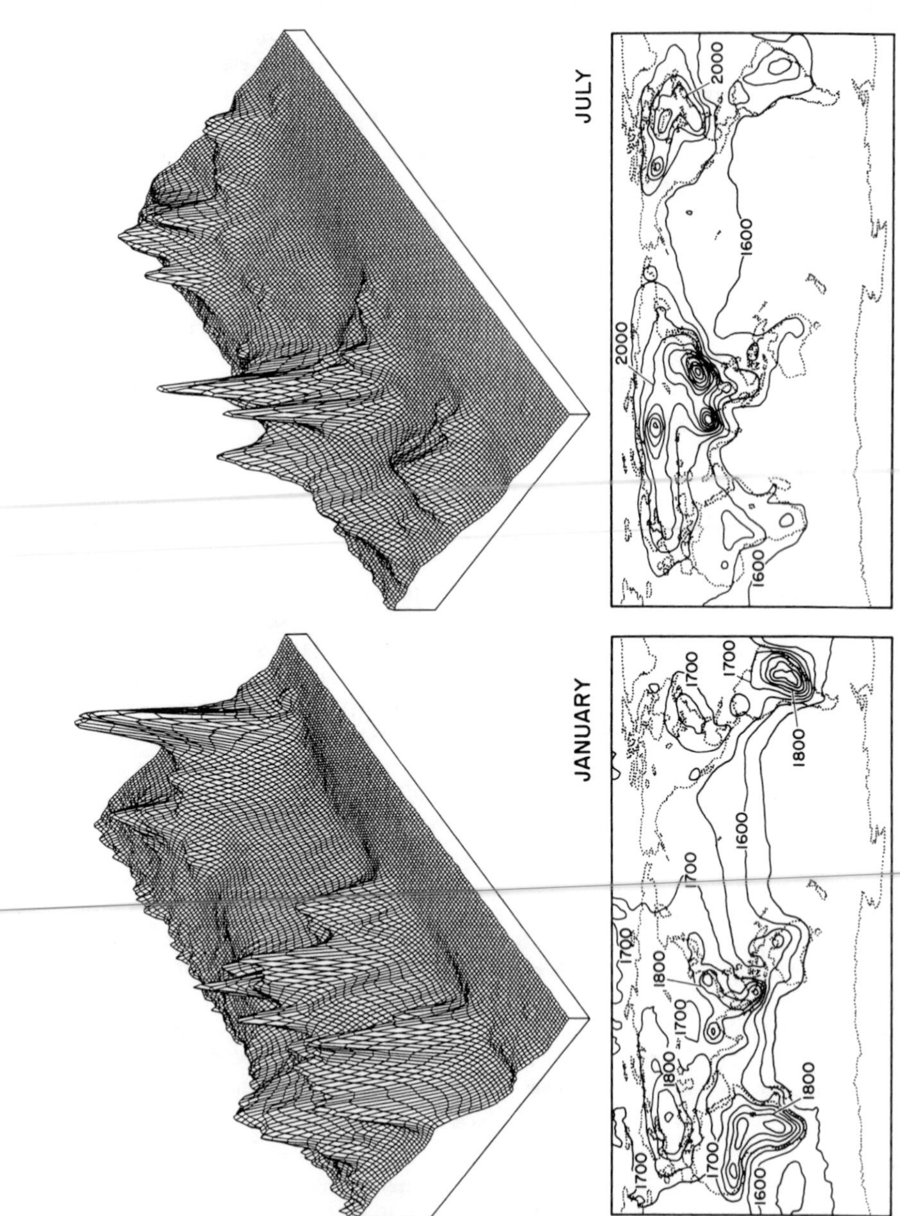
JULY
2000
2000
1600
1600
1600
JANUARY
1700
1700
1700
1800
1600
1800
1700
1800
1700
1700
1700
1800
1700
1600
1800

for all the measured sources along with the best available estimates of the absolute strengths and distribution of the individual sources to compare the computer-generated $\delta^{13}CH_4$ of background air in each hemisphere with measured values.

The data set by Stevens began in 1978, and the two-box model was used to study the trend in sources over a period from 1978 to 1987. Because the hemispheric exchange rate used (0.7 to 1.1 $year^{-1}$) is much greater than the loss rate (0.125 $year^{-1}$), the trends in the average $\delta^{13}C$ of the source fluxes in each hemisphere are much greater than the trends in the atmosphere. Among other conclusions, two features of the $\delta^{13}CH_4$ trends since 1978 stood out: (i) the decrease in the average $\delta^{13}CH_4$ of the source fluxes from 1978 to 1983 in the northern hemisphere and (ii) the increase in the $\delta^{13}CH_4$ trend after 1982–1983 in both hemispheres.

This is evidence for variability of the fluxes from sources over a relatively short time span. The decrease in average $^{13}CH_4$ of methane sources in the northern hemisphere ($-0.3‰$ $year^{-1}$), noted in the first conclusion above, apparently requires that natural source fluxes were changing, i.e., that some combination of increasing ^{13}C-depleted CH_4 (relative to the atmospheric value) and decreasing ^{13}C-enriched CH_4 fluxes existed rather than a change in anthropogenic source fluxes. Stevens argues that changes in relatively ^{13}C-depleted sources such as rice paddy methane and herbivores can be ruled out since they could not have increased so rapidly in such a short period of time. In explaining the increasing trends after 1983 ($+0.06‰$/year globally), also noted above, the source fluxes in each hemisphere must be changing such that ^{13}C-enriched fluxes are increasing. One possible explanation is that fluxes of CH_4 are increasing from an acceleration in biomass burning. If so, this effect may be more pronounced in the southern hemisphere where more deforestation of tropical rain forests is taking place. However, the increase in $\delta^{13}CH_4$ is greater in the northern hemisphere than in the southern during this time span, suggesting that the origin of the changing flux is in the north. Stevens postulates that because the global growth rate of methane concentration may be slowing down (Steele et al., 1987), the increase in $\delta^{13}CH_4$ occurring at this time may be caused by a decrease in the flux of relatively ^{13}C-depleted CH_4 sources rather than by an increase in a relatively ^{13}C-enriched source in the north such as fossil fuels.

The model by Lowe and co-workers shows that a different relative mix of methane sources in the two hemispheres leads to different $\delta^{13}C$ values for the net source flux in each hemisphere. Because of the long lifetime for methane (9 years in their model) relative to the time to mix between hemispheres (about 1 year), the interhemispheric difference in source flux $\delta^{13}C$ must be about 6‰ in order to maintain a difference of 0.6‰ in the atmosphere. Therefore, if the northern hemisphere value is $-46.6‰$ and the southern value is $-47.2‰$, it indicates that the weighted average of combined northern sources must be about 6‰ enriched in ^{13}C compared to southern sources of methane.

Figure 5. Contour plot of the predicted methane concentrations (parts per billion by volume [ppbv]) over the model surface layer for the months January and July, based on land use data source function (from Taylor et al. [1991]). Above each contour plot is a 3-D representation of the same data (courtesy J. A. Taylor).

CONCLUSION

Understanding the methane budget requires much more work in diverse areas such as field sampling and measurements, laboratory experiments, and mathematical modeling. These types of information will have to be fully integrated in the future to obtain a clearer model for atmospheric methane and its sources and sinks. In particular, understanding the methane budget requires (i) detailed chemical schemes where calculations of concentrations of OH, CH_4, CO, and nonmethane hydrocarbons are coupled; (ii) better determination of sources at the surface (including interaction with biologists to study mechanisms of methane formation and destruction); and (iii) a good transport code for models which includes advection and convection (cloud convection).

Detailed chemical schemes needed in the first requirement must properly consider the feedback effects from OH reactions with increasing CH_4, CO, and other hydrocarbons. As concentrations of these compounds rise, the ability of OH to scavenge a particular compound lessens. On the other hand, since methane reacts yielding products which include water vapor, OH levels may increase to partially mitigate the rise in concentration of the aforementioned gases.

The second requirement includes quantification of both natural and anthropogenic sources. Integration of ground, aircraft, and satellite measurements will be needed to provide accurate estimates of CH_4 flux to the atmosphere at regional scales (Harriss, 1989b). Because the boreal and arctic wetlands store much of the earth's soil carbon in wetlands, it will be especially important to quantify the temperature sensitivity of the physical and biological processes responsible for CH_4 release from these systems. Measurements of stable and ^{14}C isotopes of carbon in methane may be particularly important in these ecosystems. Climate models predict an enhanced greenhouse warming effect in these areas which may be detected in part by a relatively unique isotope signal from these biogenic sources.

The third requirement is to reproduce the interhemispheric exchange time of 1.1 years in such a way that the transport code conserves mass and does not introduce unwanted numerical diffusion. Ultimately, the models will need to have a sufficient resolution to differentiate between spatial differences on the order of a 1° grid scale. At some point, with three-dimensional transport and chemistry determined properly, inverse modeling should be able to derive sources compatible with observed concentrations of methane.

Acknowledgments. I would like to give special thanks to Ed Dlugokencky, Sue Schauffler, and Chris Ennis of the National Center for Atmospheric Research for helpful suggestions through all versions of the manuscript. I would also like to thank Guy Brasseur and Steven Schneider, both of the National Center for Atmospheric Research; Martin Manning, of the Department of Scientific and Industrial Research in New Zealand; and Ralph Cicerone, of the University of California at Irvine, for discussions regarding aspects of the manuscript including chemistry, greenhouse gases, and modeling atmospheric methane. The National Center for Atmospheric Research is funded by the National Science Foundation. This work has also been supported by a grant from the National Aeronautics and Space Administration under order W-16,184, mod. 5.

REFERENCES

Aselmann, I., and P. J. Crutzen. 1989. Global distribution of natural freshwater wetlands and rice paddies, their net primary productivity, seasonality and possible methane emissions. *J. Atmos. Chem.* **8**:307–358.

Barnola, J. M., D. Raynaud, Y. S. Korotkevich, and C. Lorius. 1987. Vostok ice core provides 160,000-year record of atmospheric CO_2. *Nature* (London) **329**:408–414.

Bartlett, K. B., P. M. Crill, D. I. Sebacher, R. C. Harriss, J. O. Wilson, and J. M. Melack. 1988. Methane flux from the central Amazonian floodplain. *J. Geophys. Res.* **93**:1571–1582.

Bingemer, H. G., and P. J. Crutzen. 1987. The production of methane from solid wastes. *J. Geophys. Res.* **92**:2181–2187.

Blake, D. R. 1983. Increasing concentrations of atmospheric methane, 1979–1983. Ph.D. thesis. University of California at Irvine, Irvine.

Blake, D. R., and F. S. Rowland. 1988. Continuing worldwide increase in tropospheric methane, 1978 to 1987. *Science* **239**:1129–1131.

Born, M., H. Dorr, and I. Levin. 1989. Methane consumption in aerated soils of the temperate zone. *Tellus* **42B**:2–8.

Burke, R. A., Jr., C. S. Martens, and W. M. Sackett. 1988. Seasonal variations of D/H and $^{13}C/^{12}C$ ratios of microbial methane in surface sediments. *Nature* (London) **322**:829–831.

Burke, R. A., Jr., and W. M. Sackett. 1986. Stable hydrogen and carbon isotopic compositions of biogenic methanes from several shallow aquatic environments, p. 297–313. *In* M. L. Sohn (ed.), *Organic Marine Geochemistry*. American Chemical Society, Washington, D.C.

Cantrell, C. A., R. E. Shetter, A. H. McDaniel, J. G. Calvert, J. A. Davidson, D. C. Lowe, S. C. Tyler, R. J. Cicerone, and J. P. Greenberg. 1990. Carbon kinetic isotope effect in the oxidation of methane by hydroxyl radicals. *J. Geophys. Res.* **95**:22455–22462.

Chanton, J. P., C. S. Martens, C. A. Kelley, P. M. Crill, and W. J. Showers. Methane transport mechanisms and isotopic fractionation in emergent macrophytes of an Alaskan tundra lake. *J. Geophys. Res.*, in press.

Chanton, J. P., G. G. Pauly, C. S. Martens, N. E. Blair, and J. W. H. Dacey. 1988. Carbon isotopic composition of methane in Florida Everglades soils and fractionation during its transport to the troposphere. *Global Biogeochem. Cycles* **2**:245–252.

Cicerone, R. J., and R. S. Oremland. 1988. Biogeochemical aspects of atmospheric methane. *Global Biogeochem. Cycles* **2**:299–327.

Cicerone, R. J., and J. D. Shetter. 1981. Sources of atmospheric methane: measurements in rice paddies and a discussion. *J. Geophys. Res.* **86**:7203–7209.

Cicerone, R. J., J. D. Shetter, and C. C. Delwiche. 1983. Seasonal variation of methane flux from a California rice paddy. *J. Geophys. Res.* **88**:11022–11024.

Coleman, D. D., J. B. Risatti, and M. Schoell. 1981. Fractionation of carbon and hydrogen isotopes by methane-oxidizing bacteria. *Geochim. Cosmochim. Acta* **45**:1033–1037.

Craig, H. 1953. The geochemistry of the stable carbon isotopes. *Geochim. Cosmochim. Acta* **3**:53–92.

Craig, H. 1957. Isotopic standards for carbon and oxygen and correction factors for mass-spectrometric analysis of carbon dioxide. *Geochim. Cosmochim. Acta* **12**:133–149.

Craig, H., and C. C. Chou. 1982. Methane: the record in polar ice cores. *Geophys. Res. Lett.* **9**:1221–1224.

Crill, P. M., K. B. Bartlett, J. O. Wilson, D. I. Sebacher, R. C. Harriss, J. M. Melack, S. MacIntyre, L. Lesack, and L. Smith-Morrill. 1988. Tropospheric methane from an Amazonian floodplain lake. *J. Geophys. Res.* **93**:1564–1570.

Crutzen, P. J. 1987. Role of the tropics in atmospheric chemistry, p. 107–130. *In* R. E. Dickinson (ed.), *The Geophysiology of Amazonia*. John Wiley & Sons, Inc., New York.

Crutzen, P. J., I. Aselmann, and W. Seiler. 1986. Methane production by domestic animals, wild ruminants, other herbivorous fauna, and humans. *Tellus* **38B**:271–284.

Crutzen, P. J., A. C. Delaney, J. Greenberg, P. Haagenson, L. Heidt, R. Lueb, W. Pollack, W. Seiler, A. Wartburg, and P. Zimmerman. 1985. Tropospheric chemical composition measurements in Brazil during the dry season. *J. Atmos. Chem.* **2**:233–256.

Crutzen, P. J., L. E. Heidt, J. P. Krasnec, W. H. Pollack, and W. Seiler. 1979. Biomass burning as source of atmospheric gases CO, H_2, N_2O, NO, CH_3Cl and COS. *Nature* (London) **282:**253–256.

Dacey, J. W. H. 1987. Knudsen-transitional flow and gas pressurization in leaves of Nelumbo. *Plant Physiol.* **85:**199–203.

Dacey, J. W. H., and M. J. Klug. 1979. Methane efflux from lake sediments through water lilies. *Science* **203:**1253–1255.

Davidson, J. A., C. A. Cantrell, S. C. Tyler, R. E. Shetter, R. J. Cicerone, and J. G. Calvert. 1987. Carbon kinetic isotope effect in the reaction of CH_4 with HO. *J. Geophys. Res.* **92:**2195–2199.

Devol, A. H., J. E. Richey, W. A. Clark, S. L. King, and L. A. Martinelli. 1988. Methane emissions to the troposphere from the Amazon floodplain. *J. Geophys. Res.* **93:**1583–1592.

Dickinson, R. E., and R. J. Cicerone. 1986. Future global warming from atmospheric trace gases. *Nature* (London) **319:**109–115.

Donner, L., and V. Ramanathan. 1980. Methane and nitrous oxide: their effects on the terrestrial climate. *J. Atmos. Sci.* **37:**119–124.

Ehhalt, D. H. 1973. Methane in the atmosphere, p. 144–158. *In* G. M. Woodwell and E. V. Pecan (ed.), *Carbon and the Biosphere*. U. S. Atomic Energy Commission, Oak Ridge, Tenn.

Ehhalt, D. H. 1974. The atmospheric cycle of methane. *Tellus* **26:**58–70.

Ehhalt, D. H. 1979. Der atmosphärische Kreislauf von Methan. *Naturwissenschaften* **66:**307–311.

Ehhalt, D. H., and A. Schmidt. 1978. Sources and sinks of atmospheric methane. *Pure Appl. Geophys.* **116:**452–464.

Fraser, P. J., R. A. Rasmussen, J. W. Creffield, J. R. French, and M. A. K. Khalil. 1986. Termites and global methane—another assessment. *J. Atmos. Chem.* **4:**295–310.

Games, L. M., and J. M. Hayes. 1975. On the mechanisms of CO_2 and CH_4 production in natural anaerobic environments, p. 51–73. *In* J. Nriagu (ed.), *Environmental Biogeochemistry*, vol. 1. Ann Arbor Science, Ann Arbor, Mich.

Gold, T. J. 1979. Terrestrial sources of carbon and earthquake outgassing. *J. Petrol. Geol.* **1:**13–19.

Gordon, S., and W. A. Mulac. 1975. Reactions of the OH ($X^2\pi$) radical produced by the pulse radiolysis of water vapor. *Int. J. Chem. Kinet. Symp.* **7:**289–299.

Greenberg, J. P., P. R. Zimmerman, L. Heidt, and W. Pollock. 1984. Hydrocarbon and carbon monoxide emissions from biomass burning in Brazil. *J. Geophys. Res.* **89:**1350–1354.

Harriss, R. C. 1989a. Historical trends in atmospheric methane concentration and the temperature sensitivity of methane outgassing from boreal and polar regions, p. 79–84. *In Ozone Depletion, Greenhouse Gases, and Climate Change*. National Academy Press, Washington, D.C.

Harriss, R. C. 1989b. Experimental design for studying atmosphere-biosphere interactions, p. 291–301. *In* M. O. Andreae and D. S. Schimel (ed.), *Exchange of Trace Gases between Terrestrial Ecosystems and the Atmosphere*. John Wiley & Sons, Chichester, Great Britain.

Harriss, R. C., E. Gorham, D. I. Sebacher, K. B. Bartlett, and P. A. Flebbe. 1985. Methane flux from northern peatlands. *Nature* (London) **315:**652–653.

Harriss, R. C., and D. I. Sebacher. 1981. Methane flux in forested freshwater swamps of the southeastern United States. *Geophys. Res. Lett.* **8:**1002–1004.

Hileman, B. 1989. Global warming. *Chem. Eng. News* **67:**25–44.

Hoefs, J. 1987. *Stable Isotope Geochemistry*, 3rd ed., p. 22–24. Springer-Verlag, New York.

Holzapfel-Pschorn, A., and W. Seiler. 1986. Methane emission during a cultivation period from an Italian rice paddy. *J. Geophys. Res.* **91:**11803–11814.

Keeling, C. D., J. A. Adams, Jr., C. A. Ekdahl, Jr., and P. R. Guenther. 1976a. Atmospheric carbon dioxide variations at the South Pole. *Tellus* **28:**552–564.

Keeling, C. D., R. B. Bacastow, A. E. Bainbridge, C. A. Ekdahl, Jr., P. R. Guenther, and L. S. Waterman. 1976b. Atmospheric carbon dioxide variations at Mauna Loa Observatory, Hawaii. *Tellus* **28:**538–551.

Keeling, C. D., R. B. Bacastow, A. F. Carter, S. C. Piper, T. P. Whorf, M. Heimann, W. G. Mook, and H. Roeloffzen. 1989. A three-dimensional model of atmospheric CO_2 transport based on observed winds. 1. Analysis of observational data, p. 165–236. *In* D. H. Peterson (ed.), *Aspects of Climate Variability in the Pacific and the Western Americas.* American Geophysical Union, Washington, D.C.

Kerr, R. A. 1989. Greenhouse skeptic out in the cold. *Science* **246:**1118–1119.

Khalil, M. A. K., and R. A. Rasmussen. 1985. Causes of increasing atmospheric methane: depletion of hydroxyl radicals and the rise of emissions. *Atmos. Environ.* **19:**397–407.

Kiehl, J. T., and R. E. Dickinson. 1987. A study of the radiative effects of enhanced atmospheric CO_2 and CH_4 on early Earth surface temperatures. *J. Geophys. Res.* **92:**2991–2998.

King, G. M. 1984. Utilization of hydrogen, acetate and "non-competitive" substrates by methanogenic bacteria in marine sediments. *Geomicrobiol. J.* **3:**275–306.

King, G. M., M. J. Klug, and D. R. Lovley. 1983. Metabolism of acetate, methanol, and methylated amines in intertidal sediments of Lowest Cove, Maine. *Appl. Environ. Microbiol.* **48:**719–725.

King, S. L., P. D. Quay, and J. M. Lansdown. 1989. The $^{13}C/^{12}C$ kinetic isotope effect for soil oxidation of methane at ambient atmospheric concentrations. *J. Geophys. Res.* **94(D15):** 18273–18277.

Koyama, T. 1963. Gaseous metabolism in lake sediments and paddy soils and the production of atmospheric methane and hydrogen. *J. Geophys. Res.* **68:**3971–3973.

Kvenvolden, K. A. 1988. Methane hydrates and global climate. *Global Biogeochem. Cycles* **2:**221–229.

Kvenvolden, K. A. 1990. Estimate of current methane release from gas hydrates. Abstr. no. 10748, p. 195. Geological Society of America National Meeting, Dallas, Tex.

Lassey, K. R., D. C. Lowe, and M. R. Manning. 1990. Sources and sinks of atmospheric methane in New Zealand: a perspective. INS Report 419. Department of Scientific and Industrial Research, Institute of Nuclear Sciences, Lower Hutt, New Zealand.

Lerner, J., E. Matthews, and I. Fung. 1988. Methane emission from animals: a global high-resolution database. *Global Biogeochem. Cycles* **2:**139–156.

Levine, J. S., C. P. Rinsland, and G. M. Tenille. 1985. The photochemistry of methane and carbon monoxide in the troposphere in 1950 and 1985. *Nature* (London) **318:**254–257.

Levy, H. 1971. II. Normal atmosphere: large radical and formaldehyde concentrations predicted. *Science* **173:**141–143.

Logan, J. A., M. J. Prather, S. C. Wofsy, and M. B. McElroy. 1981. Tropospheric chemistry: a global perspective. *J. Geophys. Res.* **86:**7210–7254.

Lowe, D. C., C. A. M. Brenninkmeijer, M. R. Manning, R. Sparks, and G. Wallace. 1988. Radiocarbon determination of atmospheric methane at Baring Head, New Zealand. *Nature* (London) **332:**522–525.

Lowe, D. C., C. Brenninkmeijer, M. Manning, S. C. Tyler, E. Dlugokencky, P. Steele, and P. Lang. 1989. Carbon isotopic composition of atmospheric methane. Presented at American Geophysical Union National Meeting, San Francisco, Calif.

Lowe, D. C., M. R. Manning, C. A. M. Brenninkmeijer, and K. R. Lassey. 1990. Proceedings of the International Clean Air Conference, p. 125–132. Auckland, New Zealand.

Mah, R. A., D. M. Ward, L. Baresi, and T. Glass. 1977. Biogenesis of methane. *Annu. Rev. Microbiol.* **31:**309–341.

Manning, M. R., D. C. Lowe, W. H. Melhuish, R. J. Sparks, G. Wallace, C. A. M. Brenninkmeijer, and R. C. McGill. 1990. The use of radiocarbon measurements in atmospheric studies. *Radiocarbon* **32:**37–58.

Martens, C. S., N. E. Blair, C. D. Green, and D. J. Des Marais. 1986. Seasonal variations in the stable carbon isotopic signature of biogenic methane in a coastal sediment. *Science* **233:**1300–1303.

Matthews, E., and I. Fung. 1987. Methane emission from natural wetlands: global distribution, area, and environmental characteristics of sources. *Global Biogeochem. Cycles* **1:**61–86.

Miller, T. L., and M. J. Wolin. 1986. Methanogens in human and animal intestinal tracts. *Syst. Appl. Microbiol.* **7:**223–229.

Oona, S., and E. S. Deevey. 1960. Carbon 13 in lake waters and its possible bearing on paleolimnology. *Am. J. Sci.* **258A:**253–272.

Oremland, R. S., L. M. Marsh, and S. Polcin. 1982. Methane production and simultaneous sulfate reduction in anoxic saltmarsh sediments. *Nature* (London) **296:**143–145.

Oremland, R. S., and S. Polcin. 1982. Methanogenesis and sulfate reduction: competitive and noncompetitive substrates in estuarine sediments. *Appl. Environ. Microbiol.* **44:**1270–1276.

Ourisson, G., P. Albrecht, and M. Rohmer. 1984. The microbial origin of fossil fuels. *Sci. Am.* **251:**44–51.

Ovsyannikov, V. M., and V. S. Lebedev. 1967. Isotopic composition of carbon in gases of biogenic origin (English translation). *Geochem. Int.* **4:**453–458.

Pearman, G. I., D. Etheridge, F. de Silva, and P. J. Fraser. 1986. Evidence of changing concentrations of atmospheric CO_2, N_2O, and CH_4 from air bubbles in Antarctic ice. *Nature* (London) **320:**248–250.

Quay, P. D., S. L. King, J. M. Lansdown, and D. O. Wilbur. 1988. Isotopic composition of methane released from wetlands: implications for the increase in atmospheric methane. *Global Biogeochem. Cycles* **2:**385–397.

Ramanathan, V., L. Callis, R. Cess, J. Hansen, I. Isaksen, W. Kuhn, A. Lacis, F. Luther, J. Mahlman, R. Reck, and M. Schlesinger. 1987. Climate-chemical interactions and effects of changing atmospheric trace gases. *Rev. Geophys.* **25:**1441–1482.

Ramanathan, V., R. J. Cicerone, H. B. Singh, and J. T. Kiehl. 1985. Trace gas trends and their potential role in climate change. *J. Geophys. Res.* **90:**5547–5566.

Raskin, I., and H. Kende. 1985. Mechanism of aeration in rice. *Science* **228:**327–329.

Rasmussen, R. A., and M. A. K. Khalil. 1981. Atmospheric methane (CH_4): trends and seasonal cycles. *J. Geophys. Res.* **86:**9826–9832.

Rasmussen, R. A., and M. A. K. Khalil. 1983. Global production of methane by termites. *Nature* (London) **301:**700–702.

Rasmussen, R. A., and M. A. K. Khalil. 1984. Atmospheric methane in the recent and ancient atmospheres: concentrations, trends, and interhemispheric gradient. *J. Geophys. Res.* **89:**11599–11605.

Raynaud, D., J. Chappellaz, J. M. Barnola, Y. S. Korotkevich, and C. Lorius. 1988. Climatic and CH_4 cycle implications of glacial-interglacial CH_4 change in the Vostok ice core. *Nature* (London) **333:**655–657.

Rice, D., and G. Claypool. 1981. Generation, accumulation and resource potential of biogenic gas. *Am. Assoc. Petrol. Geol. Bull.* **65:**5–25.

Rowland, F. S., N. R. P. Harris, and D. R. Blake. 1990. Methane in cities. *Nature* (London) **347:**432–433.

Rowland, F. S., and M. J. Molina. 1975. Chlorofluoromethanes in the environment. *Rev. Geophys. Space Phys.* **13:**1–35.

Rust, F. E. 1981. $\delta(^{13}C/^{12}C)$ of ruminant methane and its relationship to atmospheric methane. *Science* **211:**1044–1046.

Rust, F., and C. M. Stevens. 1980. Carbon kinetic isotope effect in the oxidation of methane by hydroxyl. *Int. J. Chem. Kinet.* **12:**371–377.

Sackett, W. M., and T. R. Barber. 1988. Fossil carbon sources of atmospheric methane. *Nature* (London) **334:**201.

Schneider, S. H. 1989. The changing climate. *Sci. Am.* **260:**70–79.

Schoell, M. 1980. The hydrogen and carbon isotopic composition of methane from natural gases of various origins. *Geochim. Cosmochim. Acta* **44:**649–661.

Sebacher, D. I., R. C. Harriss, and K. B. Bartlett. 1985. Methane emissions to the atmosphere through aquatic plants. *J. Environ. Qual.* **14:**40–46.

Sebacher, D. I., R. C. Harriss, K. B. Bartlett, S. M. Sebacher, and S. S. Grice. 1986. Atmospheric methane sources: Alaskan tundra bogs, an alpine fen, and a subarctic boreal marsh. *Tellus* **38B:**1–10.

Seiler, W. 1984. Contribution of biological processes to the global budget of CH_4 in the atmosphere, p. 468–477. *In* M. J. Klug and C. A. Reddy (ed.), *Current Perspectives in Microbial Ecology*. American Society for Microbiology, Washington, D.C.

Seiler, W., R. Conrad, and D. Scharffe. 1984a. Field studies of methane emission from termite nests into the atmosphere and measurements of methane uptake by tropical soils. *J. Atmos. Chem.* **1**:171–186.

Seiler, W., A. Holzapfel-Pschorn, and R. Scharffe. 1984b. Methane emission from rice paddies. *J. Atmos. Sci.* **1**:241–268.

Senum, G. I., and J. S. Gaffney. 1985. A reexamination of the tropospheric methane cycle: geophysical implications, p. 61–69. *In* E. T. Sundquist and W. S. Broecker (ed.), *The Carbon Cycle and Atmospheric CO_2: Natural Variations Archean to Present*. Geophysical Monograph Series, vol. 32. American Geophysical Union, Washington, D.C.

Silverman, S. R. 1971. Influence of petroleum origin and transformation on its distribution and redistribution in sedimentary rocks, p. 47–54. *In Proceedings of the 8th World Petroleum Congress*, vol. 2. Centre de Documentation de l'Industrie Chimique et Petroliere, Bucarest, Rumania.

Stauffer, B., E. Lochbronner, H. Oeschger, and J. Schwander. 1988. Methane concentration in the glacial atmosphere was only half that of the preindustrial Holocene. *Nature* (London) **332**:812–813.

Steele, L. P., P. J. Fraser, R. A. Rasmussen, M. A. K. Khalil, T. J. Conway, A. J. Crawford, R. H. Gammon, K. A. Masarie, and K. W. Thoning. 1987. The global distribution of methane in the troposphere. *J. Atmos. Chem.* **5**:125–171.

Steudler, P. A., R. D. Bowden, J. M. Melillo, and J. D. Aber. 1989. Influence of nitrogen fertilization on methane uptake in temperate forest soils. *Nature* (London) **341**:314–316.

Stevens, C. M. 1988. Atmospheric methane. *Chem. Geol.* **71**:11–21.

Stevens, C. M., and A. Engelkemeir. 1988. Stable carbon isotopic composition of methane from some natural and anthropogenic sources. *J. Geophys. Res.* **93**:725–733.

Stevens, C. M., and F. E. Rust. 1982. The carbon isotopic composition of atmospheric methane. *J. Geophys. Res.* **87**:4879–4882.

Taylor, J. A., G. Brasseur, P. Zimmerman, and R. Cicerone. 1991. A study of the sources and sinks of methane and methyl chloroform using a global 3-d Lagrangian tropospheric tracer transport model. *J. Geophys. Res.* **96**:3013–3044.

Thompson, A. M., and R. J. Cicerone. 1986. Possible perturbations to atmospheric CO, CH_4, and OH. *J. Geophys. Res.* **91**:10853–10864.

Tyler, S. C. 1983. Chlorinated hydrocarbons in the troposphere. Ph.D. thesis. University of California at Irvine, Irvine.

Tyler, S. C. 1986. Stable carbon isotope ratios in atmospheric methane and some of its sources. *J. Geophys. Res.* **91**:13232–13238.

Tyler, S. C. 1989a. $^{13}C/^{12}C$ ratios in atmospheric methane and some of its sources, p. 395–409. *In* P. W. Rundel, J. R. Ehleringer, and A. K. Nagy (ed.), *Stable Isotopes in Ecological Research*. Ecology Studies, vol. 68. Springer-Verlag, New York.

Tyler, S. C. 1989b. Some recent studies of atmospheric methane using stable isotopes of carbon. Presented at Isotope Ratio Mass Spectrometry Users' Forum, Geological Society of America National Meeting, St. Louis, Mo.

Tyler, S. C., D. C. Lowe, E. Dlugokencky, P. R. Zimmerman, and R. J. Cicerone. 1990. Methane and carbon monoxide emissions from asphalt pavement: measurements and estimates of their importance to global budgets. *J. Geophys. Res.* **95**:14007–14014.

Tyler, S. C., P. R. Zimmerman, C. Cumberbatch, J. P. Greenberg, C. Westberg, and J. P. E. C. Darlington. 1988. Measurements and interpretations of $\delta^{13}C$ of methane from termites, rice paddies, and wetlands in Kenya. *Global Biogeochem. Cycles* **2**:349–355.

Wahlen, M., B. Deck, R. Henry, N. Tanaka, A. Shemesh, R. Fairbanks, W. Broecker, H. Weyer, B. Marino, and J. Logan. 1989a. Profiles of $\delta^{13}C$ and δD of CH_4 from the lower stratosphere. *EOS Trans. Am. Geophys. Union* **70**:1017.

Wahlen, M., N. Tanaka, R. Henry, B. Deck, J. Zeglen, J. S. Vogel, J. Southon, A. Shemesh, R. Fairbanks, and W. Broecker. 1989b. Carbon-14 in methane sources and in atmospheric methane: the contribution from fossil carbon. *Science* **245**:286–290.

Wahlen, M., N. Tanaka, R. Henry, T. Yoshinari, R. G. Fairbanks, A. Shemesh, and W. S. Broecker. 1987. ^{13}C, D, and ^{14}C in methane. *EOS Trans. Am. Geophys. Union* **68**:1220.

Welhan, J. A., and H. Craig. 1979. Methane and hydrogen in East Pacific rise hydrothermal fluids. *Geophys. Res. Lett.* **6**:829–831.

Whiticar, M. J., E. Faber, and M. Schoell. 1986. Biogenic methane formation in marine and freshwater environments: CO_2 reduction vs. acetate fermentation-isotope evidence. *Geochim. Cosmochim. Acta* **50**:693–709.

Zimmerman, P. R., and J. P. Greenberg. 1983. Termites and methane. *Nature* (London) **302**:354–355.

Zimmerman, P. R., J. P. Greenberg, S. O. Wandiga, and P. J. Crutzen. 1982. Termites: a potentially large source of atmospheric methane, carbon dioxide and molecular hydrogen. *Science* **218**:563–565.

Chapter 3

Diversity and Physiology of Methanogens

W. Jack Jones

Methanogens represent a large and diverse group of strictly anaerobic bacteria which obligately produce methane as an end product of their energy metabolism. To date, more than 50 distinct species of methane-producing bacteria have been isolated from diverse anaerobic habitats, but primarily from those where the supply of oxygen is restricted and where degradable organic carbon is present. Among the more common methanogenic habitats are freshwater and marine sediments, the intestinal tract of humans and animals, flooded soils, and anaerobic digestors. However, methanogens have been isolated from unusual anaerobic environments such as geothermal vents and springs (Huber et al., 1982; Jones et al., 1983a; Stetter et al., 1981), decaying wetwood of living trees (Zeikus and Henning, 1975), and large dental caries (Belay et al., 1988). In most anaerobic habitats, the growth and survival of methanogenic bacteria are directly dependent on the metabolic activities of diverse eubacterial and eucaryotic microbes for supply of methanogenic substrates and/or for provision of anoxia and reducing conditions.

This report is a general review of the diversity of methanogenic bacteria with an emphasis on their physiology, biochemistry, and taxonomic status.

METHANOGENS AS ARCHAEBACTERIA

In the late 1970s, Carl Woese and associates proposed a classification scheme that placed all living organisms into three major taxonomic groups: eucaryotes, eubacteria (true bacteria), and archaebacteria. This scheme was based primarily on homologies of partial sequences of 16S rRNAs of diverse procaryotic organisms and the corresponding rRNAs of eucaryotes (Balch et al., 1979; Woese and Wolfe, 1985; Woese et al., 1978). The results revealed the following: the eucaryotic organisms remained as a coherent group, but analysis of the diverse procaryotes revealed two major divisions, (i) the eubacteria, consisting of the traditional bacterial groups such as the photosynthetic bacteria, spirochetes, actinomycetes, coliforms, and sporeformers (to name a few), and (ii) the archaebacteria, which consisted of three major bacterial groups. Together with the methanogens, the extremely halophilic bacteria and the extremely thermophilic, sulfur-respiring bacteria comprised the

W. Jack Jones • Environmental Research Laboratory, Athens, U.S. Environmental Protection Agency, College Station Road, Athens, Georgia 30613-7799.

Table 1. Substrates and energetics of methane production

Reactions		$\Delta G_0'$ (kJ/mol of methane)[a]
Hydrogenotrophic reactions		
$4 H_2 + CO_2$ ⟶	$CH_4 + 2H_2O$	− 135.6
4 Formate ⟶	$CH_4 + 3CO_2 + 2H_2O$	− 130.1
4 (2-propanol) + CO_2 ⟶	CH_4 + 4 acetone + $2H_2O$	− 36.5
Aceticlastic reaction		
Acetate ⟶	$CH_4 + CO_2$	− 31.0
Disproportionation reactions		
4 Methanol ⟶	$3CH_4 + CO_2 + 2H_2O$	− 104.9
4 Methylamine + $3H_2O$ ⟶	$3CH_4 + CO_2 + 4NH_4^+$	− 75.0
2 Dimethyl sulfide + $2H_2O$ ⟶	$3CH_4 + CO_2 + H_2S$	− 73.8

[a] Data from Thauer et al. (1977).

archaebacteria. This revolutionary scheme of classification, based on comparisons of nucleic acid (rRNA) sequences, has had a major impact on the taxonomic organization of the microbial world.

Aside from the rRNA evidence, methanogens and other archaebacteria share a wide array of unique chemical and structural features which distinguish them not only from eucaryotes, but also from the eubacteria. For instance, methanogens possess a unique mode of energy metabolism (via methane production), synthesize several novel coenzymes, contain distinctive cell envelope and lipid structures, and contain distinctive tRNAs. Methanogens and archaebacteria also share characteristics with both eucaryotes and eubacteria, including aspects of nucleic acid and genome organization, as well as distinct features of transcriptional and translational proteins (Berghofer et al., 1988; Brown et al., 1989; Kopke and Wittmann-Liebold, 1989). As an example, archaebacteria are similar to eucaryotes in that both systems are relatively insensitive to specific antibiotics and chemicals which typically affect procaryotic protein synthesis (Bock and Kandler, 1985). As a group, however, methanogens are unique among the archaebacteria because they are not restricted to "extreme" environments of high salinity or high temperature as are the extreme halophiles and the extremely thermophilic, sulfur-respiring archaebacteria. In fact, methanogens are numerous in a variety of moderate and extreme habitats of differing temperature, salinity, and pH.

METHANOGENIC SUBSTRATES

Pure cultures of methanogenic bacteria are known to utilize only a narrow array of relatively simple substrates for growth and methane production (Table 1). The most common and almost universal mode of methane production is via the H_2-mediated reduction of CO_2, although many methanogens may utilize formate and, less commonly, alcohols or CO as electron donor. In most organic-rich methanogenic habitats, only about one-third of the total methane formed is derived from the reduction of CO_2. However, this methanogenic reaction is important for maintenance of low levels of H_2 and formate which directly affect the syntrophic metabolism of organic compounds (primarily organic acids) in many anaerobic environments (Wolin and Miller, 1987). In specific ecosystems, such as the bovine rumen and in geothermal vents and springs, the H_2-mediated reduction of CO_2 is the primary source of biogenic methane, and aceticlastic methanogenesis is apparently not significant.

The major source of methane in most environments active in organic decomposition is acetate. In anaerobic digestors and freshwater sediments, approximately two-thirds of the methane produced is derived from the methyl carbon of acetate. Interestingly, only a few of the numerous methanogenic isolates can metabolize acetate. The primary "aceticlastic" methanogens are species of *Methanosarcina* and *Methanosaeta* ("*Methanothrix*"). Because of the relatively poor energetics associated with the anaerobic catabolism of acetate, the growth of aceticlastic methanogens is slow (approximately 24 h doubling time) compared with the growth of hydrogenophilic methanogens (1 to 4 h doubling time).

Methane may also be derived from the reduction of the methyl group of a variety of one-carbon compounds, including methanol, dimethyl sulfide, and mono-, di-, and trimethylamines (Table 1). In these instances, methane is usually formed by a disproportionation reaction in which some of the substrate is oxidized to generate reducing equivalents for subsequent substrate (methyl group) reduction. As with acetate catabolism, this mode of methanogenesis is found only in a few groups of methanogens but is an important source of methane in specific habitats. For example, methane production in marine sediments may occur not only from the H_2-mediated reduction of CO_2 and/or acetate catabolism, but also from the metabolism of trimethylamines which are derived from choline and glycine betaine. Glycine betaine and other osmoprotectants are produced by marine plants and bacteria to maintain osmotic balance of their cytoplasm with their environment. In other anaerobic environments, methionine and the osmoregulant dimethylsulfoniopropionate may be metabolized to dimethyl sulfide, which is a substrate for some methanogens (Kiene et al., 1986). Another non-disproportionation methanogenic reaction, the direct reduction of methanol to methane by H_2, was discovered in a methanogen isolated from the human bowel (Miller and Wolin, 1985). In the bowel, it is presumed that methanol is formed from the anaerobic transformation of methoxy groups of pectin and is subsequently reduced to methane by H_2.

Finally, it has been recently reported that secondary alcohols, including 2-propanol and 2-butanol, as well as the primary alcohols ethanol, 1-propanol, and 1-butanol, are utilized by some methanogens (Widdel, 1986; Zellner and Winter, 1987). In each instance, the alcohols are only partially oxidized and they serve as electron donors for CO_2 reduction to methane. The extent of methanogenesis from primary and secondary alcohols in natural environments is relatively unknown.

ECOLOGICAL SIGNIFICANCE/RELATION TO OXYGEN

As previously stated, methanogens have a limited substrate range and they compete directly with other microbial groups in most natural habitats. In the absence of electron acceptors such as O_2, nitrate, sulfate, and Fe^{3+}, methanogens usually proliferate in organic-rich anaerobic environments and little competition exists for the major methanogenic substrates ($H_2 + CO_2$, acetate). These conditions are more common in habitats such as anaerobic digestors, freshwater sediments, flooded soils, and the intestinal tract of animals. In general, any habitat which is devoid of oxygen or in which oxygen diffusion is limited will support methanogenesis. Even in rather dilute environments, such as groundwater supplies, which may receive organic contamination and in which alternative electron acceptors are absent or in low concentration, methanogenesis may occur. This indicates that resident aerobic microbial populations are capable of maintaining oxygen levels sufficiently low for the development of the oxygen-sensitive methanogens, even in habitats of low organic input.

The presence of alternative electron acceptors such as sulfate, nitrate, and Fe^{3+} in anoxic habitats is known to inhibit methanogenesis. In marine habitats where the concentration of sulfate is high, sulfate-reducing bacteria will outcompete methanogens for available substrates (H_2, acetate) and hydrogen sulfide production will predominate over methanogenesis. In this instance, organic carbon is oxidized to CO_2 and sulfate is reduced to hydrogen sulfide. With the exception of methylamines and possibly dimethyl sulfide, the major methanogenic substrates, including acetate, formate, alcohols, and $H_2 + CO_2$, also serve as substrates for sulfate-reducing bacteria. Thus, methane production in sulfate-rich environments is limited by substrate availability. The ability of sulfate-reducing bacteria to outcompete methanogens for these substrates is likely due to the more positive reduction potential of the electron acceptor (sulfate) than that of CO_2 (the methanogenic electron acceptor), as well as the increased affinity of sulfate-reducing bacteria for the competitive substrates (Kristjansson et al., 1982). In anaerobic zones or microenvironments where sulfate becomes limiting (either by consumption of sulfate or limitation by diffusion), methanogens may assume the role of terminal electron acceptor, and organic decomposition reactions will proceed via methane formation. The competitive mechanism (reaction) of acetate catabolism in natural environments can be readily determined by following the labeling pattern of gaseous products resulting from the metabolism of exogenously added $^{14}CH_3COOH$. The production of $^{14}CO_2$ suggests complete acetate oxidation by sulfate reducers, while the production of $^{14}CH_4$ indicates methanogenesis from acetate. This method provides a reliable estimate of aceticlastic methanogenic activity within an ecosystem (Winfrey and Zeikus, 1979).

In addition to nitrate, sulfate, and Fe^{3+} (which may also serve as a competitive electron sink), oxygen is a known inhibitor of methanogenesis. In fact, even traces of oxygen inhibit the growth of methanogenic bacteria, and strict anaerobic procedures are required for successful cultivation of pure cultures (Balch and Wolfe, 1976). In organic-rich natural environments, the physical restriction of oxygen diffusion (by overlying water or other barriers) and the rapid consumption of the available O_2 by the resident aerobic microflora usually provide conditions which allow methanogenesis to occur. The extreme sensitivity of methanogenic bacteria to oxygen is well documented and is likely due to the O_2 lability of essential enzymes involved in the methanogenic pathway. Several investigators have demonstrated the inhibitory nature of O_2 on the growth and survival of pure methanogen cultures as well as the O_2 sensitivity of specific methanogen enzymes (Zhilina, 1972; Xing and Whitman, 1987). In contrast, one report demonstrated that oxygen toxicity in methanogens was significantly reduced when growth substrates were absent, indicating that methanogens were more susceptible to oxygen during active metabolism (Kiener and Leisinger, 1983).

A more complete discussion of methanogenic ecosystems, the importance of interspecies H_2 transfer, and the ecology of methanogenesis is presented elsewhere in this volume.

CLASSIFICATION OF METHANOGENS

The taxonomic classification of methane-producing bacteria has undergone drastic reorganization since the 8th edition of *Bergey's Manual of Determinative Bacteriology* was published in 1974. At that time, 10 species of methanogens were identified, and their classification was based primarily on cell morphology and only few physiological characteristics. Because of their restricted catabolic activity and lack of distinguishing phenotypic characteristics, methanogens are difficult to characterize without discrete phylogenetic data. In 1979, Balch et al. presented a reorganization of the taxonomy of 16 isolates of methanogenic bacteria, based primarily on 16S rRNA catalogs but coupled with phenotypic and genotypic data. The use of phylogenetic methods, such as rRNA analysis and nucleic acid hybridization, coupled with physiological, immunological, and biochemical characteristics, has proven to be indispensable for the description and classification of the 50 or more species of methanogens described to date.

Additional analyses have proven to be useful for classification of methanogens, including analysis of lipid constituents (Koga et al., 1987; Mori et al., 1988; Jones et al., 1987), profiles of soluble cell proteins, molecular weight distribution of specific enzyme subunits (Berghofer et al., 1988; Rouviere and Wolfe, 1987; Thomm et al., 1986), antigenic relationships (Conway de Macario et al., 1981), cell wall composition (Kandler and Konig, 1985; Konig, 1988), and the presence and distribution of polyamines (Kneifel et al., 1986). This information has been used separately and in conjunction with phylogenetic data for assessment of methanogen taxonomic status at the family, genus, and species level of classification.

The phylogenetic data currently support the taxonomic organization of methanogenic bacteria into three major groups (orders), comprising six families (Table 2). Nineteen genera, containing more than 50 species, have been described to date. The following information describes the major characteristics of the six distinct methanogen families. A more complete description of the individual methanogen species has been reported (Jones et al., 1987) and was recently updated in a review by Whitman et al. (1991).

Table 2. Summary of major taxonomic groups of the methanogenic archaebacteria[a]

Taxa	mol% G + C (range)
Order 1. *Methanobacteriales*	
Family 1. *Methanobacteriaceae*	23–61
Genera: *Methanobacterium, Methanobrevibacter, Methanosphaera*	
Family 2. *Methanothermaceae*	33–34
Genus: *Methanothermus*	
Order 2. *Methanococcales*	
Family 1. *Methanococcaceae*	29–34
Genus: *Methanococcus*	
Order 3. *Methanomicrobiales*	
Family 1. *Methanomicrobiaceae*	39–61
Genera: *Methanoculleus, Methanogenium, Methanolacinia, Methanoplanus, Methanospirillum, Methanomicrobium*	
Family 2. *Methanocorpusculaceae*	48–52
Genus: *Methanocorpusculum*	
Family 3. *Methanosarcinaceae*	36–52
Genera: *Methanosarcina, Methanosaeta, Methanococcoides, Methanolobus,* "halophilic methanogens"	

[a] Some genera listed are not yet formally recognized.

The order *Methanobacteriales* consists of two closely related families (*Methanobacteriaceae* and *Methanothermus*) comprising four genera. With the exception of the genus *Methanosphaera*, all species are rod shaped, stain Gram positive, contain pseudomurein as the major cell wall constituent, are nonmotile, and utilize H_2 plus CO_2 and possibly formate and secondary alcohols as substrates for methane production. (Species of *Methanosphaera* are coccoid-shaped cells and grow only via the H_2-mediated reduction of methanol.) Although moderate and thermophilic species of the *Methanobacteriales* have been described (Blotevogel et al., 1985; Whitman et al., 1991), most are mesophilic and grow optimally near neutral pH. Isolates have been obtained from freshwater soils and sediments, sewage digestors, the bovine rumen, and unusual habitats such as hot springs and decaying wetwood of living trees. Species of the genus *Methanobacterium* are often isolated from anaerobic digestors, while species of the genus *Methanobrevibacter* are characteristic inhabitants of the gastrointestinal tract of animals and humans (Miller and Wolin, 1985, 1986). Organic compounds, such as vitamins, amino acids, acetate, or yeast extract, are either required for or stimulate growth of most species of the *Methanobacteriales*. A summary of characteristics of representative members of the order *Methanobacteriales* is presented in Table 3.

Table 3. Characteristics of representative members of the methanogen order *Methanobacteriales*[a]

Taxa	Morphology	Substrates	Optimal temp (°C)	pH	% G + C	Nutrition	Habitats
Family: *Methanobacteriaceae*							
Genus 1: *Methanobacterium*							
M. formicicum	Rod	H_2, formate, sec. alcohols	37	7.0	40.7	Acet stimulated	Digestors, sediments
M. alcaliphilum	Rod	H_2	37	8–9	57	Peptone or YE required	Alkaline lake sediment
M. thermoautotrophicum	Rod	H_2	65–70	7.5	49.6	Autotroph	Sewage sludge
Genus 2: *Methanobrevibacter*							
M. ruminantium	Rod	H_2, formate	37–41	6–8	30.6	Acet, aa, CoM required	Bovine rumen
M. arboriphilicus	Rod	H_2	30–37	7.5	27–31	Autotroph	Wetwood of trees
Genus 3: *Methanosphaera*							
M. stadtmaniae	Coccus	H_2 + Me	36–40	7.0	25.8	Acet, vit, aa, CO_2 required	Human colon
Family: *Methanothermaceae*							
Genus 1: *Methanothermus*							
M. fervidus	Rod	H_2	83	6.5	33	YE required	Thermal spring

[a] Data from Bryant and Boone, 1987; Miller, 1989; Miller and Wolin, 1985; Stetter et al., 1981; Zeikus and Henning, 1975; Zeikus and Wolfe, 1972; Worakit et al., 1986; and Jones et al., 1987. Abbreviations: Acet, acetate; Me, methanol; sec., secondary; YE, yeast extract; vit, vitamins; aa, amino acids.

The second order of methanogens (*Methanococcales*) consists of one family (*Methanococcaceae*) with only one genus. All species of *Methanococcus* described to date are irregular, osmotically fragile, motile cocci (1 to 3 μm) isolated solely from marine habitats (primarily sediments). All species utilize H_2 and usually formate as electron donor for CO_2 reduction to methane. Cell walls composed of protein subunits are characteristic of this order, and most species grow rapidly and autotrophically at neutral pH. *Methanococcus jannaschii, Methanococcus thermolithotrophicus*, and several unnamed isolates are thermophilic species obtained from geothermal vent habitats and are phylogenetically distinct from the more prevalent mesophilic methanococci (Jones et al., 1983a; Huber et al., 1982; Jones et al., 1989). *Methanococcus maripaludis* and *Methanococcus voltae* were reported to be the most predominant methanogens from estuarine and salt marsh sediments (Jones et al., 1983b; Whitman et al., 1986). Most species of the *Methanococcaceae* have a nutritional requirement for selenium and marine salts (NaCl, Mg^{2+}). Further, some species have been shown to assimilate externally added organic carbon, such as acetate and amino acids (Whitman, 1989; Whitman et al., 1982). A summary of characteristics of representative members of the order *Methanococcales* is presented in Table 4.

The third order of methanogens (*Methanomicrobiales*) is very diverse. More than 35 distinct species are divided among three families. The family *Methanosarcinaceae* contains at least five genera, including ecologically very important species of the genera *Methanosarcina* and *Methanosaeta* ("*Methanothrix*"). These genera include species capable of metabolizing acetate and/or methylated one-carbon substrates but not formate. The importance of aceticlastic and hydrogenophilic methanogens for the complete bioconversion of organic compounds in various anaerobic environments cannot be overemphasized. Members of *Methanosarcina* and *Methanosaeta* have rather distinctive morphologies and can be tentatively identified on the basis of this criterion. *Methanosarcina* cells are usually irregular and coccoid shaped, arranged either singly but usually in packets, while *Methanosaeta* cells are described as fat, often filamentous sheathed rods. Aceticlastic isolates of *Methanosarcina* and *Methanosaeta* are frequently obtained from anaerobic sediments, sewage digestors, and other habitats receiving simple and complex organic compounds. Most species grow at neutral pH and are mesophilic, although thermophilic isolates of both genera have been described (Patel, 1984; Patel and Sprott, 1990; Zinder et al., 1987; Zinder and Mah, 1979). Further, aspects of nitrogen fixation in methanogens have been most thoroughly studied in species of *Methanosarcina* (Lobo and Zinder, 1988; Bomar et al., 1985; Murray and Zinder, 1984; Scherer, 1989).

Other related yet distinct species of the *Methanosarcinaceae* have been described. These grow only via the disproportionation of methylated substrates such as methanol and methylamines and do not utilize H_2, formate, or acetate as substrates. For example, the recently described isolate *Methanococcoides methylutens*, obtained from marine sediments below a kelp bed, utilizes only methanol and methylamines as substrates and not acetate or H_2 plus CO_2 (Sowers and Ferry, 1983, 1985). Methanogens in this group are identified as obligately methylotrophic methanococci and are phylogenetically related to the *Methanosarcinaceae*. All species are coccoid shaped, were isolated from saline habitats, and require seawater

Table 4. Characteristics of representative members of the methanogen order *Methanococcales*[a]

Taxa	Morphology	Substrates	Optimal temp (°C)	pH	% G + C	Nutrition	Habitats
Family: *Methanococcaceae*							
Genus 1: *Methanococcus*							
M. voltae	Coccus	H_2, formate	32–40	6–7	29.6	Acet, Leu, Ile required	Estuarine mud
M. maripaludis	Coccus	H_2, formate	38	7.0	33.4	Autotroph	Marine sediment
M. jannaschii	Coccus	H_2	85	6.0	31	Autotroph	Hydrothermal vent

[a] Data from Jones et al., 1983a, 1983b; Whitman, 1989; Whitman et al., 1982; and Jones et al., 1987. Abbreviations: Acet, acetate; Leu, leucine; Ile, isoleucine.

concentrations of salts; a few species are extreme halophiles (Zhilina, 1986). A summary of the major characteristics of representative members of the family *Methanosarcinaceae* is presented in Table 5.

The second family (*Methanomicrobiaceae*) within the *Methanomicrobiales* also contains species of extreme morphological diversity, including cocci (*Methanogenium, Methanoculleus*), rods (*Methanomicrobium, Methanolacinia*), spirilli (*Methanospirillum*), and discs (*Methanoplanus*). All species are unified by the ability to utilize H_2 plus CO_2 and, with one exception, formate as substrates for methane production. A few species utilize ethanol as well as secondary alcohols, and acetate catabolism is absent. In general, most isolates within this family are mesophilic, grow optimally near neutral pH, and have organic carbon nutritional requirements (acetate, peptone, yeast extract). Further, specific metal requirements, including nickel, molybdate, and tungstate, have been demonstrated in certain strains. Species of *Methanogenium* and *Methanolacinia* were isolated from marine sediments and either require or tolerate seawater levels of NaCl (Rivard et al., 1983; Romesser et al., 1979). Other species have been isolated from freshwater sediments, various types of anaerobic digestors, and the bovine rumen. Interestingly, one isolate (*Methanoplanus endosymbiosus*) was characterized as an endosymbiont of a marine ciliate (van Bruggen et al., 1986). The only spirillum-shaped methanogen described to date (*Methanospirillum hungatei*) is included in this family (Ferry et al., 1974; Ferry and Wolfe, 1977); it was isolated from sewage sludge and was recently identified with antibody probes as a major hydrogenophilic methanogen from a lab-scale (phenol-fed), methanogenic digestor (unpublished data).

The third family of the *Methanomicrobiales* consists of one genus, *Methanocorpusculum*. All species are characterized as small, irregular cocci which may form clumps or aggregates and which utilize H_2 plus CO_2 and formate as substrates for methane production. Some species are capable of secondary alcohol metabolism, and most have complex organic nutritional requirements. Isolates have been obtained from diverse anaerobic habitats, including anaerobic digestors, lake sediments, and wastewater treatment ponds (Zellner et al., 1987).

A summary of characteristics of the *Methanomicrobiaceae* and *Methanocorpusculaceae* is presented in Tables 5 and 6.

BIOCHEMISTRY OF METHANE FORMATION

Several recent reports have addressed the details of the biochemistry of methane formation, and the reader is referred to reviews by Jones et al. (1987) and Rouviere and Wolfe (1988). The following text summarizes the major reactions of methanogenesis from CO_2, acetate, and methylated substrates. Note that most coenzymes involved in the pathway are novel to methanogens and are key components of the methanogenic reactions (see Fig. 1).

In the initial steps of methanogenesis, CO_2 is activated and reduced to the formyl level of oxidation; in addition to reducing equivalents, nonstoichiometric amounts of ATP are required, as well as a coordinated reaction with the terminal step of methane formation. Electrons for the initial and subsequent reduction

Table 5. Characteristics of representative members of the methanogen families *Methanocorpusculaceae* and *Methanosarcinaceae* within the order *Methanomicrobiales*[a]

Taxa	Morphology	Substrates	Optimal temp (°C)	pH	% G + C	Nutrition	Habitats
Family: *Methanocorpusculaceae*							
Genus 1: *Methanocorpusculum*							
M. parvum	Coccus	H_2, formate, sec. alcohols	37	7–7.5	48.5	Acet, YE, tungstate required	Whey digestor
Family: *Methanosarcinaceae*							
Genus 1: *Methanosarcina*							
M. barkeri	Coccus, packets	H_2, Me, Acet, $MeNH_4$	35	7.0	40–43	Autotroph	Digestors, sludge, sediments
M. mazei	Coccus	H_2, Me, Acet, $MeNH_4$	30–40	6–7	42		Swamp, digestors
Genus 2: *Methanosaeta*							
M. concilii	Fat rods, filaments	Acetate	37	7–7.5	49–51	Vit	Digestors, sludge
M. thermoacetophila	Rods	Acetate	65	6.8	57	None	Thermal lake sediment
Genus 3: *Methanococcoides*							
M. methylutens	Coccus	Me, $MeNH_4$	30–35	7–7.5	42	Biotin, Mg^{2+}	Kelp bed sediment
Genus 4: *Methanolobus*							
M. tindarius	Coccus	Me, $MeNH_4$	25	6.5	46	Vit. stimulated; 0.5 M NaCl	Marine sediment
Genus 5: *Halomethanococcus*							
H. mahii	Coccus	Me, $MeNH_4$	35	7.5	48.5	1–2.5 M NaCl	Saline sediment

[a] Data from Huser et al., 1982; Konig and Stetter, 1982; Mah, 1980; Patel, 1984; Patel and Sprott, 1990; Paterek and Smith, 1985, 1988; Sowers and Ferry, 1983; Sowers et al., 1984; Zehnder et al., 1980; and Jones et al., 1987. Abbreviations: sec., secondary; Acet, acetate; Me, methanol; $MeNH_4$, methylamines; vit, vitamins; YE, yeast extract.

Table 6. Characteristics of representative members of the methanogen family *Methanomicrobiaceae* within the order *Methanomicrobiales*[a]

Taxa	Morphology	Substrates	Optimal temp (°C)	pH	% G + C	Nutrition	Habitats
Family 1: *Methanomicrobiaceae*							
Genus 1: *Methanogenium*							
M. cariaci	Coccus	H_2, formate	20–25	6.8–7.3	51.6	Acet, YE required	Marine sediment
Genus 2: *Methanoculleus*							
M. marisnigri	Coccus	H_2, formate, sec. alcohols	20–25	6.2–6.6	61.2	Peptones	Marine sediment
M. thermophilicum	Coccus	H_2, formate, sec. alcohols	55	7.0	59	Autotroph	Marine sediment, digestors
Genus 3: *Methanolacinia*							
M. paynteri	Cocco-rod	H_2, sec. alcohols	40	6.5–7.0	44.9	Acet required	Marine sediment
Genus 4: *Methanomicrobium*							
M. mobile	Rod	H_2, formate	40	6.1–6.9	48.8	Complex requirements (Acet, vit)	Bovine rumen
Genus 5: *Methanoplanus*							
M. limicola	Planes	H_2, formate	40	7.0	47.5	Acet required	Swamp sediment
M. endosymbiosus	Disc	H_2, formate	37	7.0	39	ND	Marine ciliate
Genus 6: *Methanospirillum*							
M. hungatei	Curved rod	H_2, formate, sec. alcohols	35–40	6.6–7.4	45	Acet?	Sewage sludge, digestors

[a] Data from Ferry and Wolfe, 1977; Ferry et al., 1974; Paynter and Hungate, 1968; Rivard et al., 1983; Romesser et al., 1979; Tanner and Wolfe, 1988; Wildgruber et al., 1982; and Jones et al., 1987. Abbreviations: Acet, acetate; vit, vitamins; sec., secondary; ND, not determined.

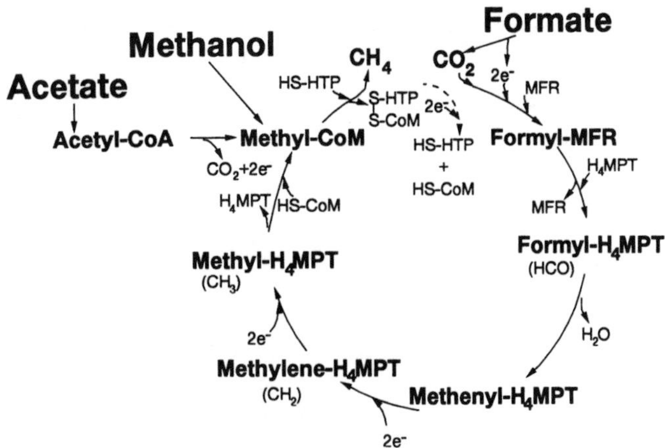

Figure 1. Generalized pathway for methane production from CO_2, acetate, methanol, and formate. Abbreviations: CoM, coenzyme M; H_4MPT, tetrahydromethanopterin; MFR, methanofuran; HS-HTP, 7-mercaptoheptanoylthreonine phosphate.

reactions are generated from H_2 (via a hydrogenase) or formate. The resulting formyl group of the initial reaction is bound by a novel methanogenic coenzyme, methanofuran, and is subsequently transferred to tetrahydromethanopterin (H_4MPT), another novel coenzyme which functions analogously to folates. The formyl group is dehydrated to methenyl-H_4MPT and sequentially reduced to methylene-H_4MPT and methyl-H_4MPT. In the final methanogenic reactions, the methyl group of methyl-H_4MPT is transferred by a methyltransferase to coenzyme M (CoM), another novel one-carbon carrier, where it is subsequently reduced to methane by the enzyme complex known as the methyl-S-CoM methylreductase. This complex consists of four protein components and three novel coenzymes, including CoM. The terminal reduction reaction results in the formation of methane and, as mentioned previously, is coupled to the activation of another molecule of CO_2.

Methane formation from acetate also proceeds through the methylreductase system. Initially, acetate is activated to acetyl coenzyme A; acetyl coenzyme A is then metabolized to form an enzyme-bound CO and a methyl group which is transferred to HS-CoM via H_4MPT. Electrons for the subsequent reduction of the CoM-bound methyl group are derived from the oxidation of the enzyme-bound CO.

Methanol is converted to methane either by direct reduction of the methyl group (via H_2) after transfer to HS-CoM via methyltransferases, or by a disproportionation reaction in which methanol is both oxidized and reduced. A cobamide moiety is involved as a methyl carrier during the methyltransferase reaction. Methylamines are also converted to methane by this reaction. In both cases, some of the methanol or methylamine is oxidized to generate reducing equivalents for subsequent reduction of the methyl group to methane by the methylreductase system. The stoichiometry of methanogenesis from methanol is 1 mol of methanol oxidized for 3 mol reduced. A generalized scheme of the methanogenic pathway is presented in Fig. 1.

SUMMARY

All methanogens are obligately methanogenic, and methane synthesis is required for growth. Substrates include H_2 plus CO_2, CO, formate, acetate, methanol, methylamines, dimethyl sulfide, and some alcohols (primarily secondary alcohols). Cells are obligately anaerobic, and growth is inhibited by traces of oxygen. Most species grow optimally at neutral pH and at mesophilic temperatures, although moderate and extreme thermophiles have been described. In natural environments, methanogenesis has been reported at neutral, acidic (peat bogs), and alkaline (soda lakes) pH and at temperatures ranging from psychrophilic to extremely thermophilic. The current classification system, based on physiological, biochemical, and genetic evidence, supports the organization of the more than 50 described methanogen species into three orders, six families, and 19 genera.

REFERENCES

Balch, W. E., G. E. Fox, L. J. Magrum, C. R. Woese, and R. S. Wolfe. 1979. Methanogens: reevaluation of a unique biological group. *Microbiol. Rev.* **43:**260–296.

Balch, W. E., and R. S. Wolfe. 1976. New approach to the cultivation of methanogenic bacteria: 2-mercaptoethanesulfonic acid (HS-CoM)-dependent growth of *Methanobacterium ruminantium* in a pressurized atmosphere. *Appl. Environ. Microbiol.* **32:**781–791.

Belay, N., R. Johnson, B. S. Rajagopal, E. Conway de Macario, and L. Daniels. 1988. Methanogenic bacteria from human dental plaque. *Appl. Environ. Microbiol.* **54:**600–603.

Berghofer, B., L. Krockel, C. Kotner, M. Truss, J. Schallenberg, and A. Klein. 1988. Relatedness of archaebacterial RNA polymerase core subunits to their eubacterial and eukaryotic equivalents. *Nucleic Acids Res.* **16:**8113–8128.

Blotevogel, K.-H., U. Fischer, M. Mocha, and S. Jannsen. 1985. *Methanobacterium thermoalcaliphilum* spec. nov., a new moderately alkaliphilic autotrophic methanogen. *Arch. Microbiol.* **142:**211–217.

Bock, A., and O. Kandler. 1985. Antibiotic sensitivity of archaebacteria, p. 525–544. *In* C. R. Woese and R. S. Wolfe (ed.), *The Bacteria*, vol. 8. Academic Press, Inc., New York.

Bomar, J., K. Knoll, and F. Widdel. 1985. Fixation of molecular nitrogen by *Methanosarcina barkeri*. *FEMS Microbiol. Ecol.* **31:**47–55.

Brown, J. W., C. J. Daniels, and J. N. Reeve. 1989. Gene structure, organization, and expression in archaebacteria. *Crit. Rev. Microbiol.* **16:**287–338.

Bryant, M. P., and D. R. Boone. 1987. Isolation and characterization of *Methanobacterium formicicum* MF. *Int. J. Syst. Bacteriol.* **37:**171.

Conway de Macario, E., M. J. Wolin, and A. J. L. Macario. 1981. Immunology of archaebacteria that produce methane gas. *Science* **214:**74–75.

Ferry, J. G., P. H. Smith, and R. S. Wolfe. 1974. *Methanospirillum*, a new genus of methanogenic bacteria, and characterization of *Methanospirillum hungatii* sp. nov. *Int. J. Syst. Bacteriol.* **24:**465–469.

Ferry, J. G., and R. S. Wolfe. 1977. Nutritional and biochemical characterization of *Methanospirillum hungatii*. *Appl. Environ. Microbiol.* **34:**371–376.

Huber, H., M. Thomm, H. Konig, G. Thies, and K. O. Stetter. 1982. *Methanococcus* *thermolithotrophicus*, a novel thermophilic lithotrophic methanogen. *Arch. Microbiol.* **132:**47–50.

Huser, B. A., K. Wuhrmann, and A. J. B. Zehnder. 1982. *Methanothrix soehngenii* gen. nov. sp. nov., a new acetotrophic non-hydrogen-oxidizing methane bacterium. *Arch. Microbiol.* **132:**1–9.

Jones, W. J., J. A. Leigh, F. Mayer, C. R. Woese, and R. S. Wolfe. 1983a. *Methanococcus jannaschii* sp. nov., an extremely thermophilic methanogen from a submarine hydrothermal vent. *Arch. Microbiol.* **136:**254–261.

Jones, W. J., D. P. Nagle, Jr., and W. B. Whitman. 1987. Methanogens and the diversity of archaebacteria. *Microbiol. Rev.* **51:**135–177.

Jones, W. J., M. J. B. Paynter, and R. Gupta. 1983b. Characterization of *Methanococcus maripaludis* sp. nov., a new methanogen isolated from salt marsh sediment. *Arch. Microbiol.* **135:**91–97.

Jones, W. J., C. E. Stugard, and H. W. Jannasch. 1989. Comparison of thermophilic methanogens from submarine hydrothermal vents. *Arch. Microbiol.* **151:**314–319.

Kandler, O., and H. Konig. 1985. Cell envelopes of archaebacteria, p. 413–457. *In* C. R. Woese and R. S. Wolfe (ed.), *The Bacteria*, vol. 8. Academic Press, Inc., New York.

Kiene, R. P., R. S. Oremland, A. Catena, L. G. Miller, and D. G. Capone. 1986. Metabolism of reduced methylated sulfur compounds in anaerobic sediments and by a pure culture of an estuarine methanogen. *Appl. Environ. Microbiol.* **52:**1037–1045.

Kiener, A., and T. Leisinger. 1983. Oxygen sensitivity of methanogenic bacteria. *Syst. Appl. Microbiol.* **4:**305–312.

Kneifel, H., K. O. Stetter, J. R. Andreesen, J. Wiegel, H. Koing, and S. M. Schoberth. 1986. Distribution of polyamines in representative species of archaebacteria. *Syst. Appl. Microbiol.* **7:**241–245.

Koga, Y., M. Ohga, M. Nishihara, and H. Morii. 1987. Distribution of a diphytanyl ether analog of phosphatidylserine and an ethanolamine-containing tetraether lipid in methanogenic bacteria. *Syst. Appl. Microbiol.* **9:**176–182.

Konig, H. 1988. Archaebacterial cell envelopes. *Can. J. Microbiol.* **34:**395–406.

Konig, H., and K. O. Stetter. 1982. Isolation and characterization of *Methanolobus tindarius* sp. nov., a coccoid methanogen growing only on methanol and methylamines. *Zentralbl. Bakteriol. Parasitenkd. Infektionskr. Hyg. Abt. 1 Orig. Reihe C* **3:**478–490.

Kopke, A. K. E., and B. Wittmann-Liebold. 1989. Comparative studies of ribosomal proteins and their genes from *Methanococcus vannielii* and other organisms. *Can. J. Microbiol.* **35:**11–20.

Kristjansson, J. K., P. Schonheit, and R. K. Thauer. 1982. Different Km values for hydrogen of methanogenic bacteria and sulfate reducing bacteria. *Arch. Microbiol.* **131:**278–282.

Lobo, A. L., and S. H. Zinder. 1988. Diazotrophy and nitrogenase activity in the archaebacterium *Methanosarcina barkeri* 227. *Appl. Environ. Microbiol.* **54:**1656–1661.

Mah, R. A. 1980. Isolation and characterization of *Methanococcus mazei*. *Curr. Microbiol.* **3:**321–326.

Miller, T. L. 1989. Genus II. *Methanobrevibacter*, p. 2178–2183. *In* J. T. Staley, M. P. Bryant, N. Pfennig, and J. G. Holt (ed.), *Bergey's Manual of Systematic Bacteriology*, vol. 3. Williams and Wilkins, Baltimore.

Miller, T. L., and M. J. Wolin. 1985. *Methanosphaera stadtmaniae* gen. nov., sp. nov.: a species that forms methane by reducing methanol with hydrogen. *Arch. Microbiol.* **141:**116–122.

Miller, T. L., and M. J. Wolin. 1986. Methanogens in human and animal intestinal tracts. *Syst. Appl. Microbiol.* **7:**223–229.

Mori, H., M. Nishihara, and Y. Koga. 1988. Composition of polar lipids of *Methanobrevibacter arboriphilicus* and structure determination of the signature phosphoglycolipid of *Methanobacteriaceae*. *Agric. Biol. Chem.* **52:**3149–3156.

Murray, P. A., and S. H. Zinder. 1984. Nitrogen fixation by a methanogenic archaebacterium. *Nature* (London) **312:**284–286.

Patel, G. B. 1984. Characterization and nutritional properties of *Methanothrix concilii* sp. nov., a mesophilic, aceticlastic methanogen. *Can. J. Microbiol.* **30:**1383–1396.

Patel, G. B., and G. D. Sprott. 1990. *Methanosaeta concilii* gen. nov., sp. nov. (*"Methanothrix concilii"*) and *Methanosaeta thermoacetophila* nom. rev., comb. nov. *Int. J. Syst. Bacteriol.* **40:**79–82.

Paterek, J. R., and P. H. Smith. 1985. Isolation and characterization of a halophilic methanogen from Great Salt Lake. *Appl. Environ. Microbiol.* **50:**877–881.

Paterek, J. R., and P. H. Smith. 1988. *Methanohalophilus mahii* gen. nov., sp. nov., a methylotrophic halophilic methanogen. *Int. J. Syst. Bacteriol.* **38:**122–123.

Paynter, M. J. B., and R. E. Hungate. 1968. Characterization of *Methanobacterium mobilis* sp. n., isolated from the bovine rumen. *J. Bacteriol.* **95:**1943–1951.

Rivard, C. J., J. M. Henson, M. V. Thomas, and P. H. Smith. 1983. Isolation and characterization of *Methanomicrobium paynteri* sp. nov., a mesophilic methanogen isolated from marine sediments. *Appl. Environ. Microbiol.* **46:**484–490.

Romesser, J. A., R. S. Wolfe, F. Mayer, E. Speiss, and A. Walther-Mauruschat. 1979. *Methanogenium*, a new genus of marine methanogenic bacteria, and characterization of *Methanogenium cariaci* sp. nov. and *Methanogenium marisnigri* sp. nov. *Arch. Microbiol.* **121:**147–153.

Rouviere, P. E., and R. S. Wolfe. 1987. Use of subunits of the methylreductase protein for taxonomy of methanogenic bacteria. *Arch. Microbiol.* **148:**253–259.

Rouviere, P. E., and R. S. Wolfe. 1988. Novel biochemistry of methanogenesis. *J. Biol. Chem.* **263:**7913–7916.

Scherer, P. A. 1989. Vanadium and molybdenum requirement for the fixation of molecular nitrogen by two *Methanosarcina* strains. *Arch. Microbiol.* **151:**44–48.

Sowers, K. R., and J. G. Ferry. 1983. Isolation and characterization of a methylotrophic marine methanogen, *Methanococcoides methylutens* gen. nov., sp. nov. *Appl. Environ. Microbiol.* **45:**684–690.

Sowers, K. R., and J. G. Ferry. 1985. Trace metal and vitamin requirements of *Methanococcoides methylutens* grown with trimethylamine. *Arch. Microbiol.* **142:**148–151.

Sowers, K. R., J. L. Johnson, and J. G. Ferry. 1984. Phylogenetic relationships among the methylotrophic methane-producing bacteria and emendation of the family *Methanosarcinaceae*. *Int. J. Syst. Bacteriol.* **34:**444–450.

Stetter, K. O., M. Thomm, J. Winter, G. Wildgruber, H. Huber, W. Zillig, D. Janecovic, H. Koing, P. Palm, and S. Wunderl. 1981. *Methanothermus fervidus*, sp. nov., a novel extremely thermophilic methanogen isolated from an Icelandic hot spring. *Zentralbl. Bakteriol. Parasitenkd. Infektionskr. Hyg. Abt. 1 Orig. Reihe C* **2:**166–178.

Tanner, R. S., and R. S. Wolfe. 1988. Nutrient requirements of *Methanomicrobium mobile*. *Appl. Environ. Microbiol.* **54:**625–628.

Thauer, R. K., K. Jungermann, and K. Decker. 1977. Energy conservation in chemotrophic anaerobic bacteria. *Bacteriol. Rev.* **41:**100–180.

Thomm, M., J. Madon, and K. O. Setter. 1986. DNA-dependent RNA polymerases of the three orders of methanogens. *Biol. Chem. Hoppe-Seyler* **367:**473–481.

van Bruggen, J. J. A., K. B. Zwart, J. G. F. Hermans, E. M. van Hove, C. K. Stumm, and G. D. Vogels. 1986. Isolation and characterization of *Methanoplanus endosymbiosus* sp. nov., an endosymbiont of the marine sapropelic ciliate *Metopus contortus* Quennerstedt. *Arch. Microbiol.* **144:**367–374.

Whitman, W. B. 1989. *Methanococcales*, p. 2185–2190. *In* J. T. Staley, M. P. Bryant, N. Pfennig, and J. G. Holt (ed.), *Bergey's Manual of Systematic Bacteriology*, vol. 3. Williams & Wilkins, Baltimore.

Whitman, W. B., E. Ankwanda, and R. S. Wolfe. 1982. Nutrition and carbon metabolism of *Methanococcus voltae*. *J. Bacteriol.* **149:**852–863.

Whitman, W. B., T. L. Bowen, and D. R. Boone. 1991. Methanogens. *In* A. Balows, H. G. Truper, M. Dworkin, W. Harder, and K. H. Schleifer (ed.), *The Procaryotes*. Springer-Verlag, New York.

Whitman, W. B., J. Shieh, S. Sohn, D. S. Caras, and U. Premachandran. 1986. Isolation and characterization of 22 mesophilic methanococci. *Syst. Appl. Microbiol.* **7:**235–240.

Widdel, F. 1986. Growth of methanogenic bacteria in pure culture with 2-propanol and other alcohols as hydrogen donors. *Appl. Environ. Microbiol.* **51:**1056–1062.

Wildgruber, G., M. Thomm, H. Loing, K. Ober, T. Ricchiuto, and K. O. Stetter. 1982. *Methanoplanus limicola*, a plate-shaped methanogen representing a noval family, the *Methanoplanaceae*. *Arch. Microbiol.* **132:**31–36.

Winfrey, M. R., and J. G. Zeikus. 1979. Microbial methanogenesis and acetate metabolism in a meromictic lake. *Appl. Environ. Microbiol.* **37:**213–221.

Woese, C. R., L. J. Magrum, and G. E. Fox. 1978. Archaebacteria. *J. Mol. Evol.* **11:**245–252.

Woese, C. R., and R. S. Wolfe (ed.). 1985. *The Bacteria*, vol. 8, *Archaebacteria*. Academic Press, Inc., New York.

Wolin, M. J., and T. L. Miller. 1987. Bioconversion of organic carbon to CH_4 and CO_2. *Geomicrob. J.* **5:**239–259.

Worakit, S., D. R. Boone, R. A. Mah, M.-E. Abdel-Samie, and M. M. El-Halwagi. 1986. *Methanobacterium alcaliphilum* sp. nov., an H_2-utilizing methanogen that grows at high pH values. *Int. J. Syst. Bacteriol.* **36:**380–382.

Xing, R. Y., and W. B. Whitman. 1987. Sulfometuron methyl-sensitive and -resistant acetolactate synthases of the archaebacteria *Methanococcus* spp. *J. Bacteriol.* **169:**4486–4492.

Zehnder, A. J. B., B. A. Huser, T. D. Brock, and K. Wuhrmann. 1980. Characterization of an acetate-decarboxylating, non-hydrogen-oxidizing methane bacterium. *Arch. Microbiol.* **124:** 1–11.

Zeikus, J. G., and D. L. Henning. 1975. *Methanobacterium arboriphilum* sp. nov. An obligate anaerobe isolated from wetwood of living trees. *Antonie van Leeuwenhoek* **41:**543–552.

Zeikus, J. G., and R. S. Wolfe. 1972. *Methanobacterium thermoautotrophicus* sp. n., an anaerobic, autotrophic, extreme thermophile. *J. Bacteriol.* **109:**707–713.

Zellner, G., C. Alten, E. Stackebrandt, E. Conway de Macario, and J. Winter. 1987. Isolation and characterization of *Methanocorpusculum parvum*, gen. nov., spec. nov., a new tungsten requiring, coccoid methanogen. *Arch. Microbiol.* **147:**13–20.

Zellner, G., and J. Winter. 1987. Secondary alcohols as hydrogen donors for CO_2-reduction by methanogens. *FEMS Microbiol. Lett.* **44:**323–328.

Zhilina, T. N. 1972. Death of *Methanosarcina* in the air. *Microbiology* **41:**980–981. (English translation.)

Zhilina, T. N. 1986. Methanogenic bacteria from hypersaline environments. *Syst. Appl. Microbiol.* **7:**216–222.

Zinder, S. H., T. Anguish, and A. L. Lobo. 1987. Isolation and characterization of a thermophilic acetotrophic strain of *Methanothrix*. *Arch. Microbiol.* **146:**315–322.

Zinder, S. H., and R. A. Mah. 1979. Isolation and characterization of a thermophilic strain of *Methanosarcina* unable to use H_2-CO_2 for methanogenesis. *Appl. Environ. Microbiol.* **38:**996–1008.

Chapter 4

Ecology of Methanogenesis

David R. Boone

The importance of methane-forming bacteria in the ecology of many natural environments and in the global cycling of carbon and methane has long been recognized. In methanogenic habitats, only a portion of the microbes can actually produce methane, and these methane-producing organisms (methanogens) can use only a very limited number of catabolic substrates. Thus, the conversion of most organic molecules to methane requires interactions between nonmethanogenic bacteria, which grow by using a wide variety of organic molecules and forming a smaller number of metabolic end products, and methanogens, which use those metabolic end products of the nonmethanogens to form methane.

Methanogens are important not only in global ecology, but also in our understandings of cell biology and the evolution of cells. Methanogens are members of the urkingdom *Archaebacteria*, which are different from "normal" bacteria (prokaryotes) and from eukaryotes (plants, animals, fungi, and protozoa) in a number of important ways. These three major cell lines have evolved along different pathways since a time early in the evolution of life on earth. Thus, the characteristics which methanogens share with prokaryotes and eukaryotes likely were developed early in the evolution of life on earth, and studies of methanogens' biochemistry and comparisons with bacteria and eukaryotes may help us to understand how early life evolved. During the past 15 years methanogens have become a focal point of studies in microbial physiology, genetics, biological membranes, and evolution.

BIOLOGY OF METHANOGENIC BACTERIA

Methanogens have a major impact on atmospheric hydrocarbons because they are the only known life forms which produce a hydrocarbon as a major catabolic product. Whereas some other microbes produce small amounts of methane as a by-product of their metabolism (Rimbault et al., 1988), methane is a requisite and major catabolic product for methanogens. Methanogens are extremely anaerobic bacteria whose substrates are limited to a few small molecules, but these molecules are supplied as the end products of the metabolic activities of other microbes. Most

David R. Boone • Environmental Science and Engineering, Oregon Graduate Institute of Science and Technology, 19600 N.W. von Neumann Drive, Beaverton, Oregon 97006-1999.

methanogens reduce CO_2 to methane with the electrons derived by oxidizing H_2 or formate, or sometimes alcohols. Other methanogens use methyl groups (acetate, methanol, trimethylamine, dimethyl sulfide) as substrate for methanogenesis. The methyl group is reduced to methane by using electrons generated either by disproportionation (oxidation of a stoichiometric portion of the substrate), by oxidation of the carboxyl group of acetate, or by oxidation of H_2 (as in the case of *Methanosphaera*).

Methanogens along with two other groups (extreme halophiles and sulfur-dependent thermophiles) make up the archaebacteria, a group of organisms which are evolutionarily distinct from true bacteria and eukaryotes (Jones et al., 1987; Woese, 1987). Archaebacteria are distinguished from bacteria and prokaryotes by a number of features, including cell membranes composed of isoprenoid hydrocarbons ether-linked to glycerol (Langworthy, 1985). The cell walls do not contain murein, although some have pseudomurein cell walls whose structure is analogous to murein (Kandler and König, 1985).

Catabolic substrates for pure cultures of methanogens are limited to a few small molecules, including acetate, formate, methanol, the methyl groups of amines, sulfides, and selenides, H_2 + CO_2, and some alcohols + CO_2. In environments where the major ultimate fate of organic matter is complete conversion to CH_4 and CO_2, acetate is the precursor to most of the methane. Acetate is usually degraded by the aceticlastic reaction, in which the methyl group is reduced to CH_4 with its substituent hydrogen atoms intact while the carboxyl group is oxidized to CO_2. Methyl groups may also be reduced without first being oxidized to the CO_2, but all of the other substrates listed above are oxidized, thereby providing electrons for reduction of CO_2 to CH_4.

BIOCHEMISTRY OF METHANE FORMATION

The biochemical pathways for microbial formation of methane (see recent reviews by Vogels et al. [1988] and Whitman [1985]) culminate in the reductive cleavage of a methyl group from the one-carbon carrier coenzyme M (CoM; 2-mercaptoethanesulfonate [Taylor and Wolfe, 1974]). The reactions leading to this final step vary depending on the substrates used, but the final reduction, which is apparently coupled to ATP synthesis, is conserved. In 1956, Barker proposed that when CO_2 is reduced to CH_4 by H_2, the one-carbon intermediates are bound to one or more carriers, and that substrates other than H_2-CO_2 feed into this pathway. Today, this hypothesis is believed to be essentially correct.

The biochemistry of the reduction of CO_2 to methane has been recently reviewed (DiMarco et al., 1990). The reduction of CO_2 begins with its condensation with methanofuran in a reaction which is stimulated by methyl-CoM, through an unknown mechanism of energy coupling to the final, exergonic step of the pathway (i.e., the release of methane from methyl-CoM) or the methyltransferase system. The formyl group is then transferred to tetrahydromethanopterin (DiMarco et al., 1990), which is dehydrated to methenyl-tetrahydromethanopterin and then reduced to methylene- and methyl-tetrahydromethanopterin by two 2-electron reductions analogous to reductions in folate biochemistry. The source of electrons for at least the first of these two reductions is coenzyme F_{420} (Cheeseman et al., 1972), a 2-electron carrier with $E_0' = -0.34$ to -0.35 V (Jacobson and Walsh, 1984; Pol et al., 1980). The methyl group of methyl-tetrahydromethanopterin is transferred to the mercapto group of CoM by a corrinoid-containing transferase (Poirot et al., 1987), and the terminal step in methanogenesis is the reductive cleavage of this methyl group to yield methane. This final reaction is catalyzed by a multiprotein, multicofactor complex (Gunsalus and Wolfe, 1978, 1980; Nagle and Wolfe, 1983).

Methanogens live in environments where their catabolic reactions yield very small amounts of energy, often less than enough to form one ATP from ADP. Some methanogens grow by splitting acetate to methane and CO_2, which yields -35.8 kJ/mol under standard conditions ($\Delta G_0'$). However, concentrations of acetate in methanogenic environments are much lower; the threshold concentration above which acetate is degraded by *Methanosaeta* (*"Methanothrix"*) is about 70 μM (Westermann et al., 1989). At the levels of CO_2 (30 kPa) and methane (2.5 kPa) in that experiment, the free-energy change for methanogenesis from acetate is -21.8 kJ/mol. The free-energy change for methanogenesis from H_2 (9 Pa) + CO_2 (30 kPa) to methane (22.3 kPa) in growing, propionate-enrichment cultures is -35.4 kJ/mol (Boone and Xun, 1987). Thus, the mechanism for coupling ATP formation to methanogenesis must allow the production of less than 1 mol of ATP per mol of methane formed.

The reduction of CO_2 to methane includes four 2-electron reductions. When catabolic substrates such as methyl compounds are used, methanogens make use of some or all of the biochemical mechanisms used for CO_2 reduction to methane. For instance, the methyl group of methanol is transferred to CoM by a corrinoid transferase, and methyl-CoM is reduced by the methyl-CoM methylreductase system described above. The source of the electrons for this reduction varies. For methanogens (*Methanosphaera* and several genera of the family *Methanosarcinaceae*) growing on H_2 plus methanol, electrons are generated from H_2 oxidation by hydrogenase. Most methanogens in the family *Methanosarcinaceae* can also grow by disproportionating methanol in the absence of H_2, and in this case, some of the methanol (approximately one-fourth) is oxidized to CO_2 to provide electrons for the reduction of methyl-CoM. The mechanism of this oxidation is not clear, but high concentrations of a special methanopterin called sarcinapterin suggest that this may be a C_1 carrier for this oxidation (van Beelen et al., 1984). The utilization of other methylated compounds such as methylated amines, sulfides, and perhaps selenides is accomplished by independent specific enzyme systems which may function in an analogous manner.

Acetate is the major precursor to methane in many environments. However, acetate can be catabolized by only two genera of methanogens, *Methanosarcina* and *Methanosaeta* (*"Methanothrix"*). Although the mechanism is not well understood, it is clear that the methyl group of acetate is transferred to CoM (making methyl-CoM), and that the carboxyl group is oxidized to CO_2, providing electrons for the reductive cleavage of methyl-CoM to methane. This is called the aceticlastic reaction. Soluble cell-free fractions of aceticlastic methanogens form methane from acetate much more slowly than in vivo rates, and only when H_2 is added (Krzycki and Zeikus, 1984). However, particulate cell-free systems form methane from acetate at rates comparable to whole cells in the absence of added H_2 (Baresi, 1984). In fact, H_2 inhibited aceticlastic methanogenesis by these particulate fractions (Baresi, 1984).

METHANOGENIC ECOSYSTEMS

The dependence of methanogens on their microbial partners in consortia is mainly a dependence for catabolic substrates. Some methanogens have requirements for growth factors which must also be supplied by nonmethanogens, but many methanogens can grow in mineral medium on H_2 + CO_2 or medium with the only organic compound added being the catabolic substrate. The most important interaction methanogens have with other microbes is that nonmethanogens release fermentation products which are the catabolic substrates for methanogens. In anaerobic environments lacking electron acceptors such as nitrate or sulfate, the catabolic substrates supplied to methanogens by other microbes are mainly acetate, formate, H_2, and CO_2.

Nonsaline Environments

Acetate is the major substrate for methanogenesis in many nonsaline environments, accounting for about two-thirds of the methane formed. Acetic acid is formed by many fermentative bacteria and then consumed by methanogens, so that acetate concentrations tend to remain low (less than 1 or 2 mM). This catabolic interaction benefits not only the methanogen, which derives its major carbon and energy source, but also the fermentative bacterium, because the removal of the catabolic product, acetic acid, benefits the energetics of its catabolic reactions and also prevents accumulation of acidic products which could inhibit growth by lowering the pH.

A less obvious interaction between methanogens and nonmethanogens in nonsaline environments has been termed "interspecies electron transfer," which is discussed in greater detail in a later section of this chapter. Interspecies electron transfer is the shuttling of reducing equivalents from nonmethanogens to methanogens via either H^+/H_2 or $HCO_3^-/HCOO^-$ couples. Nonmethanogens reduce protons to H_2 (or bicarbonate to formate), which can diffuse to a methanogen; the methanogen oxidizes H_2 back to protons (or formate to bicarbonate). The net result of this phenomenon is the transfer of reducing equivalents from reduced electron carriers of nonmethanogens to those of methanogens. The source of electrons in the nonmethanogens may be an intracellular electron carrier such as NADH (McInerney, 1986).

Saline Environments

Most saline environments contain sulfate, so sulfate reducers outcompete methanogens for most catabolic substrates. When sulfate is abundant, the sulfate-reducing bacteria grow by oxidizing acetate, formate, and H_2 and reducing sulfate to sulfide. These activities lower the concentrations of acetate, formate, and H_2 to levels lower than are utilized by the methanogens. However, sulfate reducers do not easily catabolize some methyl compounds such as trimethylamine, dimethyl sulfide, and perhaps methanol, so methanogens convert these substrates to methane. Because methanogens do not face competition with sulfate reducers for these compounds, they have been termed "noncompetitive substrates." The source of trimethylamine in saline environments is the osmoregulatory compounds glycine betaine and trimethylamine oxide, which cells use to lower the water activity of their cytoplasm and balance the lower water activity of their saline environment. Dimethyl sulfide may be formed from dimethylsulfonopropionate, another osmoregulatory compound common in algae.

Gastrointestinal Tracts

Methane formation in gastrointestinal tracts is described in another chapter of this volume (Miller). The rapid turnover of organic matter in such environments prevents complete anaerobic decomposition because some methanogens and fatty acid-degrading bacteria cannot grow rapidly enough to maintain large populations. However, H_2- and formate-using methanogens are found in many gastrointestinal tracts such as those of the rumens of ungulates (cattle, sheep, deer, etc.) and the lower digestive tracts of herbivores, some insects, and some humans.

ENVIRONMENTAL FACTORS AFFECTING MICROBIAL PRODUCTION OF ATMOSPHERIC METHANE

The rate of input of biogenic methane to the atmosphere depends on two biological processes, methanogenesis and methane oxidation. Anoxia is required for the formation of methane, so methanogenic environments must be somehow isolated from the atmosphere. Biogenic methane formed in these environments must migrate to the atmosphere, and this transit often requires the methane to first pass through oxic waters or soils, where methane oxidation can occur.

Requirements for Methane Formation: Anoxia

Of all known microbes, methanogens are the most sensitive to O_2. Thus, microbial sources of atmospheric methane require environments which are protected from the intrusion of O_2. The presence of O_2 inhibits methanogenesis by its toxic effects on methanogens and the toxic effects of by-products of O_2, such as superoxide or singlet oxygen. In addition, O_2 stimulates the activity of bacteria which can outcompete the methanogens for reduced catabolic substrates. Other electron acceptors such as nitrate, ferric, and sulfate ions can also stimulate activity of organisms which compete with methanogens. When these electron acceptors are present, concentrations of acetate, formate, and H_2 are maintained at levels so low that methanogens are unable to use them as catabolic substrates. However, when these alternate electron acceptors are exhausted, the concentrations of acetate, formate, and H_2 rise to levels which can be used for the formation of methane gas.

The boundary which separates anoxic zones from the atmosphere is often a site of high metabolic activity, and this boundary may exclude O_2 because resident microbes consume O_2 faster than it enters through diffusion or dispersion. We can use a lake to illustrate this phenomenon. The waters of an oxic lake contain organic matter which bacteria can oxidize to CO_2, creating a steady demand for O_2 as an electron acceptor. At steady state, O_2 must be delivered to all portions of the lake, otherwise the demand for O_2 exhausts the supply. The supply of O_2 must be nearly continuous, because the O_2 content of even completely aerated water is very low (pure water at 20°C in equilibrium with air contains 0.29 mM O_2).

When O_2 diffuses into a well-mixed lake, diffusion and especially dispersion (such as convection currents) distribute the O_2 throughout the water column, and most of these lakes remain oxic. Stratification of lakes, which may be caused by thermal or saline gradients, blocks the dispersion of O_2 into the lower layers of the lake. Because diffusion alone does not deliver sufficient O_2, the disruption of dispersive mechanisms results in anoxia in the lower layers.

In the sediments of lakes and rivers, the physical structure of the sediments prevents dispersion by convection, so even when very little biodegradable organic matter is present, O_2 demand may be greater than its supply. When O_2 and other electron acceptors in the interstitial water of sediments are depleted, methanogenesis occurs in the deeper sediments. In the same way that sediments block convective dispersion and bring about methanogenic conditions, the supply of O_2 in aquifers is limited by the structure of the subsurface medium. When polluted aquifers become anoxic, it may take a very long time before O_2 enters the groundwater at a rate sufficient to exceed the microbial oxygen demand and establish an oxic system. Initially, only the upper portions of such an aquifer become oxic; later, when rapidly biodegradable organic matter is exhausted, the O_2 penetrates more deeply into the aquifer.

Electron Acceptors Other than O_2

Electron acceptors other than O_2 may also inhibit methanogenesis by stimulating the activity of bacteria which outcompete methanogens for the reduced substrates. These electron acceptors include nitrate, nitrite, trimethylamine oxide, Fe(III), Mn(IV), and sulfate. These substances tend to disappear with depth in sediment, because as they diffuse downward they are reduced by the catabolic activities of resident bacteria. The order in which these compounds disappear is approximately the order listed and is generally consistent with the redox potential at which the substances function as electron acceptors (Thauer et al., 1977). Ferric ion is lower on the list than its redox potential would predict, perhaps because the insolubility of ferric ion creates kinetic limitations in its use.

Sensitive measurements of the partial pressure of H_2 reflect the metabolic state of such environments (Lovley and Goodwin, 1988). If ferric ion is the catabolic substrate, H_2 partial pressure is very low, but when ferric ion is exhausted and sulfate becomes the electron acceptor the H_2 pressure increases to 1 or 2 Pa. Likewise, when sulfate is exhausted and the ecosystem becomes methanogenic, the H_2 pressure rises to about 10 Pa.

Because of microbial activity in sediments and aquifers, the concentrations of electron acceptors other than CO_2 decrease rapidly with depth. CO_2 is usually abundant because it is produced during the oxidation of organic compounds. In some active, anoxic environments such as gastrointestinal tracts, acetigenesis (CO_2 reduction to acetate) may be a terminal electron sink. Although methanogens can outcompete these acetigens for low concentrations of their common catabolic substrates, methanogens grow slowly and may not be able to maintain significant numbers in these ecosystems. Also, acetigens can use a wider range of catabolic substrates than can methanogens.

The chemistry of an environment determines which microbes become active, and the activity of the microbes can then affect the chemistry of the environment. This process can lead to a succession of microbes and evolution of an ecosystem, but such changes are rapid relative to geological time. Thus, for example, if an ecosystem is anoxic but with excess sulfate, we might expect to find sulfate reduction as the dominant catabolic pathway (but with methane formed from "noncompetitive substrates"). The organisms which carry out these processes are ubiquitous, and inoculation appears to have little effect on the steady-state populations which develop. However, when profound changes occur in an environment, a period of time may be required to achieve steady state, and during this transitional period the relative numbers of existing flora may strongly affect the behavior of the ecosystem. For example, if a methanogenic environment suddenly accumulates substantial concentrations of sulfate, methanogenesis may continue for some time until populations of sulfate reducers decrease the concentration of catabolic substrates so they can no longer be used by methanogens. However, the time scale for such population changes is measured in days or weeks, whereas the time scale for transport of contaminants and electron acceptors into aquifers is measured in months or years. Because of the difference between these time scales, microbial adaptation and metabolism is often considered instantaneous when predicting metabolic events in the environment. Indeed, the computer model most often used for groundwater biodegradation, Bioplume (Borden and Bedient, 1986), assumes instantaneous reaction between O_2 and hydrocarbons, with no allowance for growth or adaptation of microbes. However, some organic compounds are so poorly biodegradable under methanogenic conditions that microbial metabolism is on the same time scale as transport. For instance, benzene from fuel spills and humic acids persist for long periods of time even in the presence of degradative bacteria.

Influence of Other Environmental Factors on Methanogenesis

The effects of environmental factors on methanogenesis in sediments and soils are not well documented. It is difficult to design experiments which test the effects of temperature, pH, salinity, and nutrient status in the field, so most reports have examined slurries. The interpretation of published studies is complicated by activity changes which occur when sediments or soils are slurried. Because of these difficulties, most of the available data on environmental factors affecting methanogenesis have been collected from anaerobic reactors. When considering the extrapolation of data from anaerobic reactors to natural environments, it is important to keep in mind the important differences between these two environments. In general, anaerobic reactors have metabolic rates which are much higher than those of natural soils and sediments. Also, anaerobic reactors must be considered as open systems, with a continuous or semicontinuous flow of water and organic matter into and out of the reactor. Thus, microbes in anaerobic reactors must maintain substantial growth rates in order to maintain populations within the reactors, whereas the growth rates of microbes in soils and sediments can be very low. In some reactors which retain viable biomass while the liquid flows through, the microbes may also have very slow growth rates.

A number of environmental factors affect the rates of methanogenesis in environments such as anaerobic digestors (Boone, 1985). These factors include temperature, pH, salinity, toxic substances, and the presence of nutrients. pH values near neutral are considered optimal for anaerobic digestion. Sewage sludge digestors are generally operated at pH values near neutral, and low pH values are associated with accumulation of fermentation acids and digestor failure. Low pH and accumulated acids are thought to be more inhibitory to methanogens than to fermentative bacteria, so the production of fermentation acids causes a drop in pH and inhibits the ability of methanogens to remove those acids. Fermentative bacteria are less susceptible to inhibition and continue their production of acids, resulting in a catastrophic drop in both pH and methane production by the digestors. However, even though anaerobic digestors produce less methane when the pH drops, several pure cultures of methanogens grow well at low pH. One slightly acidiphilic, H_2-using species has been described (Patel et al., 1990), and two well-known species of acetate-degrading methanogens, *Methanosarcina barkeri* and *Methanosarcina vacuolata*, grow well at low pH and even have pH optima of about 5 when grown on H_2 + methanol as a catabolic substrate (Maestrojuán and Boone, 1991). H_2-oxidizing methanogens (Boone et al., 1986) and methylotrophic methanogens (Liu et al., 1990; Mathrani et al., 1988) have been found thriving at very alkaline pH values, but no aceticlastic methanogens have been found.

Methanogenic rates of anaerobic digestors generally increase with temperature up to about 60°C. There are reports of two optimal ranges, mesophilic (near 35°C) and thermophilic (55 to 60°C), with decreased rates of methanogenesis between those zones. However, low rates measured in this zone may be due to insufficient time allowed for adaptation of the microflora (Mackie and Bryant, 1981). At temperatures at or above 70°C, methanogenic rates decrease (Zinder et al., 1984).

Salinities of 0 to 0.2 M NaCl have minimal effects on growth rates of adapted, mixed methanogenic populations, but higher salinities are inhibitory (Liu and Boone, in press).

Environmental conditions exert their effect not only on methanogens per se, but more importantly on the consortia of degradative bacteria of which methanogens are a part. The nonmethanogenic partners in these consortia, which will be described in more detail below, convert a wide range of organic molecules into the few catabolic substrates which methanogens can use. In environments with slow rates of organic matter turnover, it is typically the hydrolysis of large polymers which limits microbial activity in anoxic environments (Boone, 1982), and low-molecular-weight intermediates do not accumulate appreciably. Two notable exceptions to this generalization are sphagnum bogs and alkaline, hypersaline lakes, where soluble organic molecules accumulate to substantial concentrations. Extremely alkaline and hypersaline conditions do not support active growth of any known methanogens other than methyl-compound degraders (Liu et al., 1990; Mathrani et al., 1988).

Catabolic Substrates in Methanogenic Ecosystems

Although methanogens per se are limited to a few catabolic substrates, these microbes typically occur as members of consortia. The nonmethanogenic bacteria hydrolyze and ferment a wide range of complex organic molecules into the small-molecular-weight substrates of the methanogenic bacteria. Almost all organic molecules are biodegradable under anoxic conditions, although resistance to degradation varies. The list of compounds which are recognized to completely resist decomposition by anaerobic bacteria seems to be growing smaller as we become more skilled at finding and measuring small degradative rates. Recalcitrance is enhanced by polymerization and by the presence of substituents such as alkyl branches, halogens, or nitro- and sulfo-groups (Leisinger and Brunner, 1986). The list of anaerobically recalcitrant substrates includes saturated aliphatic hydrocarbons and some chlorinated compounds such as chlorinated biphenyls. Lignin and humic substances are extremely resistant to biodegradation, but small-molecular-weight derivatives of these materials appear to be biodegradable, and perhaps our inability to measure biodegradation of larger molecules reflects a limitation on our capacity to measure extremely slow catabolic rates. Highly branched molecules such as alkylbenzensulfonates are also extremely resistant.

Interspecies Electron Transfer

In exosystems where the turnover of bacteria is slow, the fermentation pathways of bacteria tend to lead to CO_2, H_2, and short-chained, saturated monocarboxylic acids: formic, acetic, propionic, and butyric acids. The presence of methanogens has a direct effect on the mixture of products which fermentative bacteria form (for reviews, see McInerney [1986] and Wolin [1982]). Fermentative bacteria must use stoichiometrically balanced fermentation pathways, which often include as products either reduced organic compounds such as alcohols, lactate, and formate or the reduced inorganic product H_2. The formation of alcohols and lactate is often at the expense of pathways leading to acetate and extra ATP (energy) production. Thus, it is advantageous for these organisms to favor production of H_2 or formate, but the production of these substrances from electron carriers such as NADH is thermodynamically favorable only when their concentrations are low. Thus, without methanogens, fermentative bacteria produce compounds such as lactate and ethanol, but when cocultured with methanogens they shift to the production of H_2 and formate, which do not accumulate because they are consumed by methanogens (Wolin, 1982).

The activities of fatty acid-oxidizing bacteria are crucial to complete methanogenic fermentation. Low-molecular-weight fatty acids such as propionate and butyrate are common products of fermentative bacteria from these ecosystems, and these acids cannot be catabolized by methanogens. A group of specialized organisms oxidize these acids to acetate (and CO_2 in the case of propionate), producing H_2 or formate as the reduced product. Butyrate catabolism by *Syntrophomonas wolfei* proceeds by β-oxidation, with the release of acetate (McInerney et al., 1979; McInerney et al., 1981). Electrons generated by this oxidation are released as either H_2 (McInerney et al., 1979) or formate (Boone et al., 1989). Propionate oxidation is less well understood, but it is accomplished by *Syntrophobacter wolinii* (Boone and Bryant, 1980), which may oxidize propionate via a pathway similar to the reverse of the randomizing (succinate) pathway of propionate formation by propionic acid bacteria (Houwen et al., 1987). The oxidation of propionate and butyrate is thermodynamically possible only when the reduced product (H_2 or formate) is maintained at very low concentrations by the scavenging activities of methanogenic bacteria.

Syntrophomonas wolfei LYB has both hydrogenase activity and formate dehydrogenase activity, which may allow it to produce either formate or H_2 (Boone et al., 1989). The maximum H_2 concentration which it is thermodynamically possible to produce from butyrate is very low, which leads to a substantial diffusional barrier to the rapid interspecies H_2 transfer. The interspecies transfer of H_2 in some cocultures of butyrate-oxidizing bacteria does not appear to be sufficient to account for the measured rate of methanogenesis. Calculations indicate that because the concentration of formate produced from butyrate can be substantially higher than that of H_2, formate can diffuse rapidly enough between producing and consuming organisms to account for the methanogenic rate (Boone et al., 1989) and may be the principal interspecies electron carrier in some environments. However, that study (Boone et al., 1989) examined diffusion only and did not account for enzymatic activities which may favor interspecies H_2 transfer among the tightly packed bacteria within flocs, where diffusion may not be limiting (Stams et al., 1989). In fact, extracts of *Syntrophomonas wolfei* grown on crotonate had 6×10^6-fold more hydrogenase activity than formic dehydrogenase activity (Beaty et al., 1989).

The growth of these fatty acid-oxidizing bacteria in consortia with methanogens is very slow and may limit methanogenesis in some ecosystems. The accumulation of fatty acids (especially propionic aid) in anaerobic digestors is a common problem.

Such shuttling of electrons among members of consortia was first recognized by Bryant et al. (1967) and has since come to be known as interspecies hydrogen transfer (Wolin, 1977). Formate was also recognized as a potential interspecies electron carrier, and more recently data supporting its quantitative importance have been reported (Boone et al., 1989; Thiele and Zeikus, 1988). Although the analysis of interbacterial diffusion could not directly determine the quantitative contributions of H_2 and formate, it indicates that formate may be a more active electron carrier than H_2, so the two processes are collectively termed interspecies electron transfer (Boone et al., 1989). During interspecies electron transfer the concentrations of H_2 and formate usually remain so low as to be undetectable by most analyses (H_2 concentrations in methanogenic ecosystems are about 10 Pa [60 nM] or less, and formate concentrations are usually undetectable [< 0.01 mM]), so the quantitative importance of those substances is not immediately obvious. These intermediates, which are the major electron sources for CO_2-reducing methanogens, turn over so rapidly that they often account for about one-third of the methane formed.

Factors Affecting Methane Transport to the Atmosphere

Obviously, methane formed in anoxic environments must pass through the oxic-anoxic boundary before entering the troposphere, and during that passage methane is subject to oxidation by methane-oxidizing microbes. Factors which may affect how much of the methane is oxidized before it reaches the atmosphere include the residence time of methane within a biologically active, oxic environment and the biological activity of that environment. For instance, methane which diffuses upward from an aquifer through its upper oxic layer and into the soil gas above the water table must then diffuse through the soil matrix before being released as atmospheric methane. During its passage thought these oxic environments, the methane is subject to the activities of the soil microflora, which may oxidize substantial portions of the methane. The activities of these methane oxidizers may be limited by availability of water or nutrients and by the contact time with the methane. Shorter distances between the aquifer and the atmosphere (i.e., shallow water tables) may enhance the ability of methane to escape. Landscapes in which the water table is very near to the ground surface (such as saturated soils, rice paddies, swamps, bogs, and fens) are recognized as areas with the highest productivity of atmospheric methane.

In lakes, the water table is significantly higher than the sediment surface, but here a different kind of oxic environment, namely the water column, stands between the methanogenic zone and the atmosphere. Methane moving to the surface of the lake may be somewhat protected from microbial oxidation if gas production by the sediments is active enough for bubble formation. Otherwise, dissolved methane is susceptible to microbial oxidation by bacteria in the water column.

REFERENCES

Baresi, L. 1984. Methanogenic cleavage of acetate by lysates of *Methanosarcina barkeri*. *J. Bacteriol.* **160:**365–370.

Barker, H. A. 1956. *Bacterial Fermentations*, p. 1–27. Wiley, New York.

Beaty, P. S., N. Q. Wofford, and M. J. McInerney. 1989. The metabolism of formate by the syntrophic bacterium *Syntrophomonas wolfei*. I31, p. 222. *Abstr. 89th Annu. Meet. Am. Soc. Microbiol. 1989*. American Society for Microbiology, Washington, D.C.

Boone, D. R. 1982. Terminal reactions in the anaerobic digestion of animal waste. *Appl. Environ. Microbiol.* **41:**57–61.

Boone, D. R. 1985. Fermentation reactions of anaerobic digestion, p. 41–51. *In* N. Cheremisinoff and R. P. Ouellette (ed.), *Biotechnology: Applications and Research*. Technomic Publishing Co., Lancaster, Pa.

Boone, D. R., and M. P. Bryant. 1980. Propionate-degrading bacterium, *Syntrophobacter wolinii* sp. nov. gen. nov., from methanogenic ecosystems. *Appl. Environ. Microbiol.* **40:**626–632.

Boone, D. R., R. L. Johnson, and Y. Liu. 1989. Diffusion of the interspecies electron carriers H_2 and formate in methanogenic ecosystems, and its implication in the measurement of K_m for H_2 or formate uptake. *Appl. Environ. Microbiol.* **55:**1735–1741.

Boone, D. R., S. Worakit, I. M. Mathrani, and R. A. Mah. 1986. Alkaliphilic methanogens from high-pH lake sediments. *J. Syst. Appl. Microbiol.* **7:**230–234.

Boone, D. R., and L. Xun. 1987. Effects of pH, temperature, and nutrients on propionate degradation by a methanogenic enrichment culture. *Appl. Environ. Microbiol.* **53:**1589–1592.

Borden, R. C., and P. B. Bedient. 1986. Transport of dissolved hydrocarbons influenced by oxygen-limited biodegradation. 1. Theoretical development. *Water Res.* **22:**1973–1982.

Bryant, M. P., E. A. Wolin, M. J. Wolin, and R. S. Wolfe. 1967. *Methanobacillus omelianskii*, a symbiotic association of two species of bacteria. *Arch. Mikrobiol.* **59:**20–31.

Cheeseman, P., A. Toms-Wood, and R. S. Wolfe. 1972. Isolation and properties of a fluorescent compound, factor F_{420}, from *Methanobacterium* strain M.o.H. *J. Bacteriol.* **112:** 527–531.

DiMarco, A. A., T. A. Bobik, and R. S. Wolfe. 1990. Unusual coenzymes of methanogenesis. *Annu. Rev. Biochem.* **59:**355–394.

Gunsalus, R. P., and R. S. Wolfe. 1978. ATP activation and properties of the methyl coenzyme M reductase system in *Methanobacterium thermoautotrophicum*. *J. Bacteriol.* **135:**851–857.

Gunsalus, R. P., and R. S. Wolfe. 1980. Methyl coenzyme M reductase from *Methanobacterium thermoautotrophicum*: resolution and properties of the components. *J. Biol. Chem.* **255:**1891–1895.

Houwen, F. P., C. Dijkema, C. C. H. Schoenmakers, A. J. M. Stams, and A. J. B. Zehnder. 1987. ^{13}C-NMR study of propionate degradation by a methanogenic coculture. *FEMS Microbiol. Lett.* **41:**269–274.

Jacobson, F., and C. Walsh. 1984. Properties of 7,8-didemethyl-8-hydroxy-5-deazaflavins relevant to redox coenzyme function in methanogen metabolism. *Biochemistry* **23:**979–988.

Jones, W. J., D. P. Nagel, Jr., and W. B. Whitman. 1987. Methanogens and the diversity of archaebacteria. *Microbiol. Rev.* **51:**135–177.

Kandler, O., and H. König. 1985. Cell envelopes of archaebacteria, p. 413–457. *In* C. R. Woese and R. S. Wolfe (ed.), *The Bacteria*, vol. VIII, *Archaebacteria*. Academic Press, Inc., Orlando, Fla.

Krzycki, J. A., and J. G. Zeikus. 1984. Acetate catabolism by *Methanosarcina barkeri*: hydrogen-dependent methane production from acetate by a soluble cell protein fraction. *FEMS Microbiol. Lett.* **25:**27–32.

Langworthy, T. A. 1985. Lipids of archaebacteria, p. 413–457. *In* C. R. Woese and R. S. Wolfe (ed.), *The Bacteria*, vol. VIII, *Archaebacteria*. Academic Press, Inc., Orlando, Fla.

Leisinger, T., and W. Brunner. 1986. Poorly degradable substances, p. 475–513. *In* W. Schönborn (ed.), *Biotechnology: Microbial Degradations*, vol. 8. VCH Verlagsgesellschaft, Weinheim, Federal Republic of Germany.

Liu, Y., and D. R. Boone. Effects of salinity on methanogenic decomposition. *Bioresource Technol.*, in press.

Liu, Y., D. R. Boone, and C. Choy. 1990. *Methanohalophilus oregonense* sp. nov., a methylotrophic methanogen from an alkaline, saline aquifer. *Int. J. Syst. Bacteriol.* **40:**111–116.

Lovley, D. R., and S. Goodwin. 1988. Hydrogen concentration as an indicator of the predominant terminal electron acceptor reactions in aquatic sediments. *Geochim. Cosmochim. Acta* **52:**2993–3003.

Mackie, R. I., and M. P. Bryant. 1981. Metabolic activity of fatty acid-oxidizing bacteria and the contribution of acetate, propionate, butyrate, and CO_2 to methanogenesis in cattle waste at 40 and 60°C. *Appl. Environ. Microbiol.* **41:**1363–1373.

Maestrojuán, G. M., and D. R. Boone. 1991. Characterization of *Methanosarcina barkeri* MS^T and 227, *Methanosarcina mazei* S-6^T, and *Methanosarcina vacuolata* Z-761^T. *Int. J. Syst. Bacteriol.* **41:**267–274.

Mathrani, I. M., D. R. Boone, R. A. Mah, G. E. Fox, and P. P. Lau. 1988. *Methanohalobium zhilinae* gen. nov. sp. nov., an alkaliphilic, halophilic, methyltrophic methanogen. *Int. J. Syst. Bacteriol.* **38:**139–142.

McInerney, M. J. 1986. Transient and persistent associations among prokaryotes, p. 293–338. *In* E. R. Leadbetter and J. S. Poindexter (ed.), *Bacteria in Nature*, vol. 2. Plenum Publishing Corp., New York.

McInerney, M. J., M. P. Bryant, R. B. Hespell, and J. W. Costerton. 1981. *Syntrophomonas wolfei* gen. nov. sp. nov., an anaerobic, syntrophic, fatty acid-oxidizing bacterium. *Appl. Environ. Microbiol.* **41:**1029–1039.

McInerney, M. J., M. P. Bryant, and N. Pfennig. 1979. Anaerobic bacterium that degrades fatty acids in syntrophic association with methanogens. *Arch. Microbiol.* **122:**129–135.

Nagle, D. P., Jr., and R. S. Wolfe. 1983. Component A of the methyl coenzyme M methylreductase system of *Methanobacterium*: resolution into four components. *Proc. Natl. Acad. Sci. USA* **80:**2151–2155.

Patel, G. B., G. D. Sprott, and J. E. Fein. 1990. Isolation and characterization of *Methanobacterium espanolae* sp. nov., a mesophilic, moderately acidiphilic methanogen. *Int. J. Syst. Bacteriol.* **40:**12–18.

Poirot, C. M., S. W. M. Kengen, E. Valk, J. T. Keltjens, C. van der Drift, and G. D. Vogels. 1987. Formation of methylcoenzyme M from formaldehyde by cell-free extracts of *Methanobacterium thermoautotrophicum*: evidence for involvement of a corrinoid-containing methyltransferase. *FEMS Microbiol. Lett.* **40:**7–13.

Pol, A., C. van der Drift, G. D. Vogels, T. J. H. M. Cuppen, and W. H. Laarhoven. 1980. Comparison of coenzyme F_{420} from *Methanobacterium bryantii* with 7- and 8-hydroxyl-10-methyl-5-deazaisoalloxazine. *Biochem. Biophys. Res. Commun.* **92:**255–260.

Rimbault, A., P. Niel, H. Virelizier, J. C. Darbord, and G. Leluan. 1988. L-Methionine, a precursor of trace methane in some proteolytic clostridia. *Appl. Environ. Microbiol.* **54:**1581–1586.

Stams, A. J. M., J. T. C. Grotenhuis, and A. J. B. Zehnder. 1989. Structure-function relationship in granular sludge, p. 440–445. *In* T. Hattori, Y. Ishida, Y. Maruyama, R. Y. Morita, and A. Uchida (ed.), *Recent Advances in Microbial Ecology*. Japan Scientific Societies Press, Tokyo.

Taylor, C. D., and R. S Wolfe. 1974. Structure and methylation of coenzyme M ($HSCH_2CH_2SO_3$). *J. Biol. Chem.* **249:**4879–4885.

Thauer, R. K., K. Jungermann, and K. Decker. 1977. Energy conservation in chemotrophic anaerobic bacteria. *Bacteriol. Rev.* **41:**100–180.

Thiele, J. H., and J. G. Zeikus. 1988. Control of interspecies electron flow during anaerobic digestion: significance of formate transfer versus hydrogen transfer during syntrophic methanogenesis in flocs. *Appl. Environ. Microbiol.* **54:**20–29.

van Beelen, P., J. F. A. Labro, J. T. Keltjens, W. J. Geerts, G. D. Vogels, W. H. Laarhoven, W. Guijt, and C. A. G. Haasnoot. 1984. Derivatives of methanopterin, a coenzyme involved in methanogenesis. *Eur. J. Biochem.* **139:**359–365.

Vogels, G. D., J. T. Keltjens, and C. van der Drift. 1988. Biochemistry of methane production, p. 707–770. *In* A. J. B. Zehnder (ed.), *Biology of Anaerobic Microorganisms*. John Wiley & Sons, Inc., New York.

Westermann, P., B. K. Ahring, and R. A. Mah. 1989. Threshold acetate concentrations for acetate catabolism by aceticlastic methanogenic bacteria. *Appl. Environ. Microbiol.* **55:**514–515.

Whitman, W. B. 1985. Methanogenic bacteria, p. 3–84. *In* C. R. Woese and R. S. Wolfe (ed.), *The Bacteria*, vol. 8, *Archaebacteria*. Academic Press, Inc., New York.

Woese, C. R. 1987. Bacterial evolution. *Microbiol. Rev.* **51:**221–271.

Wolin, M. J. 1977. Fermentation of cellulose by *Ruminococcus flavefaciens* in the presence and absence of *Methanobacterium ruminantium*. *Appl. Environ. Microbiol.* **34:**297–301.

Wolin, M. J. 1982. Hydrogen transfer in microbial communities, p. 323–356. *In* A. T. Bull and J. H. Slater (ed.), *Microbial Interactions and Communities*, vol. 1. Academic Press, Inc. (London), Ltd., London.

Zinder, S. H., T. Anguish, and S. C. Cardwell. 1984. Effects of temperature on methanogenesis in a thermophilic (58°C) digestor. *Appl. Environ. Microbiol.* **47:**808–813.

Chapter 5

Metabolism of Radiatively Important Trace Gases by Methane-Oxidizing Bacteria

E. Topp and R. S. Hanson

Bacteria which oxidize methane (methanotrophs) are relevant to an analysis of processes which contribute to global warming and ozone depletion. Methane is the most abundant organic gas in the atmosphere, and its concentration has been increasing at the alarming rate of over 1% per year for the last 150 to 200 years (Cicerone and Oremland, 1988; Crutzen and Graedel, 1986; Graedel and Crutzen, 1989; Pearce, 1989; Seiler, 1983; Stauffer et al., 1988). Methane-oxidizing bacteria limit the flux of methane from many sources and may be a significant sink for atmospheric methane (Abramochinka et al., 1987; Cappenberg, 1972; Cicerone and Oremland, 1988; Hanson, 1980; Harriss et al., 1982; Harriss et al., 1985; Harrits and Hanson, 1980; Heyer et al., 1984; Heyer and Suchow, 1985; Iversen and Blackburn, 1979; Iversen and Jorgenson, 1985; Iversen et al., 1987; Jannasch, 1975; Keller et al., 1983; Reeburgh, 1976, 1980, 1982, 1983; Rudd and Taylor, 1980; Ward et al., 1987; Whalen and Reeburgh, 1988, 1990). Furthermore, methanotrophs contribute to a number of ecologically important activities such as ammonium oxidation and oxygen depletion which can potentially contribute to global nitrous oxide production (Bedard and Knowles, 1989; Hanson, 1980; Hutton and Zobell, 1953; Knowles and Topp, 1988; Malashenko et al., 1979; Megraw and Knowles, 1987b; Yoshinari, 1985). This contribution will review aspects of the physiology and ecology of methanotrophs which are relevant to an understanding of their role in global cycling of radiatively important gases. Techniques useful in establishing the activity, distribution, and identity of methanotrophs in situ will be presented.

The ubiquity of methanotrophs suggests that they play a significant role in the ecology of freshwater, marine, and terrestrial environments (Bedard and Knowles, 1989; Conrad, 1984; Davis et al., 1964; Ehhalt, 1974; Gal'chenko et al., 1988; Hanson, 1980; Heyer and Suchow, 1985; Ward et al., 1987; Yavitt et al., 1990). Hutton and Zobell (1949) pointed out that the widespread occurrence of methane-oxidizing bacteria was generally appreciated by 1920. Their studies confirmed that these bacteria were present in soils and marine sediments where both methane and oxygen were present (Hutton and Zobell, 1949, 1953). Heyer and co-workers

E. Topp • Land Resources Research Center, Research Branch, Agriculture Canada, C.E.F. Ottawa, Ontario K1A 0C6, Canada. *R. S. Hanson* • Gray Freshwater Biological Institute, University of Minnesota, Navarre, Minnesota 55392.

(Heyer, 1977; Heyer et al., 1984) successfully enriched for methanotrophs from 90% of the samples taken from a variety of natural sources. Only acid soils of coniferous forests and heath failed to yield methanotrophs, and it now appears that even acidic peat bogs, tundra, and forest soils contain microbes that oxidize methane (Harriss et al., 1982; Harriss et al., 1985; Seiler, 1983; Whalen and Reeburgh, 1988, 1990; Yavitt et al., 1990). The rate of oxidation of methane in different ecosystems appears to be proportional to the concentration of dissolved methane and the rate of methane production (Heyer et al., 1984). It is now clear that methane is also oxidized at significant rates in anaerobic sediments and waters of lakes and oceans (Alperin and Reeburgh, 1985; Cicerone and Oremland, 1988; Hanson, 1980; Iversen et al., 1987; Keller et al., 1983; Lidstrom, 1983; Reeburgh, 1976, 1980, 1982, 1983). The biology and biochemistry of anaerobic methane oxidation remain to be defined.

It has been particularly difficult to relate the methane-oxidizing activity of an environment to any one species by using enrichment or isolation techniques. The association of a particular species with an environment probably reflects the enrichment and isolation procedures used more than it does the organisms present (Hanson, 1980; Whittenbury et al., 1976).

The term methylotroph refers to a diverse group of bacteria, including methanotrophs, which utilize as sources of carbon and energy a variety of reduced compounds that do not contain carbon-carbon bonds and which assimilate formaldehyde as a major source of cell carbon (Table 1; Anthony, 1982). These bacteria utilize a variety of substrates including methanol, methylamines, and halomethanes (dichloromethane and methylene-chloride). The substrates are oxidized to formaldehyde and subsequently to formate and carbon dioxide. Many are facultative or restricted facultative methylotrophs that grow on sugars, fatty acids, amino acids, inorganic substrates, and complex media as well as one-carbon compounds. Facultative methanotrophs are rare (Hanson et al., 1991). Gram-positive bacteria and several yeasts grow well on methanol, while no gram-positive bacteria grow on methane. The methylotrophic bacteria that do not utilize methane have many features in common with methanotrophs, including the assimilation of formaldehyde via the serine and ribulose monophosphate (RuMP) pathways (Anthony, 1982).

PROPERTIES OF DIFFERENT GROUPS OF METHANOTROPHIC BACTERIA

Methanotrophic bacteria have in common the ability to utilize methane as a sole source of energy and as a major carbon source (Anthony, 1982). Methane is oxidized through methanol to formaldehyde, which in turn is assimilated for the synthesis of cell material and is also further oxidized to carbon dioxide. The methanotrophic bacteria that have been isolated and characterized to date have been all gram negative, obligately aerobic, and catalase and oxidase positive and have possessed intracytoplasmic membranes (Hanson et al., 1991; Whittenbury and Dalton, 1981; Whittenbury and Krieg, 1984). They form resting structures (Anthony, 1982). The work of Whittenbury and his colleagues (Davies and Whittenbury, 1970; Whittenbury and Dalton, 1981; Whittenbury et al., 1970;

Table 1. Characteristics of different groups of methylotrophs[a]

I. **Methanotrophic bacteria.** All are able to utilize methane as a sole source of energy and a major source of carbon. All are gram-negative bacteria phylogenetically related to the proteobacteria. All form resting states and are obligate aerobes.

A. **Type I methanotrophic bacteria.** These utilize the RuMP pathway for formaldehyde assimilation. They possess DNA with a G + C content of 50 to 54 mol%. All but one species have an incomplete tricarboxylic acid cycle and lack alpha-ketoglutarate dehydrogenase. The predominant phospholipid fatty acids contain 16 carbon atoms. Intracytoplasmic membranes are arranged as disc-shaped bundles. Bacteria in this group are not known to synthesize a soluble MMO. Proposed genera include *Methylomonas* and *Methylobacter*. These bacteria are phylogenetically related to the α-subgroup of the proteobacteria (Tsuji et al., 1990).

B. **Type X methanotrophic bacteria.** Bacteria in this group also utilize the RuMP pathway for formaldehyde assimilation and have an incomplete tricarboxylic acid cycle. However, enzymes of the serine pathway and ribulose-bisphosphate carboxylase are also present, and CO_2 is fixed via the Calvin cycle. DNA from these bacteria contains 62.5 mol% G + C, the predominant phospholipids contain 16 carbon atoms, and the bacteria in this group are the only well-characterized methanotrophs that grow at 45°C or higher temperatures. Soluble MMO is present in cells grown under copper-limited conditions (Dalton and Higgins, 1987). The only recognized species in this group is *Methylococcus capsulatus* and it, like type I methanotrophs, is phylogenetically related to the α-subgroup of proteobacteria (Tsuji et al., 1990).

C. **Type II methanotrophic bacteria.** Bacteria in this group utilize the serine pathway for formaldehyde fixation, possess a complete tricarboxylic acid cycle, and do not grow at temperatures above 40°C. DNA from these bacteria contain 62.5 mol% G + C, and the predominant phospholipid fatty acids contain 18 carbon atoms. Intracytoplasmic membranes are arranged in pairs aligned parallel to the cytoplasmic membrane. Some, if not all, bacteria of this group synthesize soluble MMO when grown under carbon-limited conditions. Two genera, *Methylosinus* and *Methylocystis*, have been proposed for bacteria in this group. These bacteria are phylogenetically related to the bacteria in the α-subgroup of the proteobacteria.

D. **Methanotrophic yeasts.** Species of the genera *Rhodotorula* and *Sporobolomyces* that grow very slowly on methane have been described by Wolf and Hanson (1978).

II. **Methylotrophic bacteria that do not grow on methane as an energy source.** An extraordinary range of bacteria and yeasts that grow on methanol, methylamines, halomethanes, and other methylated substrates are known (Anthony, 1982). Yeasts, gram-positive bacteria, and gram-negative bacteria are known that utilize methanol as a source of carbon and energy. Some, like species of *Hyphomicrobium*, are able to utilize nitrate as a terminal electron acceptor for anaerobic respiration. Some methanol-utilizing bacteria are obligate methylotrophs, while others are facultative methylotrophs that are able to grow on a variety of multicarbon compounds. Several of this group are restricted facultative methylotrophs that are able to use one-carbon compounds and a very limited range of multicarbon compounds. Different one-carbon substrates are oxidized by a variety of enzymes to formaldehyde. Some bacteria, including *Methylophilus methylotrophus* and methanol-utilizing bacilli, employ the RuMP pathway for formaldehyde fixation, while others, including the pink-pigmented facultative methylotrophs belonging to the genus *Methylobacterium*, employ the serine pathway for formaldehyde fixation.

Continued on following page

Table 1—*Continued*

III. **Anaerobic methane-utilizing microbes.** The oxidation of methane in the absence of oxygen is well established (Alperin and Reeburgh, 1985; Hanson, 1980; Oremland et al., 1987; Panganiban et al., 1979; Reeburgh, 1980). However, pure cultures have not been well characterized (Hanson, 1980). It is unclear that formaldehyde is assimilated for the synthesis of cell material, and therefore, these microbes may not fit the definitions applied to methylotrophs.

[a] From Anthony (1982), Hanson (1991), Whittenbury et al. (1970), and Whittenbury and Krieg (1984).

Whittenbury et al., 1976; Whittenbury and Krieg, 1984) resulted in the characterization of three groups of methanotrophic bacteria distinguishable by several morphological and biochemical characteristics listed below. The separation of methanotrophs into these groups has been supported by subsequent studies including recent phylogenetic comparisons based on 16S rRNA analysis (Hanson et al., 1991). A knowledge of the common and unique traits of each group (Table 1) is essential for the estimation of population sizes and the identification of methanotrophs in environmental samples.

Type I methanotrophs have disc-shaped bundles of intracytoplasmic membranes, lack a complete tricarboxylic acid cycle, have DNA with a G + C content of 50 to 54 mol%, utilize the RuMP pathway for formaldehyde assimilation, and possess phospholipid fatty acids with 16 carbon atoms (Davies and Whittenbury, 1970; Gal'chenko and Andreev, 1984; Hanson et al., 1991; Whittenbury and Dalton, 1981).

Type II methanotrophs differ in all the characteristics listed above for type I methanotrophs. They contain paired intracytoplasmic membranes arranged parallel to the cytoplasmic membrane, possess a complete tricarboxylic acid cycle, utilize the serine pathway for formaldehyde fixation, and lack enzymes of the RuMP pathway, and the dominant phospholipid fatty acids contain 18 carbon atoms. The DNAs of type II methanotrophs have G + C contents of 62 to 64 mol% (Davies and Whittenbury, 1970; Gal'chenko and Andreev, 1984; Hanson et al., 1991; Whittenbury and Dalton, 1981; Whittenbury and Krieg, 1984).

Two genera, *Methylomonas* and *Methylobacter*, were proposed for type I methanotrophs and two others, *Methylosinus* and *Methylocystis*, for type II methanotrophs (Whittenbury et al., 1970). Subsequent characterizations have supported this separation of the aerobic methane-utilizing bacteria as well as the formation of a fifth genus, although the proposals have not been officially adopted at this time (Hanson et al., 1991; Whittenbury and Krieg, 1984). The single species of methanotrophs classified as type X, *Methanococcus capsulatus*, is similar to type I bacteria except that it contains enzymes of the Calvin cycle and is capable of carbon dioxide fixation. In addition, this bacterium has enzymes of the serine pathway although the major route of carbon assimilation is the RuMP pathway. The DNA of this bacterium has a G + C content of 62.5 mol%, like the type II methanotrophs. *Methylococcus capsulatus* grows at 45°C, while bacteria classified in other groups do not grow at temperatures above 37°C.

Several other characteristics (reviewed by Gal'chenko and Andreev [1984], Trotsenko [1983], Hanson et al. [1991], Whittenbury and Dalton [1981], and Whittenbury and Krieg [1984]) justify the separation of methanotrophs into three groups and indicate that each group may have evolved independently. There are reports of methane-utilizing bacteria that do not fit precisely into these groups (Hanson et al., 1991; Whittenbury and Krieg, 1984). However, those bacteria most often isolated from or detected in natural samples resemble those originally described by Whittenbury et al. (1970) in the characteristics listed above.

Other characteristics that may be important in considering the roles of each group and their detection in different ecosystems are the presence of nitrogenase in type II and X methanotrophs, its absence in type I methane-utilizing bacteria, and the presence of different phospholipid fatty acids in bacteria of each type and genus (Hanson et al., 1991; Murrell and Dalton, 1983). The fatty acid signatures, as determined by gas chromatography-mass spectrometry, have been employed to detect methanotrophs in environmental samples (Gachert et al., 1985; Nichols et al., 1986).

METHODS FOR THE DETECTION OF METHANE-UTILIZING BACTERIA AND THEIR ACTIVITIES IN SITU

Metabolism of methane by aquatic samples is most frequently determined by measuring rates of mineralization and assimilation of [^{14}C]methane (Harrits and Hanson, 1980; Patt et al., 1974; Rudd and Hamilton, 1978; Rudd and Taylor, 1980). Consumption of methane from the headspace above a sample measured by gas chromatography is a less sensitive but still useful technique (Hanson, 1980). Net production or consumption rates can be measured in flux chambers placed over soils (Conrad and Seiler, 1982). Measurements of methane utilization rates made using this method may be complicated by simultaneous methanogenesis in some soils (Whalen and Reeburgh, 1988). Methane oxidation has also been estimated by determining methane carbon isotope fractionations (Cicerone and Oremland, 1988; Reeburgh, 1976).

Methanotrophic bacteria in environmental samples have been enumerated by plating serially diluted samples onto various agar media and counting the number of colonies formed following incubation under an atmosphere of methane and air (Abramochinka et al., 1987; Gal'chenko and Andreev, 1984; Gal'chenko et al., 1988; Hanson, 1980; Hanson et al., 1991; Whittenbury et al., 1976). The numbers typically found in soils, sediments, and waters range between 10^3 and 10^6 CFU/g (Heyer et al., 1984). This procedure probably underestimates the population size and diversity of methanotrophs because of the poor plating efficiency of the methanotrophs (Bone and Balkwill, 1986; Hanson, 1980; Whittenbury et al., 1976). Recovery and enumeration of all methanotrophs by means of this method would potentially require a very large number of incubation conditions consisting of various pH values, temperatures, oxygen concentrations, substrate concentrations, and nutrient supplements (Hanson et al., 1991). Given our current knowledge, it is nearly impossible to anticipate the range of growth conditions required to cultivate all methanotrophs that may be present in soils and waters. Some methanotrophs require bacterial consorts that remove toxic products of methane metabolism or supply required growth factors (Anthony, 1982; Hanson, 1980; Lamb and Garver, 1980). These bacteria may not form colonies when diluted and spread onto agar media. Cultures of some strains are sensitive to impurities in agar media, and colonies of oligocarbophilic bacteria frequently outnumber those of methanotrophs (Hanson, 1980; Whittenbury et al., 1976). Although silica gel can be effectively substituted for agar to grow methanotrophs (Gal'chenko, 1975), the other problems mentioned above remain.

Fluorescent antibodies, prepared against killed cells from pure cultures of methanotrophs, are a potentially effective means of identifying and enumerating methanotrophs and other bacteria without culturing them (Abramochinka et al., 1987; Bohlool and Schmidt, 1980; Boxrukova et al., 1983; Gal'chenko et al., 1988; Reed and Dugan, 1978; Saralov et al., 1984). However, there is insufficient information on the number of serotypes of methane-utilizing bacteria to predict the number of antibodies necessary to enumerate all of them in a sample (Boxrukova et al., 1983). It is possible that species and strains which are dominant in some environments have not been isolated and are unavailable for the preparation of antisera (Bohlool and Schmidt, 1980). The number of strains available in individual collections is another severe limitation. Recoveries of cells which reacted with fluorescent antibodies after they were added to sediments and waters were low for at least one species examined (Reed and Dugan, 1978).

Physiological groups and species of bacteria can be identified and their population sizes can be estimated using nucleic acid probes (Amann et al., 1990; Fox et al., 1980; Giovannoni et al., 1988; Holben and Tiedje, 1988; Ogram and Saylor, 1988). These probes consist of fragments of DNA which hybridize either to genes encoding enzymes specific to a particular group of organisms, or to rRNA sequences which are conserved within a particular group and not found in other organisms.

DNA-based hybridization probes have been used to detect and identify microorganisms in a few microbial communities (Amann et al., 1990; Fox et al., 1980; Giovannoni et al., 1988; Holben and Tiedje, 1988; Ogram and Saylor, 1988). Methane monooxygenase (MMO) enzyme systems are unique to methanotrophic bacteria, and the genes that encode them would be suitable as specific probes to detect the DNA of methanotrophs isolated from natural samples (Anthony, 1982; Dalton and Higgins, 1987; Haber et al., 1983). There are two distinct enzyme systems known to catalyze the oxidation of methane to methanol (Dalton and Higgins, 1987; Dalton et al., 1984). One system is defined as particulate (pMMO) because the activity sediments during centrifugation of crude extracts (Dalton et al., 1984). The pMMO has not been purified to homogeneity, and the genes that encode its components have not been cloned. A soluble MMO (sMMO) has been purified from *Methylococcus capsulatus* (Colby and Dalton, 1978) and the type II methylotrophs *Methylosinus trichosporium* OB3b and *Methylobacterium* sp. strain CRL-26 (Patel and Savas, 1987).

The sMMO is composed of three components designated A, B, and C (Colby and Dalton, 1978; Dalton and Higgins, 1987; Fox et al., 1989). Component A, the hydroxylase or oxygen- and methane-binding component, is composed of three subunits of molecular weights 54,000, 42,000, and 17,000. Component B is a colorless protein of unknown function with a molecular weight of 15,000 to 17,000. Component C is a reductase that transfers electrons from NADH to the hydroxylase component and has a molecular weight of approximately 42,000. The composition of sMMO is similar in all three species of methanotrophs from which it has been purified. The genes encoding the three subunits of the hydroxylase component of *Methylococcus capsulatus* have been cloned and sequenced (Mullens and Dalton, 1987; Stainthorpe et al., 1990). We have recently cloned the B component gene from *Methylosinus trichosporium* OB3b (H. C. Tsien and R. S. Hanson, unpublished data). This gene has been used as a hybridization probe to detect homologous sequences on Southern blots prepared from restriction enzyme digests of DNAs isolated from a variety of methanotrophs. We have detected homologous sequences in only a few type II methanotrophs and no type I methanotrophs. Therefore, it appears that the sMMO genes are not universally conserved in methanotrophs.

Genes encoding methanol dehydrogenases and the cytochrome c_L required as an electron acceptor for this enzyme have been cloned from several methylotrophs, i.e., microorganisms grown on methanol or methane (Allen and Hanson, 1985; Bastien et al., 1989; Harms et al., 1987; Machlin and Hanson, 1988; Nunn and Anthony, 1988; Nunn and Lidstrom, 1986). This enzyme catalyzes the oxidation of methanol to formaldehyde and is found in methane oxidizers as well as methylotrophs which cannot metabolize methane. It is composed of two identical polypeptide chains or a single polypeptide chain in some species (DeVries et al., 1990). The methanol dehydrogenase structural gene from *Paracoccus denitrificans* (Harms et al., 1987) and *Methylobacterium organophilum* XX (Machlin and Hanson, 1988) has been sequenced. The genes from *Methylobacterium organophilum* XX hybridized to restriction fragments of DNAs of all gram-negative methanol-utilizing bacteria and all methane-utilizing bacteria tested on Southern blots (Hanson et al., 1989). The restriction fragments from different methylotrophs which hybridized with the cloned methanol dehydrogenase gene differed in size and could be used to identify individual isolates (Hanson et al., 1990). These restriction fragment length polymorphisms can be used to identify bacteria in DNA samples from environmental sources if all bacteria in the samples have been characterized in this fashion. However, methanol-utilizing bacteria (methylotrophs) that do not utilize methane often outnumber the methanotrophs, and the number of different species present in a sample may be large. Therefore, the data obtained by detection of restriction fragments homologous to a methanol dehydrogenase gene may be impossible to analyze. However, this gene probe will be useful for measuring population shifts and determining the fate of methylotrophs in bioreactors and environmental samples that do not have complex populations.

Studies of sequence homologies of 16S rRNAs from a variety of organisms have been used to describe evolutionary relationships among them (Woese, 1987). 16S rRNA molecules are especially suitable for phylogenetic studies because of their size (approximately 1,500 nucleotides) and their abundance in cells (Olsen et al.,

1986; Woese, 1987). Signature probes containing sequences uniquely homologous to 16S rRNA molecules of target species have been used successfully for quantifying bacteria in their natural habitats (Amann et al., 1990; Putz et al., 1990; Stahl et al., 1988). Signature probes can be developed with any specificity desired. Probes that will detect all living organisms (universal probes) or organisms in any of the three kingdoms (eucaryotic, archaebacterial, and eubacterial signature probes) have been described. Probes to detect members of a bacterial species have also been devised (Putz et al., 1990; Stahl et al., 1988). We have sequenced the 16S rRNA molecules isolated from several methanol- and methane-utilizing bacteria (Tsuji et al., 1990). A signature probe with the sequence 5′-CCCTGAGTTATTCCGAAC-3′ hybridizes only to RNAs isolated from methylotrophs and methanotrophs that contain the RuMP pathway for formaldehyde assimilation. It does not hybridize to RNAs isolated from several nonmethylotrophic procaryotic bacteria or serine pathway methylotrophs. Another probe with the sequence 5′-GGTCCGAAGATC CCCCGCTT-3′ hybridized only to RNAs isolated from serine pathway methylotrophs (Tsien et al., 1990).

We have recently shown that the type II methanotrophic bacteria are clustered separately from the other type II methylotrophs that utilize methanol, methylamine, and dichloromethane, but do not utilize methane, on a phylogenetic tree derived from 16S rRNA sequences (B. Bratina and R. S. Hanson, unpublished results). Similarly, type I methanotrophs form a cluster separate from the one type X methanotroph (*Methylococcus capsulatus*) and RuMP pathway methanol utilizers from which 16S rRNAs have been sequenced. Therefore, it should be possible to select signature probes for each group of methanotrophs that will not hybridize with rRNAs from methylotrophic bacteria which are unable to oxidize methane.

Signature probes labeled with fluorescent compounds have been used to detect individual cells of specific microbes (DeLong et al., 1989). The probes were hybridized to RNA in cells fixed to microscope slides and detected by indirect fluorescence microscopy. We have shown that the signature probes described above are specific for the detection of cells of the specific groups of methylotrophs when this procedure is employed (Tsien et al., 1990).

rRNA signature probes are attractive because rRNAs are present in relatively large amounts in all living cells and there is sufficient conservation in some sequences to develop probes with broad specificity. Yet there is sufficient diversity in sequences to develop probes that are specific for species or strains (Amann et al., 1990; Fox et al., 1980; Putz et al., 1990; Stahl et al., 1988). The use of these probes to determine the presence and amount of a target 16S rRNA in an RNA sample extracted from an environmental source and bound to filters (Stahl et al., 1988) is good evidence for the presence of a particular organism or group of organisms. However, it is difficult to determine the population size of the target organisms because the number of copies of rRNA in bacteria in their natural habitat is variable and difficult to predict or measure. On the other hand, the amount of specific rRNA provides a measure of the physiological potential or activity of a population that can be as informative as population size. The use of fluorescent labeled signature probes hybridized to RNA in cells permits the enumeration of cells containing 16S rRNAs homologous to the probe (Amann et al., 1990; DeLong et al., 1989; Putz et al., 1990; Stahl et al., 1988). A small number of the target bacteria which fluoresce can be enumerated in the presence of large number of nonfluorescent bacteria. In addition, the morphology of the fluorescent bacteria is maintained.

METHANE OXIDATION IN AQUATIC ENVIRONMENTS

During summer stratification in eutrophic, dimictic freshwater lakes, methane is produced in sediments, diffuses upward through the anoxic hyperlimnia, and is rapidly oxidized as it reaches the metalimnia where oxygen is present. In Lake Mendota, Madison, Wisconsin, the average rate of methane utilization during the summer of 1979 was 460 μmol m^{-2} h^{-1} (Harrits and Hanson, 1980). This amounted to approximately 45% of the total methane production. The total methane recycled during the three summer months was estimated to be 6×10^8 g (Hanson, 1980). The remainder of the methane produced was stored in the hypolimnion until fall turnover, when it was oxidized as the entire water column became aerobic. Approximately 8% of the methane produced during the summer stratification was evaded to the atmosphere (Fallon et al., 1980). In Lake 227, Canada, methane production rates were approximately 16% of those observed in Lake Mendota (Hanson, 1980). Methane oxidation in this lake was also confined to the metalimnion during summer stratification (Flett et al., 1976; Rudd and Hamilton, 1978; Rudd et al., 1974; Rudd and Taylor, 1980). The dissolved inorganic nitrogen concentrations in Lake 227 were low, and the methanotrophs were oxygen sensitive owing to their apparent necessity to fix atmospheric nitrogen. More than 90% of the methane oxidation evasion occurred during and shortly after fall turnover in Lake 227 (Rudd and Hamilton, 1978).

In the oligotrophic Lake Stecklin, Germany, in which the entire water column was aerobic, methane oxidation was confined to the upper aerobic layers of the sediments (Heyer and Babenzein, 1985). The rates of methane oxidation were very high (475 to 961 μmol m^{-2} h^{-1}). In Lake Kivu, Africa, a permanently stratified lake which contained approximately 50 km^3 of methane, approximately 1.2×10^8 m^3 of methane was oxidized annually at the oxic-anoxic interface in the water column (Jannasch, 1975). The methane oxidation rate approximately equaled the rate of production, and little was evaded to the atmosphere. These studies and several others described elsewhere (Cappenberg, 1972; Hanson, 1980; Heyer and Babenzein, 1985; Heyer et al., 1984; Whittenbury et al., 1976) point out the importance of methanotrophic bacteria in recycling the large amounts of methane produced in freshwater environments. For this reason, lakes are not a large source of atmospheric methane (Cicerone and Oremland, 1988).

Nearly all the methane produced in anaerobic layers of peat bogs was recycled by aerobic methanotrophs (Heyer and Suchow, 1985). Rates of methane oxidation as high as 405 μmol m^{-2} h^{-1} were observed in these bogs, where the pH varied from 3.8 to 5.3. It is surprising that obligate methanotrophs identified as *Methylosinus sporium* and *Methylosinus trichosporium* were isolated by enrichment culture from acidic peat bogs (Heyer and Suchow, 1985). In our experience, the rate of oxidation of methane by these species is very low at pH values below 5.5. It is possible that exospores of these bacteria grew elsewhere and were transported to the bogs.

Estimates of methane evasion from ocean surface waters are 1 to 7 Tg year^{-1} (Seiler, 1983). Rates of aerobic methane oxidation are much lower in oceans than in freshwater lakes because the productivity of oceans is much lower (Seiler, 1983; Ward et al., 1987). In more productive marine environments and salt lakes where anaerobic conditions required for methanogenesis exist, anaerobic methane oxidation is primarily responsible for methane recycling (Iversen and Blackburn, 1979; Iversen and Jorgenson, 1985; Iversen et al., 1987; Reeburgh, 1976, 1980, 1982; Sansone and Martens, 1978). The unidentified electron acceptors required for anaerobic methane oxidation in marine environments are presumably limiting in fresh water, where oxygen is required as a terminal electron acceptor (Whalen and Reeburgh, 1988, 1990). In at least one freshwater lake, low rates of anaerobic methane oxidation were observed (Panganiban et al., 1979), and bacteria capable of oxidizing methane with sulfate as an electron acceptor were enriched from sediment surfaces (Hanson, 1980).

A very interesting symbiosis involving the intracellular growth of methane-oxidizing bacteria in gill tissues of deep-sea mussels has been described (Cavanaugh et al., 1987; Childress et al., 1986).

METHANE OXIDATION IN TERRESTRIAL ENVIRONMENTS

Soils have not been considered as significant sinks for methane until recently, and little has been published concerning the magnitude of, and factors controlling, methane oxidation in terrestrial environments. Methane consumption has been reported in agricultural soils (Megraw and Knowles, 1987a, 1987b) and in soils of forests (Keller et al., 1983; Steudler et al., 1989), savannah and tundra (Whalen and Reeburgh, 1988, 1990), and bogs (Harriss et al., 1982; Harriss et al., 1985; Yavitt et al., 1978). Schollenberger (1930) and Harper (1939) observed that soils exposed to leaks from natural gas pipelines had higher organic matter and nitrogen content than soil not exposed to natural gas. Davis et al. (1964) isolated methane-oxidizing, nitrogen-fixing bacteria from oil field soils, garden soil, and soil exposed to natural gas.

Steudler et al. (1989) measured aerobic methane consumption rates of up to 3.2 mg of methane cm^{-2} day^{-1} in aerobic, temperate forest soils in Massachusetts. They calculated global methane consumption by temporal and boreal forests to be approximately 9.3 Tg of methane-C year^{-1}, a rate considerably higher than previous estimates. Interestingly, they found that fertilization with NH_4NO_3 significantly reduced methane consumption by these soils. Fertilization is one of a number of agricultural practices including pesticide use and tillage regimes whose impact on methane metabolism in agricultural soils should be examined.

The identity of the organisms catalyzing methane oxidation in acid soils such as those of bogs, tundra, or coniferous forest is unknown because to date there are no isolates, with the exception of a methane-oxidizing yeast (Wolf and Hanson, 1978), which are known to grow and oxidize methane at pH values below 4.4. Of all the soils tested by Heyer (1977), only acid soils of coniferous forests and heath failed to yield methanotrophs in conventional enrichments for methane-oxidizing bacteria. This result may reflect a bias of the enrichment technique rather than the abundance of methanotrophs in these environments.

Methanotrophs which have been isolated from soil and water and grown in the laboratory are ill-adapted to consuming atmospheric concentrations of methane because their K_s values for this substrate are too high (Conrad, 1984; Megraw and Knowles, 1987b). K_s values ranging from 1.7 to 75 μM have been published for various mixed and pure cultures of methanotrophs (Bedard and Knowles, 1989; Conrad, 1984; Harrits and Hanson, 1980; Lamb and Garver, 1980; Megraw and Knowles, 1987b; Rudd and Hamilton, 1978; Rudd et al., 1974; Rudd and Taylor, 1980; Seiler, 1983) and for methane oxidation in freshwater lakes and soils (Conrad, 1984). Conrad (1984) estimated the requirements for methane oxidation in soils given an atmospheric concentration of 1.7 ppm in equilibrium with soil water using the equation $N = d[K_s + S/V_{max} S]$, where d was the observed rate of methane utilization (0.5 nmol cm^{-2} day^{-1} or 80 μg cm^{-2} day^{-1}) and S was the substrate concentration in equilibrium with atmospheric methane (2.5 nM). K_s, V_{max}, and the methane-consuming biomass (N) were variables. If a lower limit of 1,000 nM for the K_s and a V_{max} of 20 mmol g^{-1} of cell dry weight h^{-1} are assumed, the biomass required to achieve the observed oxidation rate would be about 10^{-5} g of cell dry weight cm^{-2}. This represents a relatively high population of methanotrophs (10^7 to 10^8 cells per g), and the value would increase if the K_s was higher or the V_{max} lower. A cultivated humisol (Megraw and Knowles, 1987a) and a drained paddy soil (Krämer et al., 1990) each contained 10^5 CFU of methanotrophs per g. This corresponds to 10^{-7} to 10^{-8} g of cell dry weight g^{-1} of soil, assuming an average mass of 1 pg per cell (Conrad, 1984). This biomass approaches that required for the observed rates of methane oxidation.

Tundra is a large carbon reservoir, comprising 15% of global soil carbon, and an important methane source, accounting for about 10% of the global methane budget (Whalen and Reeburgh, 1990). Whalen and Reeburgh (1990) recently reported rapid equilibration of moist tundra soil with ambient concentrations of atmospheric methane and oxidation rates of 0.2 to 4.2 mg of CH_4 m^{-2} day^{-1}. In waterlogged tundra, methane oxidation was confined to the oxic-anoxic boundary near the water table and net production of methane was observed (Whalen and Reeburgh, 1988). When these soils have a reduced water content, however, they consume methane without a lag and become a methane sink (Whalen and Reeburgh, 1990). Similarly, flux measurements taken in Great Dismal Swamp over a 17-month period revealed net methane production under waterlogged conditions and net methane consumption during summer drought (Harriss et al., 1982). An agricultural humisol incubated in the laboratory produced methane under anaerobic conditions and consumed it under aerobic conditions (Megraw and Knowles, 1987b). These observations suggest that soils may generally contain both methanotrophic and methanogenic populations and that the water content, and consequently oxygen status, of the soil may determine whether it is a methane source or sink. This phenomenon may be of particular significance in the context of potential global warming. If, for example, global warming increased the depth and expanse of dry tundra soils, the rate of methane consumption in this environment would be anticipated to increase. Since tundra ecosystems make up 5% of the Earth's land surface and some climate change models predict higher temperatures and drier summer conditions in Northern latitudes, this phenomenon might significantly decrease global methane concentrations (Whalen and Reeburgh, 1988, 1990). Moreover, should dry conditions prevent the growth of vascular plants which channel methane directly from anaerobic soil to the atmosphere, methane oxidation rates would be further enhanced (Whalen and Reeburgh, 1988, 1990).

FORTUITOUS METABOLISM OF METHANOTROPHIC BACTERIA

The methanotrophs oxidize ammonia through hydroxylamine to nitrite (Hutton and Zobell, 1953; O'Neill and Wilkinson, 1977). The oxidation of both ammonia and hydroxylamine is accompanied by the production of small amounts of nitrous oxide (Knowles and Topp, 1988; Yoshinari, 1985). Nitrous oxide was also produced by *Methylosinus trichosporium* when incubated anaerobically with NO_2^- or NO_3^-, suggesting a second reductive route of N_2O production (Knowles and Topp, 1988; Yoshinari, 1985). In these and numerous other respects, the methanotrophs resemble the chemoautotrophic nitrifiers (Bedard and Knowles, 1989).

MMO almost certainly catalyzes the hydroxylation of NH_4^+ to NH_2OH, although this activity has yet to be documented with purified enzyme. This conclusion is based largely on the following observations (Dalton, 1977; Dalton and Higgins, 1987; O'Neill and Wilkinson, 1977): (i) both CH_4 and NH_4^+ metabolism respond similarly to a variety of metabolic inhibitors tested; (ii) NH_4^+ competitively inhibits CH_4 oxidation (Dalton and Higgins, 1987; O'Neill and Wilkinson, 1977); and (iii) the stoichiometry of O_2 uptake and NH_4^+ oxidation suggests a pathway identical to that catalyzed by the chemoautotrophic nitrifiers, namely $NH_4^+ + 1.5\ O_2 \rightarrow HNO_2 + H_2O$.

In vitro experiments suggest that ammonia oxidation by methane oxidizers in situ may be influenced by the availability of carbon. MMO requires a source of reductant to catalyze the hydroxylase reaction. Methane metabolism yields CH_3OH which, through subsequent oxidation reactions, provides the required reductant. In the case of NH_4^+, however, it is not known whether or not the methanotrophs generally possess the ability to recover reductant from the metabolism of the intermediate NH_2OH. If they cannot, then ammonia metabolism would be expected to be reductant limited in the absence of another source of NADH. To date, no methanotroph has been found to grow with NH_4^+ as the energy source, and NH_4^+ will not support CO_2 fixation by methanotrophs (Stanley and Dalton, 1982). In one study with membrane preparations from *Methylococcus thermophilus*, the oxidation of NH_2OH to NO_2^- was accompanied by O_2 consumption, NAD^+ reduction, cytochrome *c* reduction, and the production of ATP (Malashenko et al., 1979). Ammonia metabolism by whole cells was stimulated by the addition of formate, formaldehyde (O'Neill and Wilkinson, 1977), or low concentrations of CH_4 (10 μM) (Knowles and Topp, 1988). This stimulation was presumably due to production of NADH. Interestingly, while 10 μM CH_4 greatly stimulated the production of NO_2^- from NH_4^+, at the same time nitrous oxide production was almost completely eliminated (Knowles and Topp, 1988). At increasing CH_4 concentrations, methane increasingly inhibited NH_4^+ oxidation, presumably by competing for the active site of MMO.

Evidence to date suggests that NH_2OH is oxidized by a specific enzyme. Hydroxylamine oxidation to NO_2^- was not inhibited by agents which blocked MMO (Dalton, 1977). Methanol and NH_2OH oxidation activities in extracts of *Methylococcus thermophilus* were separated by ion-exchange chromatography, and the NH_2OH oxidoreductase resembled that from the chemoautotrophic nitrifiers in a number of respects (Sokolov et al., 1980).

Per-cell activities of NH_4^+ oxidation by the methanotrophs are generally two orders of magnitude lower than those of the chemoautotrophic nitrifiers (Bedard and Knowles, 1989). Nitrification by methanotrophs is therefore likely to predominate only where they have a large numerical advantage, such as the metalimnia of stratified lakes (Bedard and Knowles, 1989; Harrits and Hanson, 1980). Interestingly, Megraw and Knowles (1989a, 1989b) have recently observed methane-dependent nitrification in a humisol. It is also possible that methanotrophs may be significant participants in the nitrogen cycle in environments where chemoautotrophic nitrifiers are inhibited, such as highly acidic soils.

Methanotrophs have a high assimilatory requirement for nitrogen (Topp and Knowles, 1984; Megraw and Knowles, 1987a). In environments where significant CH_4 oxidation is taking place, the methanotrophs could potentially reduce the availability of inorganic N, thus reducing rates of N_2O-yielding processes such as nitrification and denitrification.

Methane-dependent O_2 consumption can result in whole-lake anoxia in lakes under ice cover (Rudd and Hamilton, 1978), thus contributing to conditions favoring denitrification. CH_4 addition to soil incubated in vitro blocked nitrification, probably by imposing an O_2 limitation (Megraw and Knowles, 1987a).

A number of methanotrophs have been found to oxidize CO to CO_2 (Ferenci et al., 1975; Hubley et al., 1974; Stirling and Dalton, 1979). This activity did not support the growth of methanotrophs (Ferenci, 1974). The high affinities for CO of the methanotrophs suggest that they may be significant agents in the oxidation of CO in situ (Bedard and Knowles, 1989).

In addition to NH_4^+, MMO can transform a wide variety of alkanes, alkenes, aromatic compounds, and more complex hydrocarbons (Anthony, 1982). These include compounds which are significant environmental pollutants such as the halogenated aliphatics including chloro- and bromomethane, dichlorovinyl, and trichloroethylene. There is considerable interest in decontaminating groundwater polluted with haloaliphatics by stimulating the methanotrophs by in situ nutrient addition. Addition of methane to soil accelerated the degradation of trans-1,2-dichloroethylene, dichloromethane, 1,2-dichloroethane, trichloroethylene, and 1,1,1-trichloroethane (Henson et al., 1988; Strand and Shippert, 1986; Wilson and Wilson, 1985). Microbial methane-oxidizing consortia degraded trichloroethylene and other chlorinated alkenes (Fogel et al., 1986; Little et al., 1988), as do pure cultures of *Methylosinus trichosporium* (Oldenhuis et al., 1989; Tsien et al., 1989; Wackett et al., 1989) and a type I isolate, strain 46-1 (Little et al., 1988). The expression of soluble MMO by whole cells is required for rapid trichloroethylene degradation (Oldenhuis et al., 1989; Tsien et al., 1989), and purified soluble MMO catalyzes the degradation of chlorinated aliphatics (Fox et al., 1991). The methanotrophs have the highest specific activity of a number of mono- and dioxygenase-producing organisms tested (Tsien et al., 1989). Evidence indicating that relatively few methanotrophs can produce soluble MMO (Brusseau et al., 1991) and that soluble-MMO synthesis is repressed by Cu in concentrations which may be encountered in situ (Dalton and Higgins, 1987; Oldenhuis et al., 1989; Prior and Dalton, 1985; Stanley et al., 1983; Tsien et al., 1989) suggests that the role of these organisms in the degradation of haloalipathics in the environment may be limited.

CONCLUSIONS

It is evident that the oxidation of methane by methanotrophic bacteria provides an important sink for methane that would otherwise escape from freshwater, soil, and marine environments to the atmosphere. Methanotrophic bacteria may also play an important role in the production and consumption of other radiatively important trace gases (N_2O and CO).

A number of key issues concerning the biology and ecology of these organisms remain to be resolved. The microorganisms catalyzing methane oxidation in acidic and anaerobic environments have yet to be identified. Kinetics of methane consumption by soils suggest that methane-utilizing organisms are very abundant or can oxidize methane at concentrations lower than laboratory isolates are capable of using. More studies should be undertaken on the kinetics of methane metabolism and growth at low substrate concentrations. Perhaps there are environmental factors which permit behaviors in situ such as the synthesis of high-affinity methane uptake systems, which have yet to be mimicked in vitro. The identity and population sizes of methanotrophs must be determined before the ecology of these organisms can be fully understood. The application of nucleic acid probes should help elucidate these questions. Finally, the magnitude and direction of methane fluxes from various terrestrial environments and the factors controlling them should be further studied.

Acknowledgments. Research performed at the Gray Freshwater Biological Institute that is reported in this manuscript was supported by grants to R.S.H. (DE-FG02-88ER 13862 from the U.S. Department of Energy and BSR 8903833 from the National Science Foundation). This is Land Resources Research Center Contribution no. 90-76.

REFERENCES

Abramochinka, F. N., L. V. Bezrukova, A. V. Koshelev, V. F. Gal'chenko, and M. V. Ivanov. 1987. Microbial oxidation of methane in a body of freshwater. *Microbiology* **56:**375–382.

Allen, L. N., and R. S. Hanson. 1985. Construction of broad-host-range cosmid cloning vectors: identification of genes necessary for growth of *Methylobacterium organophilum* on methanol. *J. Bacteriol.* **161:**955–962.

Alperin, M. J., and W. S. Reeburgh. 1985. Inhibition experiments on anaerobic methane oxidation. *Appl. Environ. Microbiol.* **50:**940–945.

Amann, R. I., L. Krumholz, and D. A. Stahl. 1990. Fluorescent-oligonucleotide probing of whole cells for determinative, phylogenetic, and environmental studies in microbiology. *J. Bacteriol.* **172:**762–770.

Anthony, C. 1982. *The Biochemistry of Methylotrophs.* Academic Press, Inc. (London), Ltd., London.

Bastien, C., S. Machlin, Y. Zhang, K. Donaldson, and R. S. Hanson. 1989. Organization of genes required for the oxidation of methanol to formaldehyde in three type II methylotrophs. *J. Bacteriol.* **55:**3124–3130.

Bedard, C., and R. Knowles. 1989. Physiology, biochemistry, and specific inhibitors of CH_4, NH_4^+, and CO oxidation by methanotrophs and nitrifiers. *Microbiol. Rev.* **53:**68–84.

Bohlool, B. B., and E. L. Schmidt. 1980. The immunofluorescence approach in microbial ecology. *Adv. Microbiol. Ecol.* **4:**203–241.

Bone, T. L., and D. L. Balkwill. 1986. Improved flotation technique for microscopy of in situ soil and sediment microorganisms. *Appl. Environ. Microbiol.* **51:**462–468.

Boxrukova, L. V., Y. I. Nikolenko, A. I. Nesterov, V. F. Galchenko, and M. V. Ivanov. 1983. Comparative serological analysis of methanotrophic bacteria. *Microbiology* **52:**639.

Brusseau, G. A., H. C. Tsien, R. S. Hanson, and L. P. Wackett. 1990. Optimization of trichloroethylene oxidation by methanotrophs and the use of a colorimetric assay to detect soluble methane monooxygenase activity. *Biodegradation* **1:**19–29.

Cappenberg, T. E. 1972. Ecological observations on heterotrophic methane-oxidizing and sulfate-reducing bacteria in a pond. *Hydrobiologia* **40:**471–485.

Cavanaugh, C. M., P. R. Levering, J. S. Maki, R. Mitchell, and M. Lidstrom. 1987. Symbiosis of methylotrophic bacteria and deep-sea mussels. *Nature* (London) **325:**346–348.

Childress, J. J., C. R. Fisher, J. M. Brooks, M. C. Kennicutt II, R. Bidigare, and A. E. Anderson. 1986. A methanotrophic marine molluscan (*bivalvia, mytilidae*) symbiosis: methane-oxidizing mussels. *Science* **233:**1306–1313.

Cicerone, R., and R. S. Oremland. 1988. Biogeochemical aspects of atmospheric methane. *Global Biogeochem. Cycles* **2:**299–327.

Colby, J., and H. Dalton. 1978. Resolution of the methane monooxygenase of *Methylococcus capsulatus* (Bath) into three components: purification and properties of component C, a flavoprotein. *Biochem. J.* **171:**461–468.

Conrad, R. 1984. Capacity of aerobic microorganisms to utilize and grow on atmospheric trace gases (H_2, CO, CH_4), p. 461–467. *In* J. J. Klug and C. A. Reddy (ed.), *Current Perspectives in Microbial Ecology.* American Society for Microbiology, Washington, D.C.

Conrad, T., and W. Seiler. 1982. Arid soils as a source of atmospheric carbon monoxide. *Geophys. Res. Lett.* **9:**1353–1356.

Crutzen, P. J., and T. E. Graedel. 1986. The role of atmospheric chemistry in environment development interactions, p. 107–130. *In* W. C. Clark and R. E. Munn (ed.), *Sustainable Development of the Biosphere.* Cambridge University Press, Cambridge.

Dalton, H. 1977. Ammonia oxidation by the methane oxidizing bacterium *Methylococcus capsulatus* strain Bath. *Arch. Microbiol.* **114:**273–279.

Dalton, H., and I. J. Higgins. 1987. Physiology and biochemistry of methylotrophic bacteria, p. 89–94. *In* H. W. Van Verseveld and J. A. Duine (ed.), *Microbial Growth on C_1 Compounds.* Martinus Nijhoff Publishers, Dordrecht, The Netherlands.

Dalton, H., S. D. Prior, D. J. Leak, and S. H. Stanley. 1984. Regulation and control of methane monooxygenase, p. 75–82. *In* R. L. Crawford and R. S. Hanson (ed.), *Microbial Growth on C_1 Compounds.* American Society for Microbiology, Washington, D.C.

Davies, S. L., and R. Whittenbury. 1970. Fine structure of methane and other hydrocarbon-utilizing bacteria. *J. Gen. Microbiol.* **61:**227–232.

Davis, J. B., V. F. Coty, and J. P. Stanley. 1964. Atmospheric nitrogen fixation by methane-oxidizing bacteria. *J. Bacteriol.* **88:**468–472.

DeLong, E. F., G. S. Wickham, and N. R. Pace. 1989. Phylogenetic stains: ribosomal RNA-based probes for identification of single cells. *Science* **243:**1360–1363.

DeVries, G. E., U. Kües, and D. Stahl. 1990. Physiology and genetics of methylotrophic bacteria. *FEMS Microbiol. Lett.* **75:**57–102.

Ehhalt, D. H. 1974. The atmospheric cycle of methane. *Tellus* **26:**58–70.

Fallon, R., S. Harrits, R. S. Hanson, and T. D. Brock. 1980. The role of methane in internal carbon cycling in Lake Mendota during summer stratification. *Limnol. Oceanogr.* **25:**357–360.

Ferenci, T. 1974. Carbon monoxide stimulated respiration in methane-utilizing bacteria. *FEBS Lett.* **41:**94–98.

Ferenci, T., T. Strom, and J. R. Quayle. 1975. Oxidation of carbon monoxide and methane by *Pseudomonas methanica*. *J. Gen. Microbiol.* **91:**79–91.

Flett, R. J., D. W. Schindler, R. D. Hamilton, and N. E. R. Campbell. 1976. Nitrogen fixation in Canadian Precambrian Shield Lakes. *Can. J. Fish Aquatic Sci.* **37:**494–505.

Fogel, M. M., A. R. Taddeo, and S. Fogel. 1986. Biodegradation of chlorinated ethenes by a methane-utilizing mixed culture. *Appl. Environ. Microbiol.* **51:**720–724.

Fox, B. G., J. G. Borneman, L. P. Wackett, and J. D. Lipscomb. 1991. Haloalkene oxidation by the soluble methane monooxygenase from *Methylosinus trichosporium* OB3b: mechanistic and environmental implications. *Biochemistry* **29:**6419–6427.

Fox, B. G., W. A. Froland, J. Dege, and J. D. Lipscomb. 1989. Methane monooxygenase from *Methylosinus trichosporium* OB3b. *J. Biol. Chem.* **264:**10023–10033.

Fox, G. E., E. Stackebrandt, R. B. Hespell, J. Gibson, J. Maniloff, T. A. Dyer, R. S. Wolfe, W. E. Balch, R. S. Tanner, L. J. Magrum, L. B. Zablen, R. Blakemore, R. Gupta, L. Bonen, B. J. Lewis, D. A. Stahl, K. R. Luehrsen, K. N. Chen, and C. R. Woese. 1980. The phylogeny of prokaryotes. *Science* **209:**457–463.

Gachert, J. B., C. P. Antworth, P. D. Nichols, and D. C. White. 1985. Phospholipid ester-linked fatty acid profiles as reproducible assays for changes in prokaryotic community structure of estuarine sediments. *FEMS Microbiol. Ecol.* **31:**147–158.

Gal'chenko, V. F. 1975. Use of silica gel for isolation of pure cultures of obligate methane-oxidizing microorganisms. *Appl. Biochem. Microbiol.* **11:**447–450. (In Russian.)

Gal'chenko, V. F., F. N. Abramochkina, L. V. Bezrukova, E. N. Sokolova, and M. V. Ivanov. 1988. Species composition of aerobic methanotrophic microflora in the Black Sea. *Microbiology* **57:**248–253.

Gal'chenko, V. F., and L. V. Andreev. 1984. Taxonomy of obligate methylotrophs, p. 269–281. *In* R. L. Crawford and R. S. Hanson (ed.), *Microbial Growth on C_1 Compounds*. American Society for Microbiology, Washington, D.C.

Giovannoni, S. J., E. F. DeLong, G. J. Olsen, and N. R. Pace. 1988. Phylogenetic group-specific oligonucleotide probes for identification of single microbial cells. *J. Bacteriol.* **170:**720–726.

Graedel, T. E., and P. J. Crutzen. 1989. The changing atmosphere. *Sci. Am.* **261:**136–143.

Haber, C. L., L. N. Allen, and R. S. Hanson. 1983. Methylotrophic bacteria: biochemical diversity and genetics. *Science* **221:**1147–1151.

Hanson, R. S. 1980. Ecology and diversity of methylotrophic organisms. *Adv. Appl. Microbiol.* **26:**3–39.

Hanson, R. S., A. I. Netrusov, and K. Tsuji. 1991. The obligate methanotrophic bacteria; *Methylococcus, Methylomonas* and *Methylosinus*, p. 661–684. *In* A. Balows, H. G. Truper, M. Dworkin, W. Harder, and K. H. Schleifer (ed.), *The Procaryotes*. Springer-Verlag, New York.

Hanson, R. S., K. Tsuji, C. Bastien, H. C. Tsien, B. Bratina, G. Brusseau, and S. Machlin. 1990. Genetic and biochemical studies of methylotrophic bacteria, p. 215–234. *In* C. Aiken and J. Smith (ed.), *Coal and Gas Biotechnology*. Institute for Gas Technology, Chicago.

Harms, N., G. E. DeVries, K. Maurer, J. Hoogendak, and A. H. Stouthamer. 1987. Isolation and nucleotide sequence of methanol dehydrogenase structural gene from *Paracoccus denitrificans*. *J. Bacteriol.* **169:**3969–3975.

Harper, H. J. 1939. The effect of natural gas on the growth of microorganisms and the accumulation of nitrogen and organic matter in the soil. *Soil Sci.* **48:**461–466.

Harriss, R. C., D. I. Sebacher, and F. P. Day. 1982. Methane flux in the Great Dismal Swamp. *Nature* (London) **297:**673–674.

Harriss, R. C., E. Gorham, D. I. Sebacher, K. B. Bartlett, and P. A. Flebbe. 1985. Methane flux from Northern peatlands. *Nature* (London) **315:**652–654.

Harrits, S., and R. S. Hanson. 1980. Stratification of aerobic methane-oxidizing organisms in Lake Mendota, Madison, Wisconsin. *Limnol. Oceanogr.* **25:**412–421.

Henson, J. M., M. V. Yates, J. W. Cochrane, and D. L. Shackleford. 1988. Microbial removal of halogenated methanes, ethanes, and ethylenes in an aerobic soil exposed to methane. *FEMS Microbiol. Ecol.* **53:**193–201.

Heyer, J. 1977. Results of enrichment experiments of methane-assimilating organisms from an ecological point of view, p. 19–21. *In* G. K. Skryabin, M. V. Ivanov, E. N. Kondratjeva, G. A. Zavarzin, Yu. A. Trotsenko, and A. I. Netrusov (ed.), *Microbial Growth on C_1-Compounds*. USSR Academy of Sciences, Puschino.

Heyer, J., and H.-D. Babenzein. 1985. Untersuchunge des Methankreislaufes in einem oligotrochen See (Stechlinsee). *Limnologia* **16:**267–276.

Heyer, J., Yu. Malashenko, U. Berger, and E. Budkova. 1984. Vertreitung methanotropher Bakterien. *Z. Allgemeine Mikrobiol.* **24:**725–744.

Heyer, J., and R. Suchow. 1985. Ökologishe Untersuchungen der Methanoxidation in einem sauren Moorsee. *Limnologica* **6:**247–266.

Higgins, I. J., D. J. Best, R. C. Hammond, and D. Scott. 1981. Methane oxidizing microorganisms. *Microbiol. Rev.* **45:**556–590.

Holben, W. E., and J. M. Tiedje. 1988. Applications of nucleic acid hybridization to microbial ecology. *Ecology* **69:**561–568.

Hubley, J. H., J. R. Mitton, and J. F. Wilkinson. 1974. The oxidation of carbon monoxide by methane-oxidizing bacteria. *Arch. Microbiol.* **95:**365–368.

Hutton, W. E., and C. E. Zobell. 1949. The occurrence and characteristics of methane-oxidizing bacteria in marine sediment. *J. Bacteriol.* **58:**463–473.

Hutton, W. E., and C. E. Zobell. 1953. Production of nitrite from ammonia by methane-oxidizing bacteria. *J. Bacteriol.* **65:**216–219.

Iversen, N., and T. H. Blackburn. 1979. Methane production and oxidation in Santa Barbara Basin sediments. *Estuarine Coastal Mar. Sci.* **8:**379–385.

Iversen, N., and B. B. Jorgenson. 1985. Anaerobic methane oxidation rates at the sulfate methane transition in marine sediments from Kattegat and Skagerrak (Denmark). *Limnol. Oceanogr.* **30:**944–955.

Iversen, N., R. S. Oremland, and M. Klug. 1987. Big Soda Lake (Nevada) pelagic methanogenesis and anaerobic methane oxidation. *Limnol. Oceanogr.* **32:**804–814.

Jannasch, H. W. 1975. Methane oxidation in Lake Kivu (Central Africa). *Limnol. Oceanogr.* **20:**860–864.

Keller, M., T. J. Goreau, S. C. Wofsy, W. A. Kaplan, and M. B. McElroy. 1983. Production of nitrous oxide and consumption of methane by forest soils. *Geophys. Res. Lett.* **10:**1156–1159.

Knowles, R., and E. Topp. 1988. Some factors affecting nitrification and the production of nitrous oxide by the methanotrophic bacterium *Methylosinus trichosporium* OB3b, p. 383–393. *In* G. Giovannozzi-Sermanni and P. Nannipieri (ed.), *Current Perspectives in Environmental Biogeochemistry*. Consiglio Nazionali delle Richerche-I.P.R.A., Rome.

Krämer, M., M. Baumgärtner, M. Bender, and R. Conrad. 1990. Consumption of NO by methanotrophic bacteria in pure culture and in soil. *FEMS Microbiol. Ecol.* **73:**345–350.

Lamb, S. C., and J. C. Garver. 1980. Batch and continuous culture studies of a methane-utilizing mixed culture. *Biotechnol. Bioeng.* **22:**2097–2118.

Lidstrom, M. E. 1983. Methane consumption in Framvaren, an anoxic marine fjord. *Limnol. Oceanogr.* **28:**1247–1251.

Little, C. D., A. V. Palumbo, S. E. Herbes, M. E. Lidstrom, R. L. Tyndall, and P. J. Gilmer. 1988. Trichloroethylene biodegradation by a methane-oxidizing bacterium. *Appl. Environ. Microbiol.* **54:**951–956.

Machlin, S. M., and R. S. Hanson. 1988. Nucleotide sequence and transcriptional start site of the *Methylobacterium organophilum* XX methanol dehydrogenase structural gene. *J. Bacteriol.* **170:**4739–4747.

Malashenko, Y. R., I. G. Sokolov, V. A. Romanovskaya, and Y. B. Shkurko. 1979. Elements of lithotrophic metabolism in the obligate methylotroph *Methylococcus thermophilus*. *Microbiology* **48:**468–474. (Translation of *Mikrobiologiya* **48:**592–598.)

Megraw, S. R., and R. Knowles. 1987a. Active methanotrophs suppress nitrification in a humisol. *Biol. Fertil. Soils* **4:**205–212.

Megraw, S. R., and R. Knowles. 1987b. Methane production and consumption in a cultivated humisol. *Biol. Fertil. Soils* **5:**56–60.

Megraw, S. R., and R. Knowles. 1989a. Methane dependent nitrate production by a microbial consortium enriched from cultivated humisol. *FEMS Microbiol. Ecol.* **62:**359–366.

Megraw, S. R., and R. Knowles. 1989b. Isolation, characterization, and nitrification potential of a methylotroph and two heterotrophic bacteria from a consortium showing methane-dependent nitrification. *FEMS Microbiol. Ecol.* **62:**367–374.

Mullens, I. A., and H. Dalton. 1987. Cloning of the gamma-subunit methane monooxygenase from *Methylococcus capsulatus*. *Biotechnology* **5:**490–493.

Murrell, J. C., and H. Dalton. 1983. Nitrogen fixation in obligate methanotrophs. *J. Gen. Microbiol.* **129:**3481–3486.

Nichols, P. D., G. A. Smith, C. P. Antworth, R. S. Hanson, and D. C. White. 1986. Phospholipid and lipopolysaccharide normal and hydroxy fatty acids as potential signatures for methane-oxidizing bacteria. *FEMS Microbiol. Lett.* **31:**327–336.

Nunn, D. N., and C. Anthony. 1988. The nucleotide sequence and deduced amino acid sequence of the genes for cytochrome cL and a hypothetical second subunit of the methanol dehydrogenase of *Methylobacterium* AM1. *Nucleic Acids Res.* **16:**7722–7723.

Nunn, D. N., and M. E. Lidstrom. 1986. Isolation of complementation analysis of 10 methanol oxidation mutant classes and identification of the methanol dehydrogenase structural gene of *Methylobacterium* sp. strain AM1. *J. Bacteriol.* **166:**581–590.

Ogram, A. V., and G. S. Saylor. 1988. The use of gene probes in the rapid analysis of natural microbial communities. *J. Ind. Microbiol.* **3:**281–292.

Oldenhuis, R., R. L. J. M. Vink, D. B. Janssen, and B. Witholt. 1989. Degradation of chlorinated aliphatic hydrocarbons by *Methylosinus trichosporium* OB3b expressing soluble methane monooxygenase. *Appl. Environ. Microbiol.* **55:**2819–2826.

Olsen, G. J., D. J. Lane, S. J. Giovannoni, N. R. Pace, and D. A. Stahl. 1986. Microbial ecology and evolution: a ribosomal RNA approach. *Annu. Rev. Microbiol.* **40:**337–365.

O'Neill, J. G., and J. F. Wilkinson. 1977. Oxidation of ammonia by methane-oxidizing bacteria and the effects of ammonia on methane oxidation. *J. Gen. Microbiol.* **100:**407–412.

Oremland, R. S., L. G. Miller, and M. J. Whitticar. 1987. Sources and flux of natural gases from Mono Lake, California. *Geochim. Cosmochim. Acta* **51:**2915–2929.

Panganiban, A. T., Jr., T. E. Patt, W. Hart, and R. S. Hanson. 1979. Oxidation of methane in the absence of oxygen in lake water samples. *Appl. Environ. Microbiol.* **37:**303–309.

Patel, R. N., and J. C. Savas. 1987. Purification and properties of the hydroxylase component of methane monooxygenase. *J. Bacteriol.* **169:**2313–2317.

Patt, T. E., G. C. Cole, J. Bland, and R. S. Hanson. 1974. Isolation of bacteria that grow on methane and organic compounds as sole sources of carbon and energy. *J. Bacteriol.* **120:**955–964.

Pearce, F. 1989. Methane: the hidden greenhouse gas. *New Sci.* **122:**37–41.

Prior, S., and H. Dalton. 1985. The effect of copper ions on membrane content and methane monooxygenase activity in methanol-grown cells of *Methylococcus capsulatus* (Bath). *J. Gen. Microbiol.* **131:**155–164.

Putz, J., F. Meinert, U. Wyss, R.-U. Ehlers, and E. Stackebrandt. 1990. Development and application of oligonucleotide probes for molecular identification of *Xenorhabdus* species. *Appl. Environ. Microbiol.* **56:**181–186.

Reeburgh, W. S. 1976. Methane consumption in Cariaco Trench waters and sediments. *Earth Planet Sci. Lett.* **28:**337–344.

Reeburgh, W. S. 1980. Anaerobic methane oxidation: rate versus depth distribution in Scan Bay sediments. *Earth Planet. Sci. Lett.* **47:**345–352.

Reeburgh, W. S. 1982. A major sink and flux control for methane in marine sediments:

anaerobic consumption, p. 203–217. *In* K. Fanning and F. T. Manheim (ed.), *The Dynamic Environment of the Ocean Floor*. Heath, Lexington, Mass.

Reeburgh, W. S. 1983. Rates of biogeochemical processes in anoxic sediments. *Annu. Rev. Earth Planet Sci.* **11:**269–298.

Reed, W. M., and P. R. Dugan. 1978. Distribution of *Methylomonas methanica* and *Methylosinus trichosporium* in Cleveland Harbor as determined by an indirect fluorescent antibody-membrane filter technique. *Appl. Environ. Microbiol.* **35:**422–430.

Rudd, J. W. M., and R. D. Hamilton. 1978. Methane cycling in a eutrophic shield lake and its effects on whole lake metabolism. *Limnol. Oceanogr.* **23:**337–348.

Rudd, J. W. M., R. D. Hamilton, and N. E. R. Campbell. 1974. Measurement of microbial oxidation of methane in lake water. *Limnol. Oceanogr.* **19:**519–524.

Rudd, J. W. M., and C. D. Taylor. 1980. Methane cycling in aquatic environments. *Adv. Aquat. Microbiol.* **2:**77–150.

Sansone, F. J., and C. S. Martens. 1978. Methane oxidation in Cape Lookout Bight, North Carolina. *Limnol. Oceanogr.* **23:**349–355.

Saralov, A. I., I. N. Krylova, E. E. Saralova, and S. I. Kusnetsov. 1984. Distribution and species composition of methane-oxidizing bacteria in lake water. *Microbiology* **53:**695–701.

Schollenberger, C. J. 1930. Effect of leaking natural gas upon the soil. *Soil Sci.* **29:**261–266.

Seiler, W. 1983. Contributions of biological processes to the global budget of CH_4 in the atmosphere, p. 468–477. *In* M. J. Klug and C. A. Reddy (ed.), *Current Perspectives in Microbial Ecology*. American Society for Microbiology, Washington, D.C.

Sokolov, I. G., V. A. Romanovskaya, Y. B. Shkurko, and Y. R. Malashenko. 1980. Comparative characterization of the enzyme systems of methane-utilizing bacteria that oxidize NH_2OH and CH_3OH. *Microbiology* **49:**142–148. (Translation of *Mikrobiologiya* **49:**202–209.)

Stahl, D. A., B. Flesher, H. R. Mansfield, and L. Montgomery. 1988. Use of phylogenetically based hybridization probes for studies of ruminal microbial ecology. *Appl. Environ. Microbiol.* **54:**1079–1084.

Stainthorpe, A. C., V. Lees, G. P. C. Salmond, H. Dalton, and J. C. Murrell. 1990. The methane monooxygenase gene cluster in *Methylococcus capsulatus* Bath. *Gene* **91:**27–34.

Stanley, S. H., and H. Dalton. 1982. Role of ribulose-1,5-bisphosphate carboxylase/oxygenase in *Methylosinus capsulatus* (Bath). *J. Gen. Microbiol.* **128:**2927–2935.

Stanley, S. H., S. D. Prior, D. J. Leak, and H. Dalton. 1983. Copper stress underlies the fundamental change in intracellular location of methane monooxygenase in methane-utilizing mechanisms: studies in batch and continuous cultures. *Biotechnol. Lett.* **5:**487–492.

Stauffer, B., E. Lochbronner, H. Oeschger, and J. Schwander. 1988. Methane concentration in the glacial atmosphere was only half that of the preindustrial holocene. *Nature* (London) **332:**812–813.

Steudler, P. A., L. O. Bowden, S. M. Melille, and J. D. Aker. 1989. Influence of nitrogen fertilization on methane uptake in temperate forest soils. *Nature* (London) **341:**314–316.

Stirling, D. I., and H. Dalton. 1979. The fortuitous oxidation and cometabolism of various carbon compounds by whole-cell suspensions of *Methylococcus capsulatus* (Bath). *FEMS Microbiol. Lett.* **5:**315–318.

Strand, S. E., and L. Shippert. 1986. Oxidation of chloroform in an aerobic soil exposed to natural gas. *Appl. Environ. Microbiol.* **52:**203–205.

Topp, E., and R. Knowles. 1984. Effects of nitrapyrin [2-chloro-6(trichloromethyl)pyridine] on the obligate methylotroph *Methylosinus trichosporium* OB3b. *Appl. Environ. Microbiol.* **47:**248–262.

Trotsenko, Yu. A. 1983. Metabolic features of methane and methanol utilizing bacteria. *Acta Biotechnol.* **3:**269–277.

Tsien, H. C., B. J. Bratina, K. Tsuji, and R. S. Hanson. 1990. Use of oligonucleotide signature probes for identification of physiological groups of methylotrophic bacteria. *Appl. Environ. Microbiol.* **56:**2858–2865.

Tsien, H.-C., G. A. Brusseau, R. S. Hanson, and L. P. Wackett. 1989. Biodegradation of trichloroethylene by *Methylosinus trichosporium* OB3b. *Appl. Environ. Microbiol.* **55:**3155–3161.

Tsuji, K., H. C. Tsien, R. S. Hanson, S. R. DePalma, R. Scholtz, and S. LaRoche. 1990. 16S ribosomal RNA sequence analysis for determination of phylogenetic relationship among methylotrophs. *J. Gen. Microbiol.* **136:**1–10.

Wackett, L. P., G. A. Brusseau, S. R. Householder, and R. S. Hanson. 1989. A survey of microbial oxygenases: trichloroethylene degradation of propane-oxidizing bacteria. *Appl. Environ. Microbiol.* **55:**2960–2964.

Ward, B. B., K. A. Kirkpatrick, P. C. Novelli, and M. I. Scranton. 1987. Methane oxidation and methane fluxes in the ocean surface layer and deep anoxic waters. *Nature* (London) **327:**226–228.

Whalen, S. C., and W. S. Reeburgh. 1988. A methane flux time series for tundra environments. *Global Biogeochem. Cycles* **2:**399–409.

Whalen, S. C., and W. S. Reeburgh. 1990. Consumption of atmospheric methane to subambient concentrations by tundra soils. *Nature* (London) **36:**160–162.

Whittenbury, R., J. Colby, H. Dalton, and H. L. Reed. 1976. Biology and ecology of methane oxidizers, p. 281–292. *In* H. G. Schlegel, G. Gottschalk, and N. Pfennig (ed.), *Microbial Production and Utilization of Gases*. E. Goltze KG, Gottingen, Germany.

Whittenbury, R., and H. Dalton. 1981. The methylotrophic bacteria, p. 894–902. *In* M. P. Starr, H. Stolp, H. G. Truper, A. Balows, and H. G. Schlegel (ed.), *The Procaryotes*. Springer-Verlag KG, Berlin.

Whittenbury, R., and N. R. Krieg. 1984. *Methylococcaceae* fam. nov., p. 256–262. *In* N. R. Krieg and J. G. Holt (ed.), *Bergey's Manual of Determinative Bacteriology*, vol. 1. Williams and Wilkins, Baltimore.

Whittenbury, R., K. C. Phillips, and J. F. Wilkinson. 1970. Enrichment, isolation and some properties of methane-utilizing bacteria. *J. Gen. Microbiol.* **61:**205–218.

Wilson, J. T., and B. H. Wilson. 1985. Biotransformation of trichloroethylene in soil. *Appl. Environ. Microbiol.* **49:**242–243.

Woese, C. R. 1987. Bacterial evolution. *Microbiol. Rev.* **51:**221–271.

Wolf, H. J., and R. S. Hanson. 1978. Alcohol dehydrogenase from *Methylobacterium organophilum*. *Appl. Environ. Microbiol.* **36:**105–114.

Yavitt, J. B., D. M. Downey, E. Lancaster, and G. E. Lang. 1990. Methane consumption in decomposing sphagnum-derived peat. *Soil Biol. Biochem.* **22:**441–447.

Yoshinari, T. 1985. Nitrite and nitrous oxide production by *Methylosinus trichosporium*. *Can. J. Microbiol.* **31:**139–144.

Chapter 6

Methane Fluxes from Terrestrial Wetland Environments

Patrick M. Crill, Robert C. Harriss, and Karen B. Bartlett

Three centuries ago atmospheric methane (CH_4) concentrations were approximately 700 ppb. Recent ice core data (Chappellaz et al., 1990) indicate that in the past 160,000 years atmospheric CH_4 varied only between 350 and 700 ppb during glacial and interglacial periods. This differs considerably from the current concentration of 1,750 ppb. Estimates of present-day atmospheric CH_4 budgets require the total source strength of CH_4 to be between 200 and 600 Tg of CH_4 year^{-1} (1 Tg = 10^{12} g) (Ehhalt, 1985; Bingemer and Crutzen, 1987; Cicerone and Oremland, 1988). Analysis of the $^{14}CH_4$ content of the troposphere (Wahlen et al., 1989) indicates that about 20% of the total is derived from production and combustion of fossil or "dead" sources, with the remaining 80% divided somewhat equally among biological sources that utilize recently fixed carbon. These sources include biomass burning, rice agriculture, ruminants, and wetlands, all of which are terrestrial. In fact, nearly the entire flux of methane (CH_4) to the atmosphere is from terrestrial environments. Ocean habitats contribute only 0.2 to 3% of the annual global flux. Freshwater natural wetlands contribute a significant proportion of the terrestrial flux of atmospheric CH_4. Estimates of the global flux of CH_4 from wetlands range from 30 to 300 Tg of CH_4 year^{-1}, with the more recent calculations settling on a value close to the middle of that range (Ehhalt, 1974; Ehhalt and Schmidt, 1978; Matthews and Fung, 1987; Cicerone and Oremland, 1988; Aselmann and Crutzen, 1989).

Because CH_4 plays a central role in the oxidation chemistry and the radiation and water budgets of the atmosphere, understanding and evaluating the global CH_4 cycle have taken on a new immediacy. Increasing concentrations of CH_4 and other trace gases can have a profound influence on the earth's atmospheric chemistry and on climate (e.g., Thompson and Cicerone, 1986; Ramanathan, 1988). What is less certain is the influence that climate change will have on emissions of trace gases from the global biosphere. Positive feedback can result if global warming enhances emissions of CH_4, with increasing concentrations of CH_4 further enhancing greenhouse warming. Major uncertainties also remain in under-

Patrick M. Crill, Robert C. Harriss, and Karen B. Bartlett • Complex Systems Research Center, Institute for the Study of Earth, Oceans and Space, University of New Hampshire, Durham, New Hampshire 03824.

standing possible negative feedbacks such as increased evaporation drying out CH_4 source regions. Increased evaporation would lead to increased atmospheric water vapor which is the main source of hydroxyl (OH), the principal sink for CH_4 in the atmosphere. More water vapor would also mean more clouds, a higher albedo, and thus less heating, but, reflecting the uncertainty in our understanding, water is a strong infrared absorber and therefore could contribute to global warming.

This discussion will focus on the flux of CH_4 from natural wetlands to the atmosphere. Wetlands, both natural and anthropogenic, are important in the global carbon cycle. They can support some of the most biologically productive ecosystems on earth (Likens, 1975; Leith, 1975; Ajtay et al., 1979). Waterlogging of biologically productive soils allows heterotrophic microbiological processes to consume oxygen (O_2) more rapidly than it can diffuse into the system. For this discussion, there are two important results. First, once the soils become anoxic in freshwater environments, CH_4 production becomes the most important terminal electron sink of anaerobic respiration (see Oremland, 1988). CH_4 removes electrons from the highly reduced environments in which it is produced by diffusing or bubbling into the atmosphere or migrating into a more oxidized environment where it might be utilized by methylotrophic organisms (e.g., Rudd and Taylor, 1979; Hanson, 1980). Transport to the atmosphere may be enhanced by plants (e.g., Dacey and Klug, 1979; Sebacher et al., 1985; Chanton and Dacey, in press), and of course, where there is an impermeable seal, biogenic CH_4 can accumulate and be buried. However CH_4 leaves the system, electrons are removed, and this role of electron sink is essential for the reoxidation of the biochemical cofactors that allow carbon remineralization to continue under anoxic conditions.

The second result is directly related to the previous. Anaerobic carbon remineralization is much less efficient energetically than aerobic processes (see Thauer et al., 1977). Once a soil becomes anoxic, remineralization rates slow and organic soils and peats can accumulate and become a sink for atmospheric carbon. This is especially marked in the arctic and boreal zones, where the combination of low temperature and poor drainage has resulted in extensive peat deposits across North America and Eurasia. However, low temperatures are not a prerequisite for peats, as evidenced by the massive peat deposits of the Indonesian archipelago or the massive coal deposits laid down during the Carboniferous age.

In the following discussion, examples will be drawn mainly from the direct flux measurements that our research group has accumulated over the past 9 years in our efforts to quantify and understand the flux of CH_4 to the troposphere. Our research effort has included sites that extend from about 5°S in the Amazon basin to the North Slope of Alaska at > 65°N. We have sampled a large number of the principal wetlands in the western hemisphere and examples of many wetland habitat types.

FLUX METHODS

Two general approaches are used to measure trace gas fluxes. The first approach involves meteorological techniques to estimate fluxes by measuring concentrations, concentration gradients, and near-surface air transport phenomena (see Baldocchi et al., 1988). Tower-based micrometeorological methods fall into two general categories: eddy correlation and concentration gradient techniques (Fowler and Duyzer, 1989). Many of the micrometeorological techniques used with towers have been adapted for use on aircraft (see Desjardins and MacPherson, 1989) to evaluate regional fluxes. Even though these methods can yield integrated flux information for large areas, the size and shape of the area a tower or aircraft "sees" are often in doubt and the costly and complex equipment that is required is not very portable, which is especially a problem when sampling remote regions. The resulting flux information is on the scale of hectares for micrometeorological towers to square kilometers in the case of aircraft, so the question of scaling factors, from the square meters of chamber measurements to the square kilometers measured by aircraft, becomes critical in evaluating regional fluxes. Also, it is only very recently that the required fast-response instruments have become available for CH_4 (McManus et al., 1989). Direct measurement of regional fluxes is an exciting and important area of future research into global CH_4 exchange.

The second approach uses enclosures and the measurement of concentration changes in the trapped headspace of static chambers or, in the case of dynamic systems, of concentration changes before and after a sweep gas flows across the surface of interest. Nearly all of the data about CH_4 exchange have been gathered with static chamber techniques. The scale with these techniques is on the order of square meters. In our case, we used static chamber methods because they are sensitive, quick, and portable. Changes greater than 0.1 mg of CH_4 m^{-2} day^{-1} can be detected with a 20-min measurement period. The short time frame minimizes temperature and humidity changes and effects on enclosed plants. Even though the area sampled is very small, usually no larger than 0.5 m^2, the portability of the system allows many fluxes to be measured over broad, ecologically diverse environments so one can arrive at an assessment of within-habitat variability. Continuous sampling techniques such as the gas filter correlation method described in detail by Sebacher and Harriss (1982) monitor any change in CH_4 concentration of the enclosed air continuously. Bubbles and diffusive CH_4 flux can be distinguished. Grab sampling techniques, in which discrete samples are taken from the enclosed headspace over a period of time, yield a total net flux and cannot quantify bubble fluxes. Rates of total CH_4 flux are determined from the slope of the concentration versus time. A recent review by Mosier (1989) discusses some of the theoretical and practical problems of chamber methods.

METHANE FLUXES FROM WETLANDS

Latitudinal Variability

Wetlands are not equally distributed across latitudinal zones (Fig. 1), and so wetlands will have a different impact on the CH_4 budget at different latitudes. Tables 1 and 2 illustrate that the variability of the measured fluxes within a climatic zone is greater than the variability between zones. We know that particular wetland types tend to dominate certain latitudes (Matthews and Fung, 1987; Aselmann and

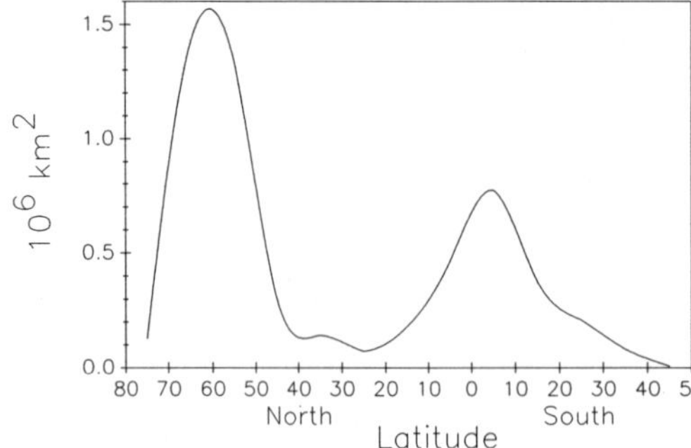

Figure 1. Latitudinal distribution of wetlands. This is the smoothed average of the data of Matthews and Fung (1987) and Aselmann and Crutzen (1989).

Crutzen, 1989) mainly because large-scale patterns of temperature and rainfall occur over broad latitudinal bands. The result is that latitudinal differences in the global CH_4 flux exist principally because of the types of wetlands, their areal extent, and the length of their active season.

High-latitude wetlands

Tundra wetlands are areally very extensive (8.8×10^{12} m^2 [Post et al., 1982]) and continuous. Their most common feature is that they are underlain by permafrost. In these regions CH_4 production and flux are strongly seasonal because the CH_4 production zone freezes during the long, severe winters. When surface organic soils become saturated with water at spring thaw, CH_4 production begins and rises rapidly to its maximum rate. Flux continues through the season until freeze-up in the fall. CH_4 production and flux can persist well into the winter (Whalen and Reeburgh, 1988) if there is an insulating layer of moss or snow. Our data and those of Whalen and Reeburgh (1988) showed very poor correlation between temperature and flux from tundra. The strongest actor on the flux was soil moisture. The highest CH_4 fluxes are measured from wet meadow sites and from tussock and low shrub tundra when the soils are saturated with water (e.g., Sebacher et al., 1986; Whalen and Reeburgh, 1988). If the surface layers drain and become drier, CH_4 fluxes are lowered because increased aeration of the soils raises the oxidation potential above that value which allows CH_4 production. Decreased substrate availability and increased CH_4 oxidation may also contribute to lowering the flux.

Between the permafrost regions in the north and approximately 45°N there is a broad expanse of wetlands that extend across large areas of North America and Eurasia with continuous areas in the Hudson/James Bay lowlands of Canada and the West Siberian Plain between the Ob and Yenisey rivers in the USSR. These and all northern wetlands are marked by large amounts of "dead" (i.e., not in living biomass) organic carbon stored as peat soils. The predominant boreal wetlands are ombrotrophic bogs and minerotrophic fens that may or may not be forested. There is also a strong seasonal signal to the CH_4 flux from these regions, but, because of the insulating effect of snow and bryophyte layers, the peats generally do not freeze and slow continuous flux of CH_4 has been measured through the winters (N. Dise,

Table 1. CH_4 fluxes measured in high latitudes[a]

Site	Period	Habitat	n	mg of CH_4 m^{-2} day^{-1}		Refence
				Mean	Range	
Stordalen, Sweden (~ 68°N)	June–Sept.	Subarctic mire:				Svensson and Rosswall, 1984
		Ombrotrophic	20	12	0.3–29	
		Intermediate	5	58	9–112	
		Minerotrophic	5	360	80–950	
		Total	30	143		
Alaska, USA (63–71°N)	Aug.	Nonforested:				Sebacher et al., 1986
		Wet tundra	44	119	34–266	
		Moist tundra	12	5	0.3–12	
		Meadow tundra	14	40	9–77	
		Alpine fen	6	289		
		Boreal marsh	7	106		
		Total	83	112		
Alaska, USA (~ 65°N)	Annual	Subarctic taiga:				Whalen and Reeburgh, 1988
		Tussock	86	22	0–164	
		Wet meadow	54	32	0–105	
		Moss	87	2	0–17	
		Intertussock	87	5	0–34	
		Total	314	15		
Schefferville, Canada (~ 56°N)	June–Aug.	Subarctic fens				Moore and Knowles, 1987
		4 sites	80	30	0–112	
Minnesota, USA (~ 47°N)	Aug.	Bogs and fens:				Harriss et al., 1985
		Forested	17	93	3–206	
		Nonforested	25	426	33–1,940	
Minnesota, USA (~ 47°N)	May–Aug.	Bogs and fens:				Crill et al., 1988a
		Forested	67	102	11–694	
		Nonforested	112	259	18–866	

[a] Midsummer fluxes.

Table 2. CH_4 fluxes from low-latitude environments

Site	Period	Habitat	*n*	mg of CH_4 m^{-2} day^{-1}		Reference
				Mean	Range	
Newport News, Va.[a] (~37°N)	Annual	Mixed hardwood swamp, 4 sites	30	155	0–463	Wilson et al., 1989
Southeast, USA (26°–33°N)	June	Cypress swamps	15	49	5–265	Harriss and Sebacher, 1981
Okefenokee Swamp, Ga. (~31°N)	Annual	Shrub swamp	42	149	−8–1,250	Bartlett et al., unpublished data
		Cypress swamp	47	40	−10–442	
		Gum/bay swamp	14	70	−8–293	
		Wet prairie	50	130	−8–1,000	
		Ponds/lakes	7	116	31–156	
Everglades, Fla. (~25°N)	Annual	Sawgrass	60	107	9–2,390	Burke et al., 1988
Everglades, Fla. (~25°N)	Annual	Salt mangrove	17	4	2–8	Harriss et al., 1988
		Swamp forest	22	59	−3–274	
		Wet prairie/sawgrass	122	61	bd[b]–624	
Amazon floodplain, Brazil (~3°N)	July–Aug.	Flooded forest	19	70	1–505	Devol et al., 1988 (corrected data)
		Floating grass	27	370	0–1,890	
		Varzea lakes	36	84	0.2–633	
Amazon floodplain, Brazil (~3°N)	July–Sept.	Flooded forest	90	192	bd–1,220	Bartlett et al., 1988
		Floating grass	55	230	bd–3,000	
		Varzea lakes	41	27	−10–111	
Amazon floodplain, Brazil (~3°N)	April–May	Flooded forest	58	126	bd–840	Bartlett et al., 1990
		Floating grass	85	201	−11–1,600	
		Varzea lakes	116	74	bd–1,160	

[a] Warm-weather fluxes.
[b] bd, below detection.

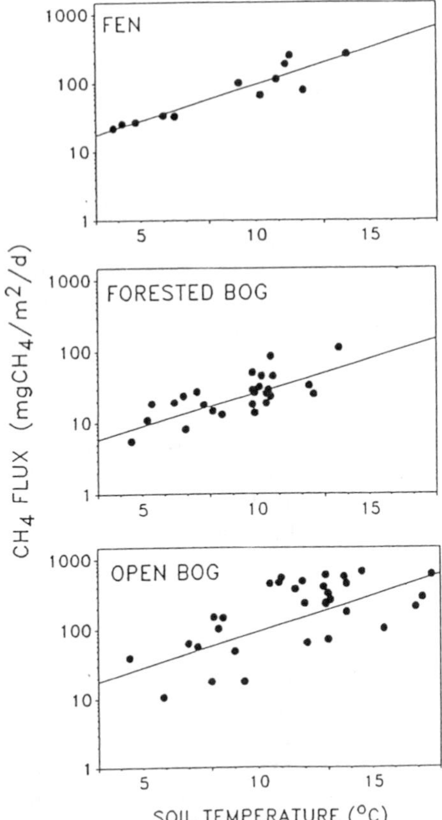

Figure 2. Temperature response of CH_4 flux from three different habitat types in the Marcell Experimental Forest, Minn. (data from Crill et al., 1988a).

personal communication). As in the tundra, water plays a central role in the CH_4 budget, with highest fluxes measured when soils are saturated to the air/soil interface. This is also observed in experimental manipulations of material from all but the most ombrotrophic, acidic boreal peats (Moore and Knowles, 1989). The residence time of the water in the surface soils may also affect the flux. Unlike the tundra, a strong temperature correlation was observed in these regions, especially when soil temperatures from at least 10-cm depth were used (Fig. 2; Crill et al., 1988a). The implications of this observation in light of changing climate are interesting given that the largest temperature and moisture variations related to "greenhouse" changes are predicted to occur in the boreal region (e.g., Hansen et al., 1988).

Low-latitude wetlands

Recent evaluations of wetland areas show a bimodal distribution (Matthews and Fung, 1987; Aselmann and Crutzen, 1989) with a minimum observed in the temperate zones (Fig. 1). Even though temperate regions have the least extensive wetlands, probably the largest variety of ecosystems is found in the mid-latitudes. Marshes and wooded swamps become areally more predominant as one moves to lower latitudes, with extensive alluvial wetlands found only in the tropics, mainly the Congo and Amazon River basins (Gore, 1983). Only 10% of the global CH_4 flux is estimated to come from the mid-latitudes; nevertheless, they are important habitats for study. The relative ease of access because of their proximity to population centers allows consideration of seasonal changes and anthropogenic influences on flux. The often restricted watersheds provide well-defined sites for mass balance studies and for the intercomparison of discrete habitats.

There is less seasonal temperature fluctuation in the lower latitudes, and, at least in the western hemisphere, more of the organic carbon pool is found in the living biomass rather than stored in the soils. The seasonal cycle of CH_4 flux is more closely associated with the seasonal cycle of inundation and plant growth and senescence. Plants affect the substrate supply and degree of aeration of water-saturated soils (e.g., Barber, 1961). Plants also become more important in the physical transport of CH_4 from anaerobic soils to the atmosphere as macrophytes with large internal lacunae occur more frequently (Dacey and Klug, 1979; Sebacher et al., 1985). Ebullition also becomes more important as a mechanism for CH_4 release in lower latitudes (e.g., Chanton and Martens, 1988; Bartlett et al., 1988).

Long growing seasons, plentiful organic substrate, and abundant shallow fresh water combine to produce high CH_4 fluxes. On the other hand, low CH_4 fluxes are always measured in highly saline wetlands such as salt marshes (Bartlett et al., 1987) or coastal mangroves (Harriss et al., 1988). Lower CH_4 fluxes in freshwater wetlands can also result from lower organic substrate concentrations, such as in the calcareous marls in the Everglades (Harriss et al., 1988). CH_4 fluxes may also be depressed by flowing water and intense light keeping the surface soil more oxidized. Temperature, water, substrate supply, and degree of oxidation of the soil are the principal controls on CH_4 flux from all soils found in both high and low latitudes.

Controls on Small-Scale Variability

There are also fine-scale changes in flux that may produce order-of-magnitude changes in the flux of CH_4 over distances on the scale of meters or less within a habitat type. For example, heat and water affect both the macro- and microbiological processes that determine the extent of anaerobiosis and the physical mechanism of gas flux. The processes responsible for this small-scale, within-habitat variability in CH_4 flux are beginning to be understood and have been the subject of recent reviews (e.g., Oremland, 1988; Conrad, 1989), but this is still an important area for future research. All this will have to be quantified if we are to understand the responses of our life-bearing ecosystems to the dynamic global climate.

Temperature

Methanogenic bacteria, like all catalysts of chemical reactions, respond to increased temperature with increased CH_4 production in both laboratory and soil environments (Baker-Blocker et al., 1977; Zeikus and Winfrey, 1976). Despite the fact that narrow ranges of activation energies are often reported in laboratory studies (e.g., 60 to 90 kJ mol^{-1} [Conrad et al., 1987]), very broad ranges of the temperature response of the net flux of CH_4 from wetland soils to the atmosphere have been observed (see Crill et al., 1988a). This is because most of the processes that affect CH_4 production and flux are also directly affected by temperature, e.g., moisture balance and rates of CH_4 oxidation and substrate supply. However, Fig. 2 illustrates that the flux of CH_4 from different habitats with similar moisture, carbon, and temperature regimes can respond to temperature in ways similar to the laboratory production CH_4 studies in that very different habitats can demonstrate a narrow range of activation energies.

Soil moisture

In every wetland we have studied, the degree of inundation of a particular habitat has been a primary factor correlated with methane flux. Wetlands, by definition, are characterized by standing water during at least part of the year. Figure 3 illustrates the average fluxes measured along a moisture gradient in Alaska. The wetter sites support higher CH_4 fluxes (the *Carex* sp. and *Arctophila fulva* sites are very wet or flooded). If the water table falls below the soil surface, then the CH_4 flux drops to low values, possibly changing from efflux to consumption (Harriss et al., 1982). Dry upland temperate and tropical soils have been found to consume atmospheric CH_4 (Keller et al., 1983; Goreau and deMello, 1988; Born et al., 1990).

Transport processes

The transport of CH_4 from anaerobic sites of production in wetland soils and sediments to the troposphere involves a number of mechanisms including diffusion, ebullition, and transport by rooted macrophytes. Each of these transport mechanisms presents challenges to the quantification of annual or longer-term CH_4 flux from a particular ecosystem. The effects of bubbles and an emergent macrophyte on the CH_4 flux from the Okefenokee Swamp are shown in Table 3. Highest fluxes are associated with bubbling events, and plant-mediated transport supports fluxes higher than those observed when diffusion is the sole transport mechanism. Diffusive flux is influenced by wind velocity (Sebacher et al., 1983), by surface roughness, and by limnological factors such as density stratification dynamics (e.g., the density effects of diel heating and cooling of the surface layers) which can limit dissolved CH_4 transport to the interface (Rudd and Hamilton, 1978; Crill et al., 1988b). Bubble fluxes can vary by orders of magnitude in both space and time at a single site (Chanton and Martens, 1988). Factors such as animal burrowing in soils and variations in hydrostatic loading have been shown to influence ebullition from subsurface environments that are supersaturated with CH_4 (Martens and Klump, 1980). Transport of CH_4 by rooted aquatic plants has been studied in detail in rice paddies, where pronounced temporal variations have been observed related to the growth stage of the plants (Cicerone et al., 1983; Holzapfel-Pschorn and Seiler, 1986). Because the relative importance of various transport mechanisms is variable

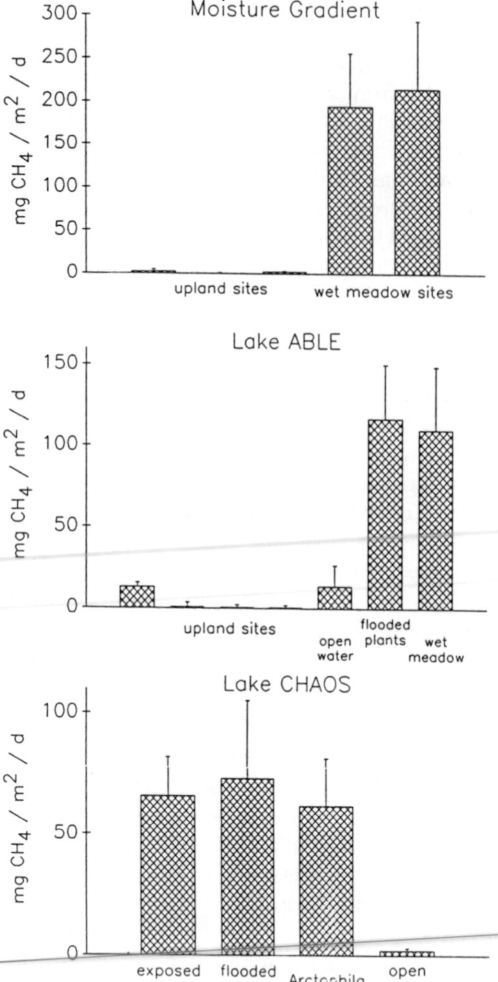

Figure 3. Average and standard error of the mean CH_4 flux measured across tundra moisture gradients in the Yukon-Kuskokwim delta, Alaska. The upland sites are dry and the others are either wet, flooded, or open water.

in both time and space, CH_4 flux studies should always include sampling schemes to test for diurnal and seasonal variations.

Substrate quality and quantity

As Conrad (1989) points out, there are three broad classes of substrates that affect CH_4 production and flux: organic matter, electron acceptors, and nutrients. If the rate of supply of organic matter is faster than the rate of supply of O_2 to oxic heterotrophic organisms that remineralize the organic matter, then anoxia will result in the soil or sediment—a necessary precondition for methanogenesis.

Table 3. Influence of gas exchange mechanisms on methane flux from the Okefenokee Swamp, Ga.

Habitat	CH_4 flux (mg of CH_4 m^{-2} day^{-1})	Flux mechanism
Flooded cypress	422	Ebullition and diffusion
Flooded cypress	59	*Nuphar* sp. and diffusion
Flooded cypress	29	Diffusion
Open lake	99	*Nuphar* sp. and diffusion
Open lake	31	Diffusion

Though a supply of organic matter is essential, only a small portion of the organic substrate pool can be directly utilized by methanogens, mainly acetate and CO_2 (see Oremland, 1988), so the seemingly paradoxical condition of carbon limitation in peats is possible.

The simple occurrence of anoxia is not the only prerequisite for CH_4 production. Methanogens cannot compete for organic substrates in the presence of mineral electron acceptors (i.e., iron, manganese, nitrate, and sulfate) that yield more energy when their reduction is coupled to carbon remineralization. Hence, CH_4 fluxes are usually lower from high-sulfate environments such as salt marshes (e.g., Bartlett et al., 1987). Even at sulfate concentrations typically found in freshwater environments, sulfate-reducing bacteria can outcompete methanogens for organic substrates (Lovley and Klug, 1983). Therefore, anthropogenic loading, especially of nitrate and sulfate, on wetlands may have potentially serious effects on patterns of organic carbon remineralization and possibly even on patterns of peat accumulation and ablation.

Direct nutrient (e.g., P, K, trace elements) effects on methanogenesis and flux are difficult to assess because of effects on the local vegetation which will adapt to the prevailing nutrient regime. A change in nutrient status will change the vegetation, which will have an effect on the CH_4 exchange rates. Vegetation may affect the aeration of the surface soils, the availability of labile organic exudates in the available soil carbon pool, and rates of gas exchange and CH_4 release from the zone of production within the soil.

pH

Low pH appears to inhibit CH_4 production even though substantial CH_4 fluxes have been measured from acidic bogs (e.g., Svensson and Rosswall, 1984). Higher CH_4 concentrations and fluxes have been observed in the more alkaline lagg or edge areas (pH 4.6 to 5.0) than in the more acidic central regions (pH 3.7 to 4.5) of the same bogs in Minnesota (Fig. 4; Crill et al., 1988a). Also, rates of CH_4 production measured from peat slurries (Williams and Crawford, 1984) and from acidophilic microbial communities isolated from bog peats (Goodwin and Zeikus, 1987) were higher when the pH was closer to neutrality than at pH below 4.5.

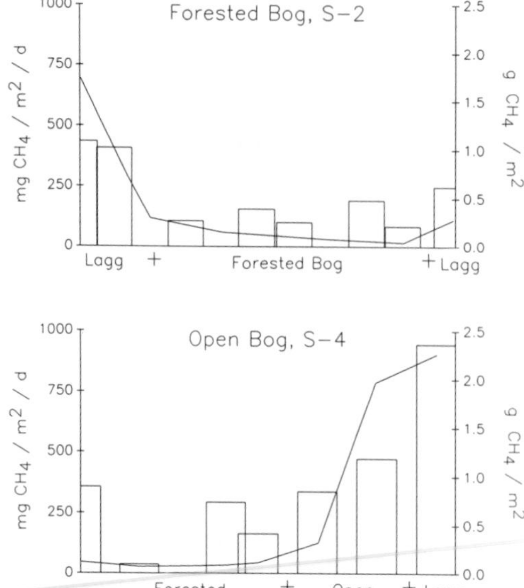

Figure 4. Transects of CH_4 flux (lines) and near-surface (top 30 cm) dissolved CH_4 concentration (bars) across drier, acidic, ombrotrophic central regions to wetter, more alkaline, minerotrophic lagg (edge) areas of bogs in the Marcell Forest, Minn. (redrawn from Crill et al., 1988a).

A CASE STUDY: THE OKEFENOKEE SWAMP, GEORGIA

How does one go about quantifying the flux of CH_4, or any other gas for that matter, from a complex landscape at a given point during the active flux season? One is always faced with the problem of scales, from the 0.5 m^2 of the flux chambers to the tens of thousands of hectares of the landscape feature of interest. Our approach has been to reduce the complex mosaic of a large ecosystem unit into readily identifiable subunits, generally using plant communities as the key, sampling those subunits until we can determine a statistically reasonable mean flux and then calculate an areally weighted average flux. Plant communities are important to the success of this survey approach because they represent long-term integrations of much less easily measured variables such as nutrient, water, and temperature regimes.

Figure 5 shows that both the mean and the standard area of the mean tend towards an asymptotic value in large data sets. After a certain number of measurements (about 20 or 30) you are at a point where taking even more measurements has less and less effect on the mean or standard error of the mean. For example, in the "cypress" and "prairie" habitats, after 20 samples the standard error is about 48 and 25% of the mean, respectively. Doubling the number of samples reduces the standard error to 40 and 21% of the mean without affecting the value of the mean very much at all. In both cases, after 40 samples in each habitat type the magnitude and the variability of the CH_4 flux have been determined.

The Okefenokee Swamp of southeastern Georgia and northeastern Florida is a large ($\sim$ 200,000-ha), low wet area impounded behind a north-south sand ridge

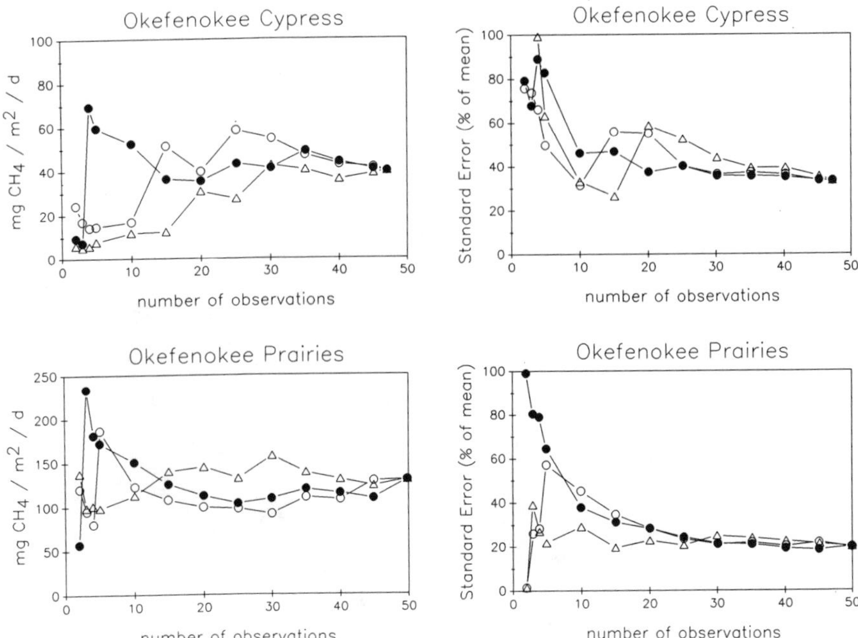

Figure 5. Change in calculated average CH_4 flux and in calculated standard error (as percent of the mean) with increasing sample size. Values are calculated from observations randomly selected from the pool of samples in each habitat. Each graph represents three individual runs of the calculation routine. See Table 4 for the total sample size (*n*).

along its eastern edge. Drainage is to the southwest via the Suwannee River and to the south and east by the St. Mary's River (Fig. 6). A complex mosaic of vegetation has developed across the swamp. Refer to Schlesinger (1978), Duever and Riopelle (1983), and Christensen (1989) to complete the following cursory discussion of the vegetation. The first step in quantifying the CH_4 flux from this environment was to identify easily recognizable habitat types. We then measured the fluxes within a habitat type using the gas filter correlation technique described above. We selected five broad habitat classes in the swamp so that we could areally weight the mean fluxes from each to arrive at a regional flux from the Okefenokee ecosystem (Fig. 6; Tables 4 and 5).

"Shrub" covers the largest area. It is a mix of evergreen or deciduous broad-leaved shrubs (e.g., *Lyonia lucida, Cyrilla racemiflora, Ilex* spp.) and needle-leaved or broad-leaved deciduous scrub (e.g., young *Taxodium ascendens*, young *Nyssa sylvatica*), probably established by the heavy logging of the cypress. The highest mean fluxes are measured in this vegetation class (Table 4) mainly because the peaty soils are wet to the air/soil interface and it is covered by a very productive plant community.

"Prairie" is broad, shallow flooded area (Fig. 6) covered by emergent macro-

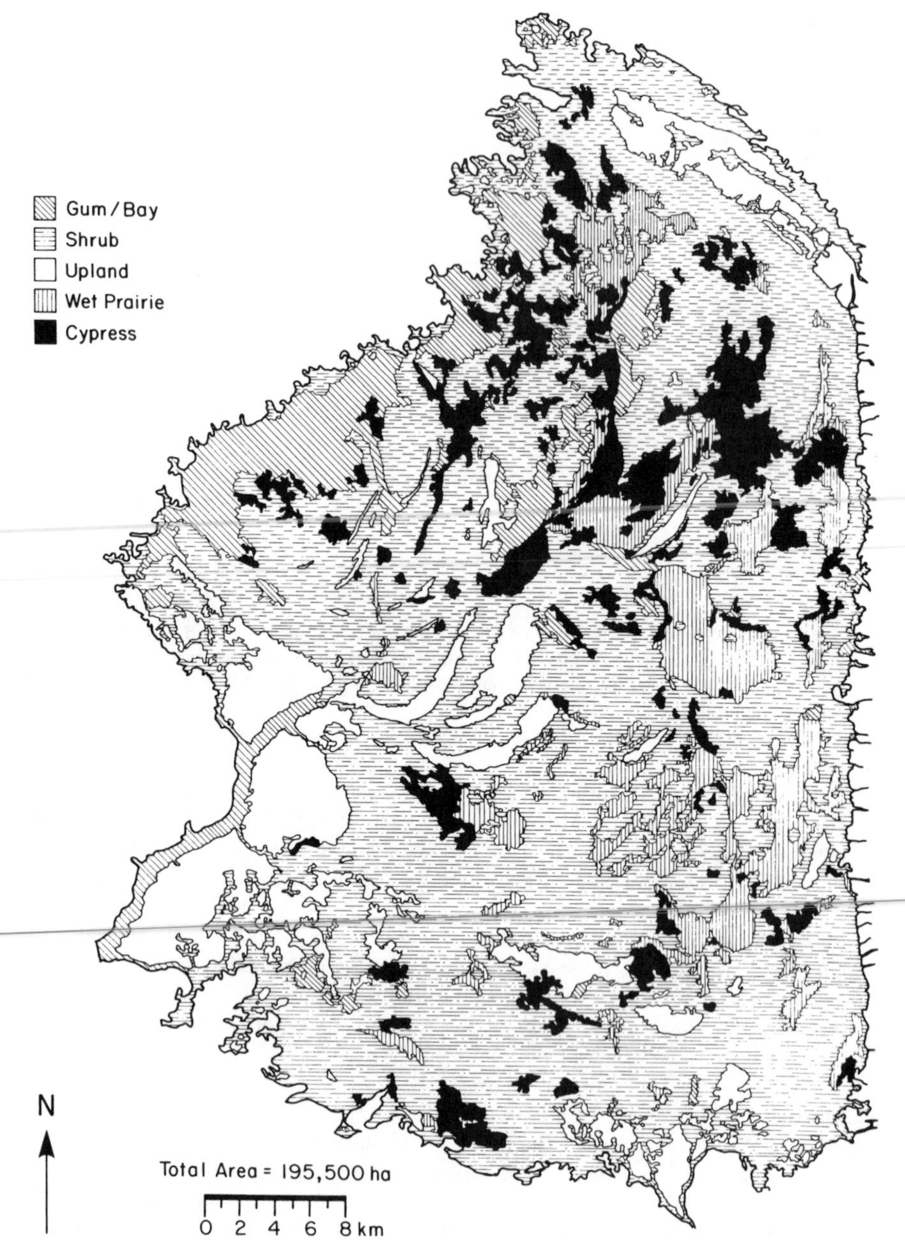

Figure 6. A vegetation map of the Okefenokee Swamp, Ga.

Table 4. Methane flux from the wetlands of the Okefenokee Swamp, Ga.

Habitat	Flux (range) (mg of CH_4 m^{-2} day^{-1})	*n*
Shrub	149.0 (−7.5–1,250)	42
Prairie	130.4 (−7.9–1,000)	50
Cypress	39.8 (−10.1–442)	47
Gum/bay	69.6 (−7.5–293)	14
Open lake	115.9 (30.8–217)	7

phytes (e.g., *Nuphar luteum* and *Nymphaea adorata*) and graminoids (e.g., *Panicum* spp. and *Carex* spp.). It is maintained by fire during dry periods (Duever and Riopelle, 1983). CH_4 flux is significantly lower than that measured in shrub habitats because of three reasons: (i) there is only a narrow zone of CH_4 production in a thin layer of organic sediments over mineral soils; (ii) there is a shallow layer of oxic water that the CH_4 must diffuse through; and (iii) even though the emergent macrophytes provide conduits for CH_4 release from the soil, they also transport O_2 to oxidize their root zone.

"Cypress" are the forested wetlands dominated by *Taxodium ascendens*. Because they form domes or wet hammocks on higher points of land or on peat accumulations in the midst of the prairies, cypress heads are often hydrologically isolated and much more nutrient limited, especially when compared with floodplain cypress swamps (Schlesinger, 1978). The result of the nutrient limitation and the input of more recalcitrant lignified organic matter into the substrate pool is a lower CH_4 flux (Table 4).

The "gum/bay" category is a combination of areas dominated by broad-leaved deciduous and evergreen species such as *Nyssa sylvatica*, *Gordonia lasianthus*, and *Magnolia virginiana*. CH_4 fluxes from these habitats are probably controlled by nutrients, moisture, and substrate because, like the cypress domes, gum/bay communities are found on slightly higher elevations. Lowered CH_4 fluxes were also observed from forested habitats in boreal forested bogs (compared with unforested

Table 5. Methane flux from the Okefenokee Swamp, Ga., ecosystem

Habitat	CH_4 flux (g m^{-2} $year^{-1}$)	% Total area	Area (ha)	Total annual CH_4 flux (10^{10} g $year^{-1}$)
Shrub	149.0	56.3	110,200	5.99
Prairie	130.4	10.5	20,500	0.98
Cypress	39.8	9.7	19,020	0.28
Gum/bay	69.6	8.0	15,630	0.40
Open lake	115.9	0.9	1,600	0.07
Uplands	0	14.6	28,560	0
Total		100	195,500	7.72

bogs [Crill et al., 1988a]) and from the flooded forests of the Amazon floodplain (compared with open floating grass mats [Bartlett et al., 1988; Bartlett et al., 1990]).

A few flux measurements were also made in open lake areas where a strong and consistent CH_4 flux was observed (Table 4). Our efforts were limited in this habitat since open water comprised less than 1% of the total area of the swamp (Table 5). The other habitat type included for completeness in Table 5 is uplands. No CH_4 flux measurements were made in this habitat since we were interested in wetland efflux. No dry upland soil has ever been shown to be a source of atmospheric CH_4. In fact, the measurements that have been made indicate that upland forest soils are a weak sink for CH_4 (Keller et al., 1983; Goreau and deMello, 1988). Therefore for our regional efflux calculation we assigned a value of zero (Table 5) for this community.

Using the vegetation map shown in Fig. 6 to estimate the areal extent of specific habitats, the total regional annual CH_4 flux is derived by multiplying the mean flux by the area and the length of the annual season of active CH_4 flux. Because the Okefenokee has a mean annual temperature of about 21°C and monthly precipitation exceeds evaporation throughout the year (see Christensen, 1989), an active CH_4 flux season of 365 days per year is assumed. The resulting 0.08 Tg year^{-1} is only 0.03 to 0.3% of the total wetland CH_4 flux to the atmosphere but could be as much as 2% of the total annual flux from wetlands between 30 and 40°N latitude (Aselmann and Crutzen, 1989).

The assumptions used to derive the annual regional flux illustrate both the difficulties of making regional and global extrapolations by CH_4 flux as well as the utility of studies of restricted, well-defined ecosystems such as the Okefenokee Swamp system. The variabilities of fluxes within habitats are usually high, with standard errors of measured fluxes approaching 20% only in the best cases. The length of the active seasons is often in doubt as well as being variable from season to season, and the areal extent of specific habitats is often in question. But we can use areally restricted systems that are well defined in terms of their hydrology, vegetation, geology, and climate to investigate the processes that control CH_4 flux and to design sampling schemes that will enable us to make better extrapolations as our understanding of the controls on flux and the data of the areal extent of environments improves.

REFERENCES

Ajtay, G. L., P. Ketner, and P. Duvigneaud. 1979. Terrestrial primary production and phytomass, p. 129–181. *In* B. Bolin, E. T. Degens, S. Kempe, and P. Ketner (ed.), *The Global Carbon Cycle, SCOPE*, vol. 13. Wiley, New York.

Aselmann, I., and P. J. Crutzen. 1989. Global distribution of natural freshwater wetlands and rice paddies, their net primary productivity, seasonality and possible methane emissions. *J. Atmos. Chem.* **8:**307–358.

Baker-Blocker, A., T. M. Donahue, and K. H. Mancy. 1977. Methane flux from wetland areas. *Tellus* **29:**245–250.

Baldocchi, D. D., B. B. Hicks, and T. P. Meyers. 1988. Measuring biosphere-atmosphere exchanges of biologically related gases with micrometeorological methods. *Ecology* **69:**1331–1340.

Barber, D. A. 1961. Gas exchange between *Equisetum limosum* and its environment. *J. Exp. Bot.* **12:**243–251.

Bartlett, K. B., D. S. Bartlett, R. C. Harriss, and D. I. Sebacher. 1987. Methane emissions along a salt marsh salinity gradient. *Biogeochemistry* **4:**183–202.

Bartlett, K. B., P. M. Crill, J. A. Bonassi, J. E. Richey, and R. C. Harriss. 1990. Methane flux from the Amazon River floodplain: emissions during rising water. *J. Geophys. Res.* **95:**16773–16788.

Bartlett, K. B., P. M. Crill, D. I. Sebacher, R. C. Harriss, J. O. Wilson, and J. M. Melack. 1988. Methane flux from the central Amazonian floodplain. *J. Geophys. Res.* **93:**1571–1582.

Bingemer, H. G., and P. J. Crutzen. 1987. The production of methane from solid wastes. *J. Geophys. Res.* **92:**2181–2187.

Born, M., H. Dörr, and I. Levin. 1990. Methane consumption in aerated soils of the temperate zone. *Tellus* **42B:**2–8.

Burke, R. A., T. R. Barbor, and W. M. Sackett. 1988. Methane flux and stable hydrogen and carbon isotope composition of sedimentary methane from the Florida Everglades. *Global Biogeochem. Cycles* **2:**329–340.

Chanton, J. P., and J. W. H. Dacey. Effects of vegetation on methane flux, reservoirs and carbon isotopic composition. *In* T. D. Sharkey, E. A. Holland, and H. A. Mooney (ed.), *Trace Gas Emissions from Plants*. Academic Press, Inc., San Diego, Calif., in press.

Chanton, J. P., and C. S. Martens. 1988. Seasonal variations in ebullitive flux and carbon isotopic composition of methane in a tidal freshwater estuary. *Global Biogeochem. Cycles* **2:**289–298.

Chapellaz, J., J. M. Barnola, D. Raynaud, Y. S. Korotkevich, and C. Lorius. 1990. Ice-core record of atmospheric methane over the past 160,000 years. *Nature* (London) **345:**127–131.

Christensen, N. L. 1989. Vegetation of the southeastern coastal plain, p. 317–364. *In* M. G. Barbour and W. D. Billings (ed.), *North American Terrestrial Vegetation*. Cambridge University Press, New York.

Cicerone, R. J., and R. S. Oremland. 1988. Biogeochemical aspects of atmospheric methane. *Global Biogeochem. Cycles* **2:**299–327.

Cicerone, R. J., J. D. Shetter, and C. C. Delwiche. 1983. Seasonal variation of methane from a California rice paddy. *J. Geophys. Res.* **88:**11022–11024.

Conrad, R. 1989. Control of methane production in terrestrial ecosystems, p. 39–58. *In* M. O. Andreae and D. S. Schimel (ed.), *Exchange of Trace Gases between Terrestrial Ecosystems and the Atmosphere*. Dahlem Workshop Reports, Life Sciences Research Report 47. Wiley, New York.

Conrad, R., H. Schütz, and M. Babbel. 1987. Temperature limitation of hydrogen turnover and methanogenesis in anoxic paddy soil. *FEMS Microbiol. Ecol.* **45:**281–289.

Crill, P. M., K. B. Bartlett, R. C. Harriss, E. Gorham, E. S. Verry, D. I. Sebacher, L. Madzar, and W. Sanner. 1988a. Methane flux from Minnesota peatlands. *Global Biogeochem. Cycles* **2:**371–384.

Crill, P. M., K. B. Bartlett, J. O. Wilson, D. I. Sebacher, R. C. Harriss, J. M. Melack, S. MacIntyre, L. Lesack, and L. Smith-Morrill. 1988b. Tropospheric methane from an Amazonian floodplain lake. *J. Geophys. Res.* **93:**1564–1570.

Dacey, J. W. H., and M. J. Klug. 1979. Methane efflux from lake sediments through water lilies. *Science* **203:**1253–1254.

Desjardins, R. L., and J. I. MacPherson. 1989. Aircraft-based measurements of trace gas fluxes, p. 135–154. *In* M. O. Andreae and D. S. Schimel (ed.), *Exchange of Trace Gases between Terrestrial Ecosystems and the Atmosphere*. Dahlem Workshop Reports, Life Sciences Research Report 47. Wiley, New York.

Devol, A. H., J. E. Richey, W. A. Clark, S. L. King, and L. A. Martinelli. 1988. Methane emissions to the troposphere from the Amazonian floodplain. *J. Geophys. Res.* **93:**1583–1592.

Duever, M. J., and L. A. Riopelle. 1983. Successional sequences and rates on tree islands in the Okefenokee Swamp. *Am. Midl. Nat.* **110:**186–193.

Ehhalt, D. H. 1974. The atmospheric cycle of methane. *Tellus* **26:**58–70.

Ehhalt, D. H. 1985. Methane in the global atmosphere. *Environment* **27:**6–33.

Ehhalt, D. H., and U. Schmidt. 1978. Sources and sinks of atmospheric methane. *Pure Appl. Geophys.* **116:**452–464.

Fowler, D., and J. H. Duyzer. 1989. Micrometeorological techniques for the measurement of trace gas exchange, p. 189–208. *In* M. O. Andreae and D. S. Schimel (ed.), *Exchange of Trace Gases between Terrestrial Ecosystems and the Atmosphere*. Dahlem Workshop Reports, Life Sciences Research Report 47. Wiley, New York.

Goodwin, S., and J. G. Zeikus. 1987. Ecophysiological adaptations of anaerobic bacteria to low pH: analysis of anaerobic digestion in acidic bog sediments. *Appl. Environ. Microbiol.* **53:**57–64.

Gore, A. J. P. (ed.). 1983. *Ecosystems of the World*, vol. 4B, *Mires: Swamp, Bog, Fen and Moor*. Elsevier, New York.

Goreau, T. J., and W. Z. deMello. 1988. Tropical deforestation: some effects on atmospheric chemistry. *Ambio* **17:**275–281.

Hansen, J., I. Fung, A. Lacis, D. Rind, S. Lebedeff, R. Ruedy, G. Russel, and P. Stone. 1988. Global climate changes as forecast by the Goddard Institute for Space Studies three-dimensional model. *J. Geophys. Res.* **93:**9341–9364.

Hanson, R. S. 1980. Ecology and diversity of methylotrophic organisms. *Adv. Appl. Microbiol.* **26:**3–39.

Harriss, R. C., E. Gorham, D. I. Sebacher, K. B. Bartlett, and P. A. Flebbe. 1985. Methane flux from northern peatlands. *Nature* (London) **315:**652–654.

Harriss, R. C., and D. I. Sebacher. 1981. Methane flux in forested swamps of the southeastern United States. *Geophys. Res. Lett.* **8:**1002–1004.

Harriss, R. C., D. I. Sebacher, K. B. Bartlett, D. S. Bartlett, and P. M. Crill. 1988. Sources of atmospheric methane in the south Florida environment. *Global Biogeochem. Cycles* **2:**231–243.

Harriss, R. C., D. I. Sebacher, and F. P. Day, Jr. 1982. Methane flux in the Great Dismal Swamp. *Nature* (London) **297:**673–674.

Holzapfel-Pschorn, A., and W. Seiler. 1986. Methane emission during a cultivation period from an Italian rice paddy. *J. Geophys. Res.* **91:**11803–11814.

Keller, M., T. J. Goreau, S. C. Wofsy, W. A. Kaplan, and M. B. McElroy. 1983. Production of nitrous oxide and consumption of methane by forest soils. *Geophys. Res. Lett.* **10:**1156–1159.

Leith, H. 1975. Primary production of the major vegetation units in the world, p. 203–215. *In* H. Leith and R. H. Whittaker (ed.), *Primary Productivity of the Biosphere*. Ecological Studies, vol. 14. Springer, New York.

Likens, G. E. 1975. Primary production of inland aquatic ecosystems, p. 185–202. *In* H. Leith and R. H. Whittaker (ed.), *Primary Productivity of the Biosphere*. Ecological Studies, vol. 14. Springer, New York.

Lovley, D. R., and M. J. Klug. 1983. Sulfate reducers can outcompete methanogens at freshwater sulfate concentrations. *Appl. Environ. Microbiol.* **45:**187–192.

Martens, C. S., and J. V. Klump. 1980. Biogeochemical cycling in an organic-rich coastal marine basin. I. Methane sediment-water exchange processes. *Geochim. Cosmochim. Acta* **44:**471–490.

Matthews, E., and I. Fung. 1987. Methane emission from natural wetlands: global distribution, area, and environmental characteristics of sources. *Global Biogeochem. Cycles* **1:**61–86.

McManus, J. B., P. L. Kebabian, and C. E. Kolb. 1989. Atmospheric methane measurement instrument using a Zeeman-split He-Ne laser. *Appl. Opt.* **28:**5016–5023.

Moore, T. R., and R. Knowles. 1987. Methane and carbon dioxide evolution from subarctic fens. *Can. J. Soil Sci.* **67:**77–81.

Moore, T. R., and R. Knowles. 1989. The influence of water table levels on methane and carbon dioxide emissions from peatland soils. *Can. J. Soil Sci.* **69:**33–38.

Mosier, A. R. 1989. Chamber and isotope techniques, p. 175–188. *In* M. O. Andreae and D. S. Schimel (ed.), *Exchange of Trace Gases between Terrestrial Ecosystems and the Atmosphere*. Dahlem Workshop Reports, Life Sciences Research Report 47. Wiley, New York.

Oremland, R. S. 1988. Biogeochemistry of methanogenic bacteria, p. 641–705. *In* A. J. B. Zehnder (ed.), *Biology of Anaerobic Microorganisms*. Wiley, New York.

Post, R. M., W. R. Emanuel, P. J. Zinke, and A. G. Stangenberger. 1982. Soil carbon pools and world life zones. *Nature* (London) **298:**156–159.

Ramanathan, V. 1988. The greenhouse theory of climate change: a test by an inadvertent global experiment. *Science* **240:**293–299.

Rudd, J. W., and R. D. Hamilton. 1978. Methane cycling in a eutrophic shield lake and its effects on whole lake metabolism. *Limnol. Oceanogr.* **23:**337–348.

Rudd, J. W., and C. D. Taylor. 1979. Methane cycling in aquatic environments. *Adv. Aquat. Microbiol.* **2:**77–150.

Schlesinger, W. H. 1978. Community structure, dynamics and nutrient cycling in the Okefenokee cypress swamp-forest. *Ecol. Monogr.* **48:**43–65.

Sebacher, D. I., and R. C. Harriss. 1982. A system for measuring methane fluxes from inland and coastal wetland environments. *J. Environ. Qual.* **11:**34–37.

Sebacher, D. I., R. C. Harriss, and K. B. Bartlett. 1983. Methane flux across the air-water interface: air velocity effects. *Tellus* **35B:**103–109.

Sebacher, D. I., R. C. Harriss, and K. B. Bartlett. 1985. Methane emissions to the atmosphere through aquatic plants. *J. Environ. Qual.* **14:**40–46.

Sebacher, D. I., R. C. Harriss, K. B. Bartlett, S. M. Sebacher, and S. S. Grice. 1986. Atmospheric methane sources: Alaska tundra bogs, an alpine fen, and a subarctic marsh. *Tellus* **38B:**1–10.

Svensson, B. H., and T. Rosswall. 1984. In situ methane production from acid peat in plant communities with different moisture regimes in a subarctic mire. *Oikos* **43:**341–350.

Thauer, R. K., K. Jungermann, and K. Decker. 1977. Energy conservation in chemotrophic anaerobic bacteria. *Bacteriol. Rev.* **41:**100–180.

Thompson, A. M., and R. J. Cicerone. 1986. Possible perturbations to atmospheric CO, CH_4 and OH. *J. Geophys. Res.* **91:**10853–10864.

Wahlen, M., N. Tanaka, R. Henry, B. Deck, J. Zeglen, J. S. Vogel, J. Southon, A. Shemesh, R. Fairbanks, and W. Broecker. 1989. Carbon-14 in methane sources in atmospheric methane: the contribution from fossil carbon. *Science* **245:**286–290.

Whalen, S. C., and W. S. Reeburgh. 1988. A methane flux time series for tundra environments. *Global Biogeochem. Cycles* **2:**399–409.

Williams, R. T., and R. L. Crawford. 1984. Methane production in Minnesota peatlands. *Appl. Environ. Microbiol.* **47:**1266–1271.

Wilson, J. O., P. M. Crill, K. B. Bartlett, D. I. Sebacher, R. C. Harriss, and R. L. Sass. 1989. Seasonal variation of methane emissions from a temperate swamp. *Biogeochemistry* **8:**55–71.

Zeikus, J. G., and M. Winfrey. 1976. Temperature limitation of methanogenesis in aquatic sediments. *Appl. Environ. Microbiol.* **31:**99–107.

Chapter 7

Production and Consumption of Methane in Aquatic Systems

Ronald P. Kiene

There is compelling evidence that atmospheric concentrations of CH_4 have increased dramatically over the last century and that the rise is continuing today at a rate of approximately 1% per year (Rasmussen and Khalil, 1981; Khalil and Rasmussen, 1987, 1990; Blake and Rowland, 1988). This effect appears to be closely tied to human activities throughout the biosphere. The increased methane levels are of considerable concern because this trace gas plays an important role in the chemistry of the atmosphere (Cicerone and Oremland, 1988) and is also a potent absorber of infrared radiation. Recent estimates indicate that current levels of CH_4 contribute about 10 to 15% to the atmospheric greenhouse forcing (Lashof and Ahuja, 1990; Rodhe, 1990). Methane has both natural and anthropogenic sources, and this molecule is involved in complex biogeochemical cycles. The alarming changes that are detectable for CH_4, as well as other trace gases, have prompted questions about how these changes have already affected and how they will continue to affect the Earth's climate system and global biogeochemical cycles. If we are to predict with any certainty what role trace gases will play in the future, it will be necessary to understand the potential feedbacks that global changes in climate may have on the biogeochemical cycling of these compounds.

This paper will concentrate on the biogeochemical cycling of CH_4 in natural aquatic environments and the factors determining release of CH_4 from these habitats. Large amounts of CH_4 are produced in aquatic habitats by the activities of methanogenic bacteria. However, in most instances only a fraction of the gross CH_4 production is actually released to the atmosphere. Much of the CH_4 produced in these systems is intercepted by bacteria which are capable of oxidizing it. Nonetheless, natural emissions of CH_4 from aquatic habitats play a significant role in the global atmospheric CH_4 budget (Seiler, 1984; Mathews and Fung, 1987; Cicerone and Oremland, 1988; see Table 1). Because a far greater percentage of CH_4 is produced than actually reaches the atmosphere, a small perturbation in the natural CH_4 cycle, due perhaps to climate change, could significantly alter the atmospheric CH_4 budget. It is therefore critical that we understand how CH_4 is cycled in these

Ronald P. Kiene • University of Georgia Marine Institute, Sapelo Island, Georgia 31327.

Table 1. Annual methane release rates for identified sources[a]

Emission source	CH_4	
	Annual release (10^{12} g)	Range (10^{12} g)
Enteric fermentation	80	65–100
Natural wetlands (forested and nonforested bogs, forested and nonforested swamps, tundra, and alluvial formations)	115	100–200
Rice paddies	100	60–170
Biomass burning	55	50–100
Termites	40	10–100
Landfills	40	30–70
Oceans	10	5–20
Fresh waters	5	1–25
Methane hydrate destabilization	5?	0–100 (future)
Coal mining	35	25–45
Gas drilling, venting, transmission	45	25–50
Total	540	400–640

[a] Data from Cicerone and Oremland (1988).

environments and that we gain a greater comprehension of how processes might be affected by climate change.

To date, the cycling of CH_4 in the biosphere has received considerable attention. Several excellent reviews on this subject already exist. The review by Rudd and Taylor (1980), although somewhat dated, still stands as a thorough and lucid overview of how CH_4 is cycled in aquatic environments. More recent treatments (King, 1984b; Winfrey, 1984; Ward and Winfrey, 1985; Capone and Kiene, 1988; Conrad and Schütz, 1988; Oremland, 1988; Cicerone and Oremland, 1988; Oremland and King, 1989) have brought this field mostly up to date. Despite considerable advances in our knowledge over the last 20 years, there remain many significant questions about the factors which control the net ecosystem production of CH_4. Here I would like to focus attention on the fundamental processes which are involved in the net production of CH_4 from various aquatic habitats, i.e., that which is available for release to the atmosphere. Emphasis will be placed on the factors which control CH_4 production and its oxidation and the balance struck between these two processes. Out of necessity I will give a brief overview of the basic aspects of CH_4 biogeochemistry. Finally, I would like to evaluate the current state of knowledge of CH_4 cycling in the oceans.

METHANOGENESIS AND METHANOGENIC HABITATS

It has been shown that the majority of CH_4 in the atmosphere has a relatively young ^{14}C age and is therefore probably of biogenic origin (Ehhalt, 1974; Ehhalt and

Schmidt, 1978). Several papers are available which discuss isotope geochemistry of CH_4 and the role of thermogenic CH_4 in biogeochemical cycling (Winfrey, 1984; Oremland, 1988; Cicerone and Oremland, 1988). These aspects of CH_4 geochemistry will not be dealt with further here.

Biogenic methane is produced almost exclusively by the activities of a unique group of archaebacteria known as the methanogens (see Zeikus, 1977; Zehnder, 1978; Balch et al., 1979; Winfrey, 1984; Jones et al., 1987). Several cases of trace methane production by nonmethanogens have been observed (e.g., Rimbault et al., 1988), but this is considered to be insignificant in the biogeochemical cycle of methane.

Methanogenic bacteria are strict anaerobes and are inhibited by even traces of O_2 (Jones et al., 1987). Thus, CH_4 is produced in environments where organic matter accumulates and where O_2 is absent. Such environments abound in nature, and examples include organic-rich soils and sediments, stratified water bodies, the forestomachs of ruminants, the hindguts of termites, and the interior of certain trees (Zeikus, 1977). Selected rates of CH_4 emissions from various aquatic habitats are given in Table 2. In each of these environments, organic matter is plentiful, diffusion of O_2 is restricted, and anaerobic conditions prevail. When methanogenic activity is evident in habitats where O_2 is present in the bulk environment, i.e., surface ocean waters (see below), it must almost certainly originate in microenvironments which are sufficiently lacking in molecular oxygen. With respect to aquatic environments, it is the sediments which are the primary sites of methanogenesis, although in some cases the water column supports significant methanogenesis when stratification causes anoxia (Rudd and Taylor, 1980; Iverson et al., 1987).

FACTORS GOVERNING THE DISTRIBUTION AND RATES OF METHANOGENESIS

The fundamental controlling factors on the extent of CH_4 production in a given environment are related to (i) the ecological interactions among various microbial groups in the habitats of interest, (ii) the availability of electron acceptors, (iii) the quality and quantity of organic matter supply, (iv) temperature, and, to some extent, (v) pH and (vi) salinity. These factors will be discussed below.

Microbial Interactions and Anaerobic Food Chains

The anaerobic food chain requires the concerted efforts of both fermentative and respiratory organisms. In the anaerobic food chain model (Fig. 1), complex polymeric organic matter is first degraded into monomers which are then fermented to low-molecular-weight fatty acids, alcohols, H_2, and methyl compounds. These latter products are oxidized to CO_2 and H_2O with the concomitant reduction of various electron acceptors, or through methanogenic fermentation. Bacteria which specialize in the use of specific electron acceptors have evolved (i.e., SO_4^{2-}

Table 2. Methane emissions from various aquatic environments[a]

Environment	CH_4 emission rate	Reference
Temperate coastal marshlands		
Barataria Basin, La.		
Salt marsh	0.36 mol · m^{-2} · year^{-1}	DeLaune et al., 1983
Brackish marsh	6.1 mol · m^{-2} · year^{-1}	
Freshwater marsh	13.3 mol · m^{-2} · year^{-1}	
Virginia salt marsh		
Salt marsh	0.35 mol · m^{-2} · year^{-1}	Bartlett et al., 1987
Intermediate	1.4 mol · m^{-2} · year^{-1}	
Freshwater marsh	1.13 mol · m^{-2} · year^{-1}	
California salt marshes	8.4–49.3 mol · m^{-2} · year^{-1}	Cicerone and Shetter, 1981
Georgia salt marsh		
Short Spartina zone	3.3 mol · m^{-2} · year^{-1}	King and Wiebe, 1978
Tall Spartina zone	0.025 mol · m^{-2} · year^{-1}	
Northern wetlands		
Alaskan tundra		
Moist tundra	0.31 μmol · m^{-2} · day^{-1}	Sebacher et al., 1986
Waterlogged tundra	7.4 μmol · m^{-2} · day^{-1}	
Wet tussock meadow	2.5 μmol · m^{-2} · day^{-1}	
Alpine fen	18.1 μmol · m^{-2} · day^{-1}	
Boreal marsh	6.6 μmol · m^{-2} · day^{-1}	
Minnesota peatlands	0.19–119 mmol · m^{-2} · day^{-1}	Harriss et al., 1985
Swedish acid peat	0.02–59.4 mmol · m^{-2} · day^{-1}	Svensson and Rosswall, 1984
Rice paddies		
Italian rice field		
Rice plants	1.5–39 mmol · m^{-2} · day^{-1}	Holzapfel-Pschorn et al., 1986
Weeds	1.5–22.5 mmol · m^{-2} · day^{-1}	
Unplanted	1.5–25 mmol · m^{-2} · day^{-1}	
California rice field		
Unfertilized	2 mmol · m^{-2} · day^{-1}	Cicerone and Shetter, 1981
Fertilized	9.4 mmol · m^{-2} · day^{-1}	
Open ocean		
Cariaco Trench		
Surface water	0.23 μmol · m^{-2} · day^{-1}	Ward et al., 1987
Atmospheric exchange		
Southern Sargasso Sea	2.3 μmol · m^{-2} · day^{-1}	Scranton and Brewer, 1978
Lakes		
Wintergreen Lake, Mich.		
Diffusive flux	10–46 mmol · m^{-2} · day^{-1}	Strayer and Tiedje, 1978
Ebullition	35–37 mmol · m^{-2} · day^{-1}	
California lakes	8.1 mmol · m^{-2} · day^{-1}	Cicerone and Shetter, 1981
Big Soda Lake, Nev.	0.036 mmol · m^{-2} · day^{-1}	Iverson et al., 1987

[a] Data from original papers have been converted to units of moles of CH_4 per unit area where necessary.

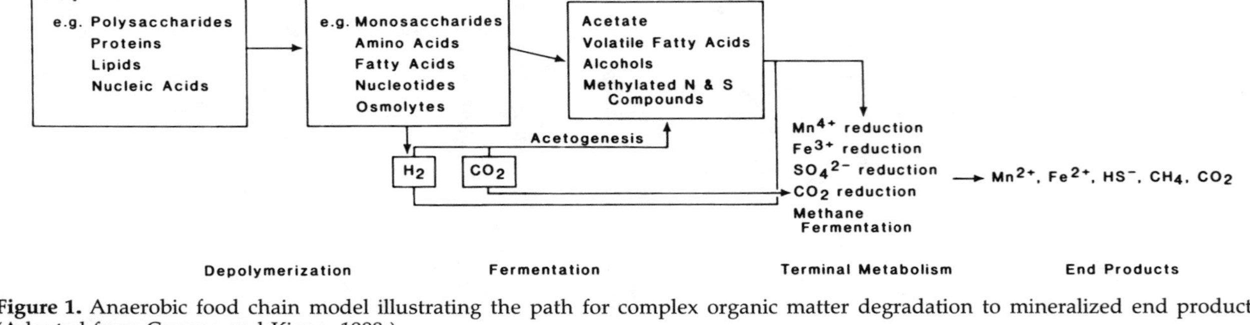

Figure 1. Anaerobic food chain model illustrating the path for complex organic matter degradation to mineralized end products. (Adapted from Capone and Kiene, 1988.)

reducers and CO_2-reducing methanogens), although some organisms are versatile and use several electron acceptors.

With oxygen or nitrate as the electron acceptor it is thermodynamically favorable for organisms to completely oxidize complex substrates without release of intermediary products. However, for all other electron acceptors, the energy yield per electron transferred is greater if complex substrates are first fermented to low-molecular-weight products (fatty acids, H_2, and alcohols), with the subsequent respiration of these products (McInerney and Beaty, 1988; Lovley and Phillips, 1989). Removal of intermediary products is required to maintain thermodynamically favorable conditions for the complete oxidation of organic matter to mineralized end products.

Methanogens cannot directly utilize complex organic molecules for energy generation and growth and therefore are dependent on the activities of other microorganisms to provide low-molecular-weight substrates. The growth substrates known to be utilized by methanogenic bacteria are restricted to a very few low-molecular-weight compounds; these include H_2 (linked to CO_2 reduction), carbon monoxide, acetate, formate, methanol, dimethyl sulfide (DMS), tri-, di-, and monomethylamines, and ethyl methylamine (see Winfrey [1984] and Oremland [1988] for reviews). Additionally, Widdel (1986) has recently reported that a pure culture of a methanogen could grow using 2-propanol as a hydrogen donor for the reduction of CO_2, provided that acetate was present as a carbon source. Acetate and H_2 have been identified as the two major substrates for methanogenesis in natural environments (Rudd and Taylor, 1980).

Of the existing species of methanogens, none are known to be capable of metabolizing all of the substrates listed above. Some species such as *Methanobacterium thermoautotrophicum* appear to be obligate H_2 oxidizers (Balch et al., 1979), while others such as *Methanococcoides methylutens* appear to require methylated substrates (Sowers and Ferry, 1983) and are considered to be obligate methylotrophs. Still others, like *Methanosarcina barkeri*, have a more versatile metabolism and are capable of metabolizing a variety of the known methanogenic substrates.

Availability of Electron Acceptors and Competition for Substrates

The thermodynamic energy yield of the oxidation of organic matter coupled to reduction of various electron acceptors decreases in the order $O_2 > NO_3^- > MnO_2 > FeOOH > SO_4^{2-} > CO_2$, and in general the most favored electron acceptor will be used preferentially. The theoretical basis for this lies in the greater energy available, and therefore greater growth yield, to the organisms which can use the most favorable electron acceptors (Thauer et al., 1977). In reality, though, other factors come into play such as kinetic competition for substrates and biochemical substrate-use limitations. Methanogenesis from the reduction of CO_2 with molecular hydrogen is the least energetically favorable mode of anaerobic respiration.

Numerous studies have shown that addition of more energetically favorable electron acceptors to methanogenic sediments results in inhibition of methanogen-

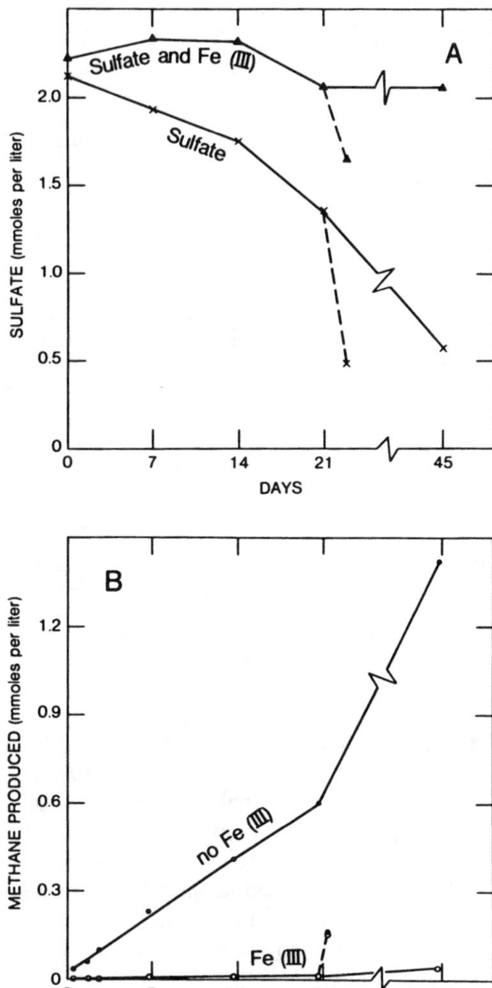

Figure 2. Effect of amorphic iron(III) oxyhydroxide-coated clay on sulfate reduction and methane production. (A) Sulfate depletion in sediments that received additions of sulfate and clay suspensions with or without Fe(III) coating. (B) Methane production in sediments that received additions of a clay suspension with or without Fe(III) coating. The dashed lines indicate samples (at ca. 21 days) which were incubated under excess hydrogen. The data presented are from one bottle of each sediment treatment and are representative of triplicates of each treatment. (From Lovley and Phillips, 1987.)

esis and diversion of the electron flow to the favored electron acceptor (Winfrey and Zeikus, 1977; Abram and Nedwell, 1978; Lovley and Klug, 1983a; Lovley and Phillips, 1987). For example, when either NO_3^{2-}, SO_4^{2-}, or amorphous iron oxides are added to methane-producing sediments, methanogenesis is inhibited (Winfrey and Zeikus, 1977; Lovley and Phillips, 1987). Figure 2 illustrates these kinds of interactions in a freshwater sediment in which SO_4^{2-} or amorphous Fe(III) was added (Lovley and Phillips, 1987). In this case methanogenesis is inhibited by both SO_4^{2-} and Fe(III), and Fe(III) inhibits sulfate consumption.

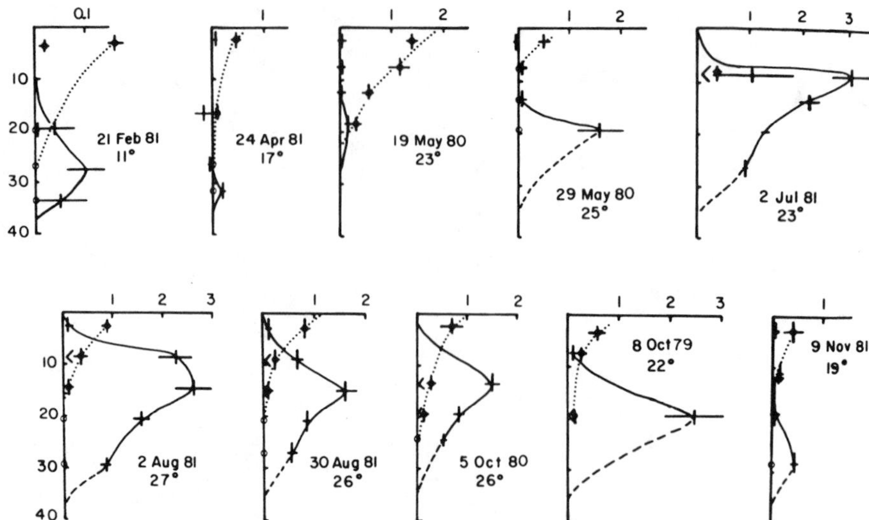

Figure 3. Rates of sulfate reduction and methane production plotted versus depth in Cape Lookout Bight, North Carolina, sediments. The seasonal pattern of methanogenesis moving closer to the interface is illustrated. Horizontal lines through data points are the standard errors of the regression coefficients; vertical lines are the depth intervals of the samples. (From Crill and Martens, 1983.)

These same authors showed that pulse additions of excess substrates (H_2 in Fig. 2) allowed methanogenesis to proceed despite the presence of more favorable electron acceptors, indicating that the electron acceptors are not directly toxic to the methanogens. In addition, many studies have also shown that when sulfate reduction is inhibited by molybdate or by fluorolactate, methanogenesis is stimulated by the increased availability of substrates (Abram and Nedwell, 1978; Oremland and Taylor, 1978; Banat et al., 1983; Capone et al., 1983). The conclusion from the above-mentioned studies is that sulfate-reducing bacteria outcompete methanogenic bacteria for common substrates such as H_2 and acetate. The only exceptions to this general rule appear to be the methylated amines, for which methanogens can compete successfully (see below).

Sulfate reducers are recognized as having more efficient uptake systems (lower K_s) for H_2 and acetate compared to methanogens (Kristjansson et al., 1982; Schonheit et al., 1982; Robinson and Tiedje, 1984) and are capable of maintaining concentrations of these substrates at levels low enough to exclude methanogens (Lovley et al., 1982; Lovley, 1985). In sulfate-rich environments (i.e., marine sediments), sulfate-reducing bacteria utilize the majority of available H_2 and acetate (see review by Capone and Kiene [1988]). This competitive interaction results in low rates of methanogenesis in the upper zone of marine sediments, where sulfate is plentiful (Oremland and Taylor, 1978; Crill and Martens, 1983) (Fig. 3). Below the zone of sulfate depletion, methanogenesis often increases and may be significant with respect to carbon mineralization (Martens and Berner, 1977; Mountfort and Asher, 1981; Crill and Martens, 1983; Martens and Klump, 1984). This is in contrast to the pattern observed in low-sulfate fresh waters, where methanogenesis is highest near the very surface of the sediments and dominates carbon flow (Fig. 4) (Kelly and Chynoweth, 1981; Lovley and Klug, 1983b; Frenzel et al., 1990).

Because of the higher affinity of nonmethanogenic organisms for hydrogen and acetate, these substrates are maintained at concentrations too low for methanogens to effectively utilize (Lovley and Klug, 1986; Lovley and Goodwin, 1988). It appears that for hydrogen and perhaps acetate there are characteristic steady-state concentrations of the substrates when a given electron acceptor is the dominant electron sink (Lovley and Phillips, 1987; Lovley and Goodwin, 1988; Michener et al., 1988; Novelli et al., 1988). For example, under iron-reducing conditions, steady-state H_2 concentrations are 0.2 nM, whereas they are 1 to 1.5 nM under sulfate-reducing conditions and 7 to 10 nM under methanogenic conditions (Lovley and Goodwin, 1988). In the presence of other electron acceptors, hydrogen concentrations are well below the K_m of methanogenic populations. Because a threshold concentration may exist for methanogens to take up H_2 (Lovley, 1985), methanogens are unable to compete for this substrate. It is interesting that the steady-state concentrations of H_2 are independent of organic matter availability.

At some concentrations each electron acceptor will become limiting and the next favorable process will begin to compete successfully. The exact concentrations at which these limitations occur are not very well known and could be dependent on organic matter availability. This would be worth investigating. Worth mentioning with respect to the use of Fe(III) as an electron acceptor is the fact that not all Fe(III) is available to microorganisms for dissimilatory reduction (Jones et al., 1983; Lovley and Phillips, 1986a, 1986b). The most easily reduced Fe(III) compounds are the amorphous iron oxides, whereas crystalline forms of Fe(III) are much less labile. Some pools of Fe(III) appear not to be reduced at all, when viewed over time scales of a few years, and therefore persist at depth in sediments.

Because of the competition for common substrates, significant methanogenesis proceeds only when other electron acceptors are absent or exhausted. This situation often arises in natural environments when the supply of more favorable electron acceptors is cut off or limited by diffusion. In sediments, the source of all electron acceptors (except CO_2) is the water column. Thus, at depth in the sediments, electron acceptors are depleted sequentially and methanogenesis is limited to the lower layers of such environments. The supply of electron acceptors varies greatly in different environments; it exerts a significant control on the rate and extent of methanogenesis (Capone and Kiene, 1988) and is responsible for the zonation and segregation of various respiratory activities.

Organic Matter Supply

The most rapid oxidation of organic matter typically occurs in the upper layers of sediments, resulting in depletion of electron acceptors. The higher metabolism in the aerobic surface layers is not the result of the greater efficiency of O_2 respiration,

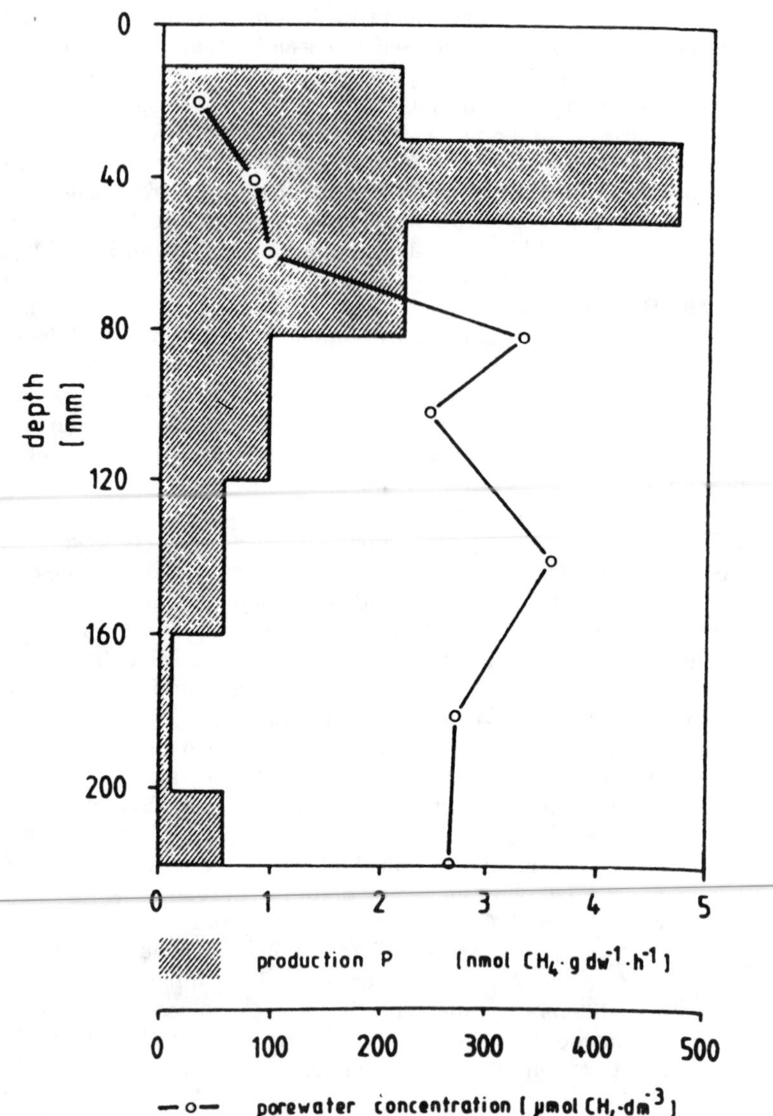

Figure 4. Vertical profile of production rates and pore water concentrations of CH_4 in profundal sediments of Lake Constance. (From Frenzel et al., 1990.)

but rather results from the fact that the most labile organic matter is present in this region (i.e., newly settled material). Henrichs and Reeburgh (1987) have recently pointed out that anaerobic decomposition rates are not intrinsically lower than aerobic rates. Thus, the rate of microbial metabolism at depth in sediments is likely to depend on the ability of the existing consortium to break down the organic matter which persists at that given depth. With respect to methanogenesis, this has several ramifications. Because methanogens are last in line in the anaerobic food chain, the methanogenic consortium is most likely to be severely limited by the refractory nature of the organic matter passing through the other respiratory zones. This limitation is overcome when other electron acceptors such as nitrate or sulfate are scarce in the system, such as occurs in most fresh waters, or when organic inputs are sufficiently high so that the supply of electron acceptors such as oxygen, nitrate, or sulfate cannot keep up with the sedimentation of newly deposited material. The latter situation occurs in organic-rich marine systems (Mountfort and Asher, 1981; Martens and Klump, 1984), and in these cases, depth-integrated methane production accounts for a major part of the carbon flow.

Temperature

Temperature effects on methanogenesis are of obvious importance with respect to climate change. As with other biological processes, methanogenesis is strongly temperature dependent. Zeikus and Winfrey (1976) have found clear temperature optima for methanogenesis in Lake Mendota sediments (Fig. 5). For the most part, these optima were the same for several sites within the lake and at different depths within sites, even though absolute rates differed considerably (Fig. 5). The optimum temperature for methanogenesis was always well above the maximum temperatures experienced by organisms in the lake. Similar findings were reported for methanogenesis in a northern wetland (Svensson, 1984), and in addition, different temperature optima for methanogenesis from either H_2/CO_2 or acetate were observed (Svensson, 1984). This indicates that considerable potential exists for increased rates of methanogenesis if temperatures were to rise.

Temperature effects on the whole methanogenic consortium need to be considered. In this regard, Conrad et al. (1987b) found that lower temperatures slowed H_2 turnover in paddy soil more than they did H_2-dependent methanogenesis. It is also important to consider that over an annual cycle in a system such as a lake or paddy soil, even if rates were to rise as a result of temperature increase, it does not necessarily mean that more total methane will be formed. If organic matter supply remains constant, then increasing temperature may just accelerate the consumption of the labile fraction of the organic matter, resulting in the same amount of end product (CH_4) (Kelly and Chynoweth, 1981). A change in the rate of formation, though, may have a significant effect on the exchange of CH_4 with the atmosphere if, for example, more bubbling occurs at the higher methane production rates (see below). To further complicate this picture, it is possible that higher temperatures could affect the total organic carbon pool which is available for degradation within a system. Westrich and Berner (1988) have recently found that

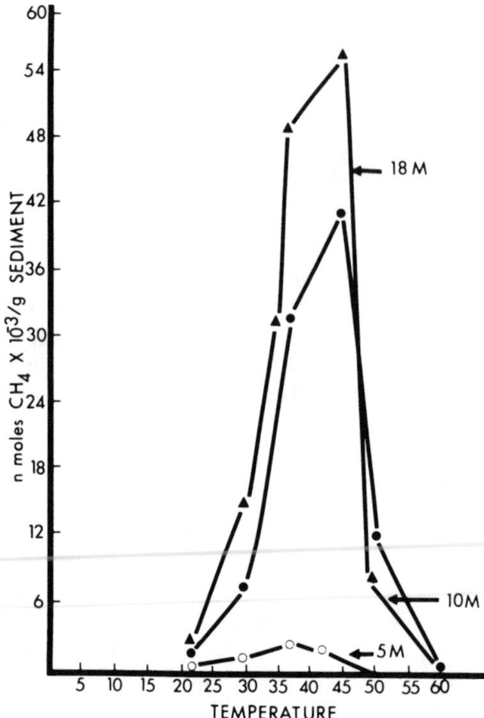

Figure 5. Effect of temperature on sediment methanogenesis in samples collected on 17 January 1975 from Lake Mendota, Wisconsin. Results represent the average of duplicate experiments. (From Zeikus and Winfrey, 1976.)

while sulfate reduction is temperature dependent, the exact dependence is a function of the lability of the organic matter undergoing decomposition. In other words, they found that the energy of activation was higher for degradation of more refractory organic matter. Thus, at higher temperatures a larger pool of organic matter may be available for degradation which would otherwise be refractory at lower temperatures. This is an intriguing concept which could have important implications and should be investigated further.

pH

Methanogenesis is known to occur in both acidic and alkaline environments (King et al., 1981; Oremland et al., 1982a; Phelps and Zeikus, 1984; Svensson and Rosswall, 1984; Iversen et al., 1987; Westermann and Ahring, 1987; Oremland and King, 1989). In general, the rates of methanogenesis reported from these environments are similar to neutral pH marine and freshwater habitats. These findings suggest that CH_4 production is not limited by pH. However, some of the ecological interactions and degradation pathways may be influenced by the pH regime. Phelps and Zeikus (1984) found that at lower pHs in lake sediments most of the CH_4 was formed from acetate fermentation rather than CO_2 reduction with H_2. In these sediments, H_2 metabolism was dominated by homoacetogenic bacteria. Furthermore, it has been found that low pH values in acidic sediments of Knaack Lake inhibited H_2 production (Conrad et al., 1987a). These findings may explain why H_2 is less important to methanogenesis at low pH.

Salinity

As mentioned above, sulfate exerts a significant control on the distribution of methanogenesis in aquatic environments because of the competition between sulfate-reducing bacteria and methanogens. Because seawater contains a large supply of sulfate (24 to 28 mM), methanogenesis is relatively unimportant with respect to carbon flow in most marine systems (Capone and Kiene, 1988). Salinity effects on CH_4 fluxes have also been observed (DeLaune et al., 1983; Bartlett et al., 1987) (Table 2), and in general, CH_4 fluxes are greater in the freshwater regions of marshes and estuaries.

In salt marshes, methane production is very heterogeneous (Senior et al., 1982; Harvey et al., 1989), which is consistent with the large variety of environmental conditions present in these systems. Highest methane production rates are found in the upper marsh zones, which are poorly drained (King and Wiebe, 1978, 1980; Harvey et al., 1989) and in which sulfate is depleted at relatively shallow depths. Because of the high sulfate content of surficial sediments in salt marshes, the bulk of methanogenesis may be due to use of methylated substrates (Oremland et al., 1982b; Senior et al., 1982; Banat et al., 1983). Franklin et al. (1988) have recently found that a large fraction of the isolable methanogens from a Georgia salt marsh were methylotrophic.

METHANOGENESIS AND METHYLATED SUBSTRATES

A look at the recent literature reveals that new species of methanogenic bacteria are being isolated at an unprecedented rate. This is due to several factors including improved isolation techniques, intense current interest in biogenic methane production, and the recognition of a broader ecological niche for methanogens than was previously believed. Several of the new isolates, particularly those from saline environments, are capable of metabolizing methylated substrates (Sowers and Ferry, 1983; Zhilina, 1983; Paterek and Smith, 1985; Mathrani and Boone, 1985; Blotevogel et al., 1986; Oremland et al., 1989). Interestingly, certain of these methylotrophic isolates are incapable of growth with H_2/CO_2 or acetate. Methanogens from saline environments may have adapted to methylotrophic growth in order to avoid competition from sulfate-reducing bacteria for H_2 and acetate.

Recently, several reports have indicated active methanogenesis in sulfate-containing environments (Oremland and Taylor, 1978; Oremland et al., 1982b; Crill and Martens, 1983; King et al., 1983; King, 1984a, 1984b; Kiene, 1988; Oremland and King, 1989). Methanogens in aquatic sediments can utilize substrates such as methanol (Oremland et al., 1982b; Oremland and Polcin, 1982), methylated sulfur compounds, and methylated amines (Zinder and Brock, 1978b; Oremland and Polcin, 1982; King et al., 1983; King, 1984a, 1984b; Kiene et al., 1986; Kiene, 1988) when sulfate is present. King (1984b) has recently reviewed the use by methanogens of "noncompetitive" substrates. It appears that when methanol is present at low concentrations it is utilized primarily by sulfate-reducing bacteria, while at high concentrations methanogens metabolize the majority of methanol (King, 1984b). A similar observation was made by Kiene and co-workers (Kiene et al., 1986; Kiene, 1988) for the metabolism of DMS by methanogenic pathways in saline sediments. At low concentrations, DMS was metabolized mostly by oxidative pathways, presumably sulfate reduction, whereas at higher concentrations a greater fraction of the DMS was converted to CH_4. Likewise, the conversion of the S-methyl group of methionine (a precursor of DMS and methanethiol) to CH_4 in sulfate-containing sediments displayed a similar concentration dependence to that of DMS and methanol (Banat et al., 1983; Winfrey and Ward, 1983; Kiene and Visscher, 1987). In apparent contrast to the above-mentioned studies, Zinder and Brock (1978b) reported that 10 mM sulfate did not affect conversion of DMS or methane thiol to CH_4 in freshwater lake sediments.

Most of the methylamines in marine sediments are consumed by methanogenic bacteria (King et al., 1983; King, 1988a). However, there is evidence that some fraction of the methylamines in marine sediments may be utilized by nonmethanogenic pathways (King et al., 1983; King, 1984b; Lovley and Phillips, 1986a). Thus, the frequently used term "noncompetitive" methanogenic substrate appears to be somewhat misleading, since some degree of competition exists between sulfate reducers and methanogens for all known methanogenic substrates. Methylated amines appear to be the only substrates that methanogens can successfully compete for at in situ substrate concentrations which are typically in the low micromolar range.

The predominance of methylotrophic methanogens in saline environments may be due to the availability of precursors for the immediate methanogenic substrates. Methanol may be formed in sediments from the decomposition of pectin (Schink and Zeikus, 1982), while methylamines arise from the degradation of more complex amines such as choline, glycine betaine, and trimethylamine oxide (King, 1984a). Choline is present in all organisms, and trimethylamine oxide is known to occur in the tissues of marine fish. Glycine betaine is utilized by a large number of aquatic organisms as an osmotic solute (Yancey et al., 1982; King, 1988b). Similarly, most of the DMS in marine systems may arise from the metabolism of dimethylsulfoniopropionate (Kiene and Visscher, 1987; Kiene and Capone, 1988), which is also an osmolyte in halophytic plants. Thus, these precursors may be the sources of methanol, methylamines, methylated sulfur compounds, and eventually methane in marine sediments. In freshwater sediments, the turnover of methylamines contributes only a small fraction to total CH_4 production due to the fact that H_2 and acetate are being channeled mostly to CH_4 (Winfrey et al., 1977; Lovley and Klug, 1982; Lovley et al., 1982), and methylamines are of far less quantitative significance in terms of carbon flow than are other sources of CH_4. In contrast, King et al. (1983) found that trimethylamine contributed a large fraction to total methanogenesis in estuarine sediments from Lowes Cove and in a hypersaline pond from the West Indies (King, 1988a). Kiene (1988) also observed that while methanogens contributed only a small amount to the total DMS consumption in salt marsh sediment slurries, the DMS contributed about 30% of the total CH_4 production in these experimental slurries. The use of methylated sulfur compounds for methanogenesis is not restricted to saline sediments, as Zinder and Brock (1978a, 1978b) have shown their conversion to CH_4 in lake sediments and sewage sludge. However, the quantitative contribution of these compounds in fresh waters remains to be determined.

Although it is clear that methylotrophic methanogenesis is important in some systems, particularly saline environments, it is not clear what consequences this has for atmospheric exchange of CH_4. The methylated precursors of CH_4 originate primarily from marine plants and therefore are expected to be most prevalent near the surface of the sediment or at root interfaces. Methane formed near the sediment/water interface could conceivably be more easily exchanged with the overlying water or to the atmosphere. This is also near the zone of CH_4 oxidation, and it is possible that these two processes could coexist in microniches within the same sediment horizon. It is clear that methanogenesis from methylamines is largely free from competition by sulfate reduction, but it is not known whether methanogens can successfully compete for these substrates under NO_3^-, Fe(III), or Mn(IV) reducing conditions, which presumably occur nearer the oxic/anoxic interface.

METHANE OXIDATION

In most cases only a small percentage of the gross methane production escapes from methanogenic habitats into the surrounding or overlying environments. A large fraction of the methane formed in aquatic habitats is consumed by methane-oxidizing bacteria before it can leave the system (Reeburgh and Heggie, 1977; Rudd and Taylor, 1980) (Tables 3 and 4). Thus, methane oxidation is a critical process governing the amount of methane reaching the atmosphere. Methane oxidation occurs both aerobically and anaerobically.

Aerobic Oxidation

Methanotrophic bacteria are a unique group possessing the ability to utilize CH_4 as an energy and carbon source during growth. Aspects of the biology and biochemistry of methane-oxidizing organisms have been reviewed by Whittenbury et al. (1970), Anthony (1975), and Hanson (1980).

In natural systems, aerobic methane oxidation activity is most active at or near the interface of oxic and anoxic conditions (Rudd and Taylor, 1980). This is a zone where CH_4 diffuses from the anoxic environment into the oxygenated zone and where its concentration is the highest. However, CH_4 concentration is not the only factor involved in governing the distribution of oxidation activity. Oxygen and combined nitrogen concentrations are important controlling factors as well (Rudd

Table 3. Rates of aerobic methane oxidation and percentage of methane oxidized on passage through the oxic layers of various aquatic environments[a]

Environment	Oxidation rate	% of CH_4 oxidized on passage through oxic zone	Comments/methods	Reference
Lake Constance, profundal sediments	447 µmol · m^{-2} · day^{-1}	93	Core incubation with and without O_2 in water	Frenzel et al., 1990
Lake Mendota during summer stratification	23.3 mmol · m^{-2} · day^{-1}	48	8% of total CH_4 was evaded to atmosphere; remainder accumulated	Fallon et al., 1980
Lake Washington, deep sediments	250–350 µmol · m^{-2} · day^{-1}	50	Based on calculation of diffusion across sediment water interface and what was actually measured	Kuivila et al., 1988
Lake 227	0.02–32 mmol · m^{-2} · day^{-1}	100	100% of hypolymnetic CH_4 flux diffusing across thermocline	Rudd and Hamilton, 1978
		60	60% during overturn of lake	
Italian rice field, laboratory and field experiments				
Rice	38 mmol · m^{-2} · day^{-1}	77	CH_4 production and emission measured, % oxidized by difference	Holzapfel-Pschorn et al., 1986
Weeds	11.7 mmol · m^{-2} · day^{-1}	95		
Everglades peat	0.91–2.3 mmol · m^{-2} · day^{-1}	74–85	Oxic/anoxic core incubation method; strong O_2 dependence	King et al., 1990
Danish wetland		16–96	Oxic/anoxic core incubation method; strong dependence on light because of benthic photosynthetic O_2 production	King, 1990
Cariaco Trench, oxic water column	0.0004 nM · day^{-1} (subsurface CH_4 maxima)	10	Used radiolabeled CH_4 and diffusive flux calculations. Oxidation at the interface probably underestimated due to low depth resolution of samples	Ward et al., 1987
	0.02 nM · day^{-1} (across oxic/anoxic zone)	5		
Cape Lookout Bight, organic-rich coastal marine basin	0.084 mmol · m^{-2} · day^{-1}	1.1	Summer CH_4 budget	Martens and Klump, 1984
Big Lagoon, Calif., coastal saline pond	0–110 nM · day^{-1}	4.4–7.0	Used radiolabeled CH_4 for oxidation; diffusive flux calculated for CH_4 input	Butler et al., 1988

[a] Data were converted to units of moles per unit area or per liter of water.

Table 4. Rates of anaerobic methane oxidation and percentage of methane oxidized on passage through sulfate-reducing zone

Environment	CH_4 concn (μM)	Rate of oxidation (per day)	% of CH_4 oxidized on passage through sulfate reducing zone	Reference
Water column				
Cariaco Trench, anoxic bottom water	2–13	0.4 nM	—[a]	Ward et al., 1987
Framvaren Fjord, water column	1–25	0–11 μM	—	Lidstrom, 1983
Big Soda Lake, Nevada, water column (meromictic, alkaline, hypersaline lake)	5–10	Mixolimnion, 2–6 nM	50% total in anoxic zone	Iversen et al., 1987
	10–53	Monimolimnion, 49–85 nM		
Sediments				
Jutland, Denmark, sediments	2–85	0.2–0.28 μM	—	Iversen and Blackburn, 1981
Skagerrak, 200-m sediments	20–1,000	0–11 μM	100	Iversen and Jørgensen, 1985
Skan Bay, Alaska	250–2,500	0–8.2 μM	> 97	Alperin and Reeburgh, 1984
Sannich Inlet, sediment	100–2,800	0.5–34 μM	—	Devol, 1983

[a] —, not possible to calculate with the data given.

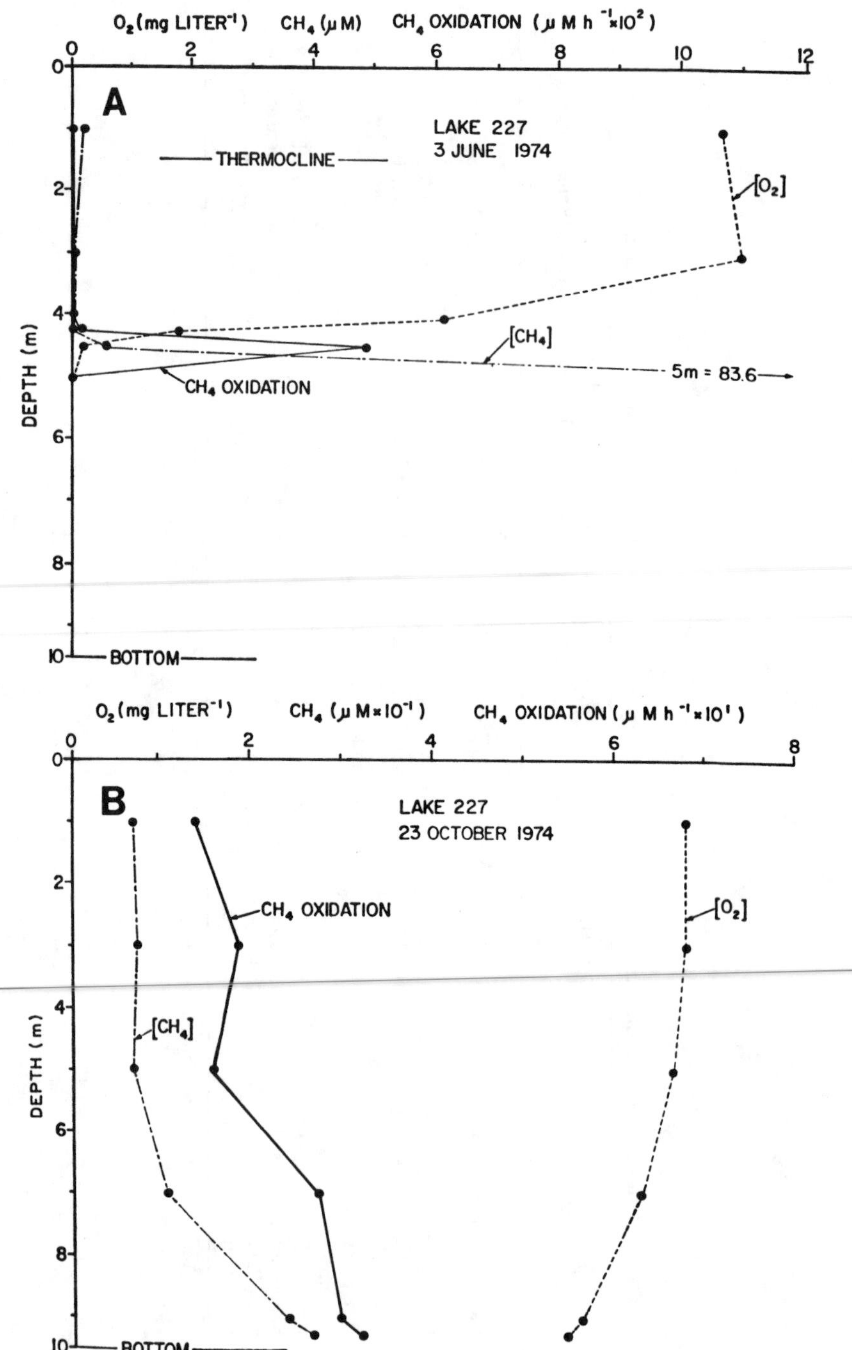
O₂ (mg LITER⁻¹) CH₄ (μM) CH₄ OXIDATION (μM h⁻¹ ×10²)
A
LAKE 227
3 JUNE 1974
THERMOCLINE
[O₂]
[CH₄]
5m = 83.6
CH₄ OXIDATION
DEPTH (m)
BOTTOM
O₂ (mg LITER⁻¹) CH₄ (μM ×10⁻¹) CH₄ OXIDATION (μM h⁻¹ ×10¹)
B
LAKE 227
23 OCTOBER 1974
CH₄ OXIDATION
[O₂]
[CH₄]
DEPTH (m)
BOTTOM

et al., 1976; Harrits and Hanson, 1980). Because their substrate supplies no nitrogen, methane oxidizers require an external source of combined nitrogen. Most methane oxidizers are capable of fixing dinitrogen (DeBont, 1976), but they can do this only when oxygen levels are relatively low (they have no specialized structures to protect their nitrogenase, such as the heterocysts found in some cyanobacteria). Nitrogen-fixing activity is not expressed when sufficient nitrogen levels are present (Harrits and Hanson, 1980). It should be noted that many methane-oxidizing bacteria may also be involved in oxidizing ammonia (Harrits and Hanson, 1980; Jones and Morita, 1983).

Working in eutrophic Lake 227, Rudd and his co-workers (Rudd et al., 1976) found that, in the summer when O_2 levels were high and dissolved inorganic nitrogen (DIN) was low in surface waters, CH_4 oxidation was restricted to the oxic/anoxic interface (Fig. 6). In contrast, during the fall overturn of the lake, when both DIN and O_2 were high, CH_4 oxidation activity was present throughout the water column. Greater than 95% of the annual CH_4 oxidation occurred during overturn periods, and this accounted for removal of 60% of the CH_4 from the lake during this time (Rudd and Hamilton, 1975). Thus, methane oxidizers are not strictly limited to microaerophilic zones, but may be prevalent there because of other factors. In general these observations were consistent among lakes of various trophic states (Rudd and Taylor, 1980).

At the boundary of the anoxic hypolimnion and the oxic metalimnion in African rift lakes, methane oxidation consumed virtually all of the upward-diffusing CH_4, and the carbon flow through CH_4 was equivalent to $\sim 10\%$ of the primary production occurring in these lakes (Rudd, 1980; Rudd and Taylor, 1980). Several lakes in this region, with considerably different methane concentration profiles, had roughly equivalent rates of methane oxidation. Since the rate of CH_4 oxidation is a function of its supply to the zone of oxidation (typically the oxic/anoxic boundary), lakes with different CH_4 concentrations must have had different physical mixing characteristics. Lake Kivu, with bottom water CH_4 concentrations of up to 20 mM, had a lower coefficient of vertical diffusion than did Lake Tanganyika (2 mM bottom water CH_4). The contrasting CH_4 concentrations and diffusion coefficients offset one another, resulting in a similar CH_4 supply to the oxidation zone. Similarly, Fallon et al. (1980) argued that greater wind-driven mixing in Lake Mendota accounted for the larger percentage of CH_4 oxidation during summer stratification there (45%) as compared to Lake 227 (11%) (Rudd and Hamilton, 1978). King et al. (1990) have discussed the possibility that V_{max} or the capacity of sediment to oxidize methane is a function of CH_4 supply to the oxidation zone. Actual rates, though, are a function of edaphic factors.

In wetland habitats where the soil interface is overlain by only a thin film of water or none at all, light may also be an important factor controlling CH_4 oxidation

---

Figure 6. Typical profile of methane oxidation, methane concentration, and oxygen concentration during summer (A) and after fall overturn (B) in Lake 227. Note sharp lens of oxidation activity during summer stratification and oxidation throughout the water column after overturn. (From Rudd et al., 1976.)

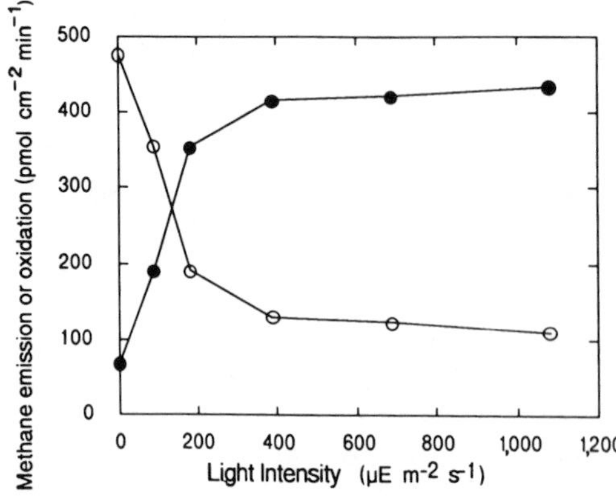

Figure 7. Methane oxidation (●) or emission (○) from a Danish wetland soil as a function of light intensity. Incubation carried out at 25°C. (From King, 1990.)

and CH_4 release to the atmosphere. King (1990) recently showed that oxygen distribution in the surface of a Danish wetland soil was greatly affected by incident light, presumably because of photosynthetic production (Fig. 7). Oxygen microprofiles are clearly affected by benthic photosynthesis and in turn affect CH_4 oxidation (King et al., 1990). Methane oxidation activity was limited by shallow oxygen penetration in the dark, causing greater CH_4 release from the sediments. In the light, increased oxygen production allowed greater CH_4 oxidation and much less escaped. Oremland and Taylor (1977) have observed similar effects of photosynthetically produced O_2 on CH_4 production and efflux from tropical seagrass sediments.

Considerable spatial variability was evident in Everglades sediments with respect to CH_4 oxidation (King et al., 1990). These authors found that marl sediments had low oxidation rates whereas peat sediments had very high rates, accounting for removal of 60 to 91% of the CH_4 flux. Similar variability was reported for both CH_4 fluxes and oxidation in tundra soils (Whalen and Reeburgh, 1988, 1990). Large spatial and temporal variations in CH_4 emissions may therefore be related to changes in CH_4 oxidation rates.

The physical mixing controls on methane oxidation mentioned above for the water columns of lakes may also play an analogous role in sediments. Bottom water currents, sediment transport, bioturbation, and tidal pumping may increase the rate of supply of methane in an oxidation zone. These factors may enhance the apparent diffusion coefficient of pore water solutes (including CH_4 and O_2). Whether this increase in supply affects the net loss of CH_4 from the system to the atmosphere is largely unknown but should be considered.

Anaerobic Methane Oxidation

Anaerobic methane oxidation is an important process accounting for removal of a large quantity of CH_4 in saline environments (Reeburgh and Heggie, 1977; Henrichs and Reeburgh, 1987). Some rates of anaerobic CH_4 oxidation are presented in Table 4. Whether anaerobic CH_4 oxidation influences the release of CH_4 to the atmosphere remains an open question. In almost all cases, any CH_4 passing the anaerobic methane oxidation zone must also pass through an aerobic oxidation zone before reaching the atmosphere. If, however, the rate of methane supply to the aerobic zone governs the amount of CH_4 oxidized there, then the amount of CH_4 oxidized anaerobically becomes important. In addition, removal of CH_4 anaerobically could also influence its transport via processes other than diffusion, e.g., ebullition (see below).

The oxidation of CH_4 in the absence of O_2 has been a subject of some debate and mystery. Reeburgh and Heggie (1977), Martens and Berner (1977), and Alperin and Reeburgh (1984) have convincingly argued from geochemical evidence (CH_4 profiles and $\delta^{13}CO_2$ distributions) that significant anaerobic methane oxidation occurs in marine sediments but not in freshwater sediments. They argued that the difference was due to the fact that sulfate was the most likely oxidant for CH_4 and that it was plentiful in seawater but not fresh water. In some marine sediments, anaerobic oxidation removes virtually all of the CH_4 before it reaches the aerobic zone. More recent studies have confirmed the earlier work of Reeburgh and his colleagues (Iversen and Blackburn, 1981; Devol, 1983; Alperin and Reeburgh, 1984; Iversen and Jørgensen, 1985; Iversen et al., 1987) and strengthened the link between anaerobic CH_4 oxidation and sulfate reduction. Near the base of the sulfate reduction zone, a peak in sulfate reduction rates has been observed (Devol, 1983; Devol et al., 1984; Iversen and Jørgensen, 1985) which may be due to the use of CH_4 as a substrate by sulfate reducers. However, in some environments, e.g., Cape Lookout Bight, anaerobic CH_4 oxidation does not seem to be important (Martens and Klump, 1984). Unfortunately, the microbial processes involved in anaerobic CH_4 oxidation remain a perplexing mystery.

While there are several reports of anaerobic methane oxidation by enrichment cultures and pure cultures of sulfate-reducing bacteria (Davis and Yarbrough, 1966; Panganiban et al., 1979), there remains very little direct microbiological information about the organisms involved. Furthermore, studies of anaerobic CH_4 oxidation in freshwater environments have never shown net CH_4 consumption (Zehnder and Brock, 1980). Attempts to resolve the microbial processes involved with inhibitor experiments have led to the conclusion that sulfate reducers are not directly involved in CH_4 oxidation, but that it occurs via a consortium of activities (Zehnder and Brock, 1980; Alperin and Reeburgh, 1985). The possibility that some as yet undiscovered metal oxide serves as the electron acceptor for anaerobic CH_4 oxidation remains a possibility, since this would be a more favorable redox couple than sulfate (Zehnder and Brock, 1980). Recently, Sorensen (1988) has postulated that methanethiol (CH_3SH) may somehow be an intermediate in the oxidation of CH_4 since it has a maximum at sediment depths where CH_4 oxidation is known to occur. Methanethiol is known to be converted to CH_4 in sediments, but it may also be metabolized via sulfate reduction (Kiene and Visscher, 1987; Kiene, 1988). The steps leading from CH_4 to methanethiol are not known but could involve reaction of CH_4 with HS^- or possibly a reversal of the methyl coenzyme M reductase system with subsequent cleavage of methanethiol from methyl coenzyme M. These mechanisms are highly speculative, and considerable work needs to be done to examine the ecology and biochemistry of anaerobic methane oxidation.

OTHER FACTORS CONTROLLING THE EMISSION OF CH_4 FROM NATURAL AQUATIC SYSTEMS

Transport of Methane through Vascular Plants

Methane emissions in vegetated wetlands such as rice paddies, marshes, swamps, bogs, and littoral zones of lakes may be significantly influenced by the vegetation growing in these areas. Plants may enhance emission of CH_4 through root-leaf transport (Dacey and Klug, 1979; Dacey, 1980; Cicerone and Shetter, 1981), as well as by releasing substrates for methanogens (Holzapfel-Pschorn et al., 1986; Schütz et al., 1989). Alternatively, the plants could retard release of CH_4 by providing a greater inflow of O_2 (and oxygenated surface area) to sediments, thereby increasing the extent of CH_4 oxidation (Schütz et al., 1989). These potential interactions are illustrated schematically for a rice paddy in Fig. 8. Because the plants themselves play a significant role in CH_4 fluxes, it is important to consider the physiological state and growth stage of the plants.

Dacey (1980) found that heat-driven, pressurized transport of gases through water lilies (*Nuphar luteum*) could account for atmospheric levels of O_2 found in the rhizomes which grow in anoxic mud. Gas flow was from the younger leaves down to the rhizomes and then back out of the older leaves to the atmosphere. Diffusion of CH_4 from the surrounding methane-rich sediments into the rhizomes resulted in enhanced transport of this gas to the atmosphere; during one period in summer, approximately 46% of the total CH_4 lost from Duck Lake, Michigan, was emitted through the leaves of water lilies (Dacey and Klug, 1979). A marked diurnal pattern was observed for CH_4 transport in this species and was undoubtedly due to patterns of solar heating. In some plants gas transport occurs via diffusion alone, which might explain why individual plant species can have vastly different transport characteristics (Oremland and Taylor, 1978; Sebacher et al., 1985; Holzapfel-Pschorn et al., 1986).

The possibility that vascular plants could affect trace gas fluxes is significant with respect to questions about the magnitude of biogenic fluxes, as well as to responses of these fluxes to global change. This is true to some extent for all radiatively important trace gases, but could be especially important for CH_4, which has a large global source in vegetated wetlands such as rice paddies and marshes. Vascular plants are likely to be greatly affected by changes in climate and atmospheric CO_2 (e.g., Idso et al., 1990). In most cases, elevated CO_2 results in increased production and biomass (above and below ground). The influence of these changes on the rhizosphere microbial flora and transport of gases is not currently known, but could be significant. Therefore plant physiology and microbial interactions deserve attention.

Bubble Ebullition

Another mechanism for CH_4 transport to the atmosphere which largely bypasses the microbial oxidation barriers is bubble ebullition from sediments. In

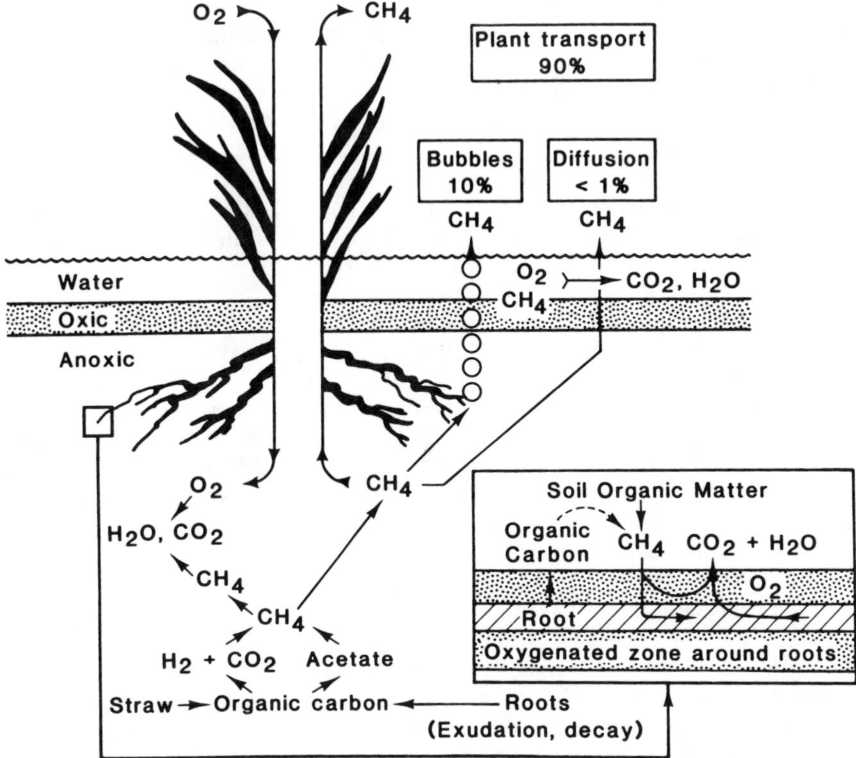

Figure 8. Scheme of production, reoxidation, and emission of CH_4 in a vegetated paddy field. Inset illustrates processes which occur near the root/soil interface. Whether O_2 is present in the roots and soil is a function of organic content of sediments and soil, light regime, plant stage, and physiology. (Redrawn from Schütz et al., 1989.)

sediments with active methanogenesis and sufficient labile organic matter, methane concentrations may reach saturating levels. This occurs in shallow lakes (Strayer and Tiedje, 1978), rice paddies (Holzapfel-Pschorn et al., 1986), and organic-rich marine sediments and rivers (Hammond et al., 1975; Martens and Klump, 1980). In rice paddies it has been found that bubble ebullition is greater in fields without vegetation. This is because the plants either facilitate CH_4 transport (in the case of rice) or foster greater oxidation in the soils (in the case of weeds) (Schütz et al., 1989).

The concentration at which CH_4 becomes saturated is a function of a variety of factors including temperature, salinity, and ambient pressure. Methane ebullition is most important in shallow water systems because of the relatively low hydrostatic pressures. Martens and co-workers have studied methane losses from Cape Lookout Bight sediments in great detail. Ebullition occurred only in the summer months when CH_4 generation was high (Crill and Martens, 1983), and release of bubbles was induced by decreases in hydrostatic pressure at low tide. Approximately 85% of the total CH_4 formed in the sediments of this site is released as bubbles. As the bubbles ascend, they redissolve and a small fraction is subsequently oxidized (Sansone and Martens, 1978; Martens and Klump, 1980), but the majority is lost via lateral transport to adjacent waters. Similar methane losses in lakes have been observed, and in these cases the loss of CH_4-carbon plays an important role in the carbon budget of these systems. The strong temperature dependence of the onset of bubbling in both marine sediments and rice fields suggests that future climate warming could increase the number of environments where bubbling occurs and prolong the duration of these bubbling periods.

THE OCEANIC METHANE PARADOX

There is ample evidence that CH_4 concentrations in the surface oceans, in areas well away from the coast and from sediment influences, are supersaturated with respect to equilibrium with the atmosphere (Atkinson and Richards, 1967; Lamontagne et al., 1973; Scranton and Brewer, 1977; Brooks et al., 1981; Burke et al., 1983; Ward et al., 1987). These observations have been made in widely different environments, both spatially and biogeographically. Indeed, the oceanic subsurface maximum of CH_4 is a world-wide phenomenon (Conrad and Seiler, 1988). The observed supersaturations make the oceans a net source of CH_4 to the atmosphere, and because of the enormous surface area of the oceans, their contribution to the global CH_4 budget cannot be ignored. However, current estimates put the oceanic course strength at 5 to 20 Tg year^{-1}, which is only 1 to 4% of the total global atmospheric flux (Seiler, 1984; Cicerone and Oremland, 1988; see Table 1).

The supersaturations observed in oceanic areas are no larger than 3× saturation, and maxima are usually located at the pycnocline well below the air/sea interface. Existing data on CH_4 concentration in the very surface waters indicate only 30 to 70% supersaturation with the atmosphere. Most of these data were collected in the 1970s or early 1980s, when CH_4 partial pressure in the atmosphere was lower than it is now. With the trend of increasing atmospheric CH_4 it has been suggested (Cicerone and Oremland, 1988) that these supersaturations could possibly disappear and that the oceans could become a sink for CH_4, unless of course the natural biogeochemical processes in the surface ocean maintain supersaturation in spite of the higher atmospheric levels. Unfortunately, our understanding of oceanic CH_4 cycling is still very limited, and it is not clear whether the oceanic methane flux will be positive or negative in the future. Therefore this area needs further attention. The question also arises as to how methane is formed in an environment in which O_2 is present in the bulk fluid.

In some cases either anthropogenic or coastal shelf sources of CH_4 could contribute to the observed supersaturations; however, in at least three cases (Scranton and Brewer, 1977; Brooks et al., 1981; Burke et al., 1983) the authors have discounted lateral transport as the sole source of CH_4 in the subsurface maximum, and several studies have indicated a net in situ source of CH_4 both in the subsurface maxima as well as in the surface mixed layer (Scranton and Brewer, 1977). Burke et al. (1983) and Brooks et al. (1981) found no consistent correlation of CH_4 concentrations and other water column characteristics, although at several stations there were significant relationships between CH_4 and chlorophyll *a*, particulate organic carbon, and total ATP. In addition, higher CH_4 concentrations were generally associated with low oxygen levels, lending credence to the idea that CH_4 is formed in microzones completely lacking O_2. Methane could originate in the guts of fish (Oremland, 1979) or from the intestinal tracts of larger zooplankton. Traganza et al. (1979) found that CH_4 concentration profiles in the Pacific Ocean were correlated with those of zooplankton biomass. Suspended particulate matter such as marine snow could also be sites of methanogenesis.

There is now convincing evidence that zones of reduced oxygen do exist in natural particulate matter in the surface ocean (Alldredge and Cohen, 1987; Paerl and Prufert, 1987; Paerl and Carlton, 1988). Certainly, if CH_4 is produced in anaerobic microzones within the surface ocean, would not energetically more favorable dissimilatory sulfate reduction also occur in these microzones? This has not, to my knowledge, been tested. However, sulfide has recently been observed in aerobic ocean waters (Luther and Tsamakis, 1989). This alone, however, does not prove that dissimilatory sulfate reduction is occurring, since other sources of HS^- exist such as hydrolysis of carbonyl sulfide (Elliot et al., 1989) and the metabolism of DMS (Kiene and Bates, 1990). In any event, with oceanic SO_4^{2-} concentrations of about 25 mM, there seems little chance for complete consumption of this electron acceptor in microzones, and from the arguments presented above, it seems clear that methanogenesis (if it occurs) must proceed through use of "noncompetitive" substrates. The most likely substrates for this would be formaldehyde, methylamines, and possibly DMS. The methylated compounds would be derived from various osmotica, namely, betaines, trimethylamine oxide, and dimethylsulfoniopropionate, whereas formaldehyde has a significant photochemical source (Mopper and Stahovec, 1986) and may also be produced from metabolism of methyl substrates. The production of CH_4 from only a selected group of precursors might explain why CH_4 concentrations do not consistently correlate with measurements of bulk seawater characteristics such as chlorophyll *a*, particulate organic carbon, etc. Recent measurements (Kiene and Bates, 1990) of the turnover of DMS in surface waters of the eastern tropical Pacific Ocean indicated rates ranging from 1 to 18 nM day^{-1}. These rates would be more than enough to sustain the methane production needed to produce the maxima in profiles thus far observed. Similar turnover rates of formaldehyde have recently been measured (Mopper and Kieber, in press). There is little known about the turnover of methylated amines in surface seawater, although there is one report of their existence (Van Neste et al., 1987). Furthermore, methylamines and betaines are likely to occur in algae as well (Sieburth, 1988), and their use by microbes is likely (Sieburth and Keller, 1991). Whether CH_4 is produced from these compounds in surface seawater has yet to be established.

Riverine inputs of CH_4 to the ocean are not very well known, but could be significant (De Angelis and Lilley, 1987). Methane concentrations in rivers vary greatly but may be nearly 3 orders of magnitude higher than in oceanic seawater

(Lamontagne et al., 1973; De Angelis and Lilley, 1987). Whether this CH_4 is largely oxidized by microbial processes or whether it is dispersed by physical processes and eventually exchanged with the atmosphere remains to be determined.

If the origin of at least some of the CH_4 in the surface ocean is lateral transport due to physical processes, then this would argue for a relatively long lifetime of CH_4 in seawater. Direct measurements of methane oxidation rates have confirmed that the biological turnover of CH_4 in seawater is relatively slow (Griffiths et al., 1982; Ward et al., 1987) (Table 5). Turnover times are generally months to years. Despite slow in situ turnover rates, CH_4 oxidation in seawater is linearly related to the concentration of added CH_4 (Fig. 9), and this activity appeared to be unsaturated even at CH_4 levels 4 orders of magnitude greater than that naturally present.

Because of the low turnover rates, a significant fraction of the CH_4 formed in the surface waters could be exchanged with the atmosphere. The reasons for slow CH_4 oxidation are not yet clear, but may be related to the lack of significant CH_4 sources to support a large methanotrophic population, and to the biological characteristics of the organisms involved. At elevated O_2 levels, methanotrophs are known to require significant concentrations of combined nitrogen (DIN). In fact, CH_4 oxidation could be carried out by ammonia-oxidizing bacteria (Jones and Morita, 1983). In the absence of high concentrations of DIN, methanotrophs require low oxygen tensions, which allows them to fix dinitrogen. Neither condition is prevalent in the surface ocean, except perhaps at the pycnocline or in microzones. Thus, CH_4 oxidation is likely to occur in or near zones which are most likely to harbor methane production. A close coupling of these activities cannot be ruled out. However, there are few direct measurements of CH_4 cycling in ocean waters.

Ward et al. (1987) have measured CH_4 oxidation rates in the water column in the Cariaco Trench, which is an anoxic basin off the coast of Venezuela. Methane oxidation activity was distributed throughout the water column (Fig. 10), with the highest rates occurring in the anoxic bottom waters. Another peak of oxidation activity was evident near the oxic/anoxic transition (~ 260-m depth), and relatively low activity was found throughout the surface waters. In the case of the Cariaco Trench, only 1.5% of the surface primary productivity was cycled through CH_4. This is probably an upper limit for oceanic systems since very few oceanic basins with anoxic bottom waters exist.

The low apparent CH_4 turnover in surface seawater is in contrast to the conclusions presented in several papers (Johnson et al., 1983; Sieburth et al., 1987). These authors have also reported the isolation of a methanotrophic bacterium from the Sargasso Sea (Sieburth et al., 1987). Based on data from changes in O_2 and total CO_2 in surface seawater, Sieburth and co-workers have hypothesized a large chemosynthetic fixation of carbon, primarily through methane production. Methane production and subsequent oxidation of this magnitude cannot be accounted for in the existing data (see Table 5). However, a contribution of C_1 compounds to oceanic bacterial production cannot be ruled out. Recent evidence suggests significant photochemical conversion of refractory dissolved organic carbon, of which there is an abundance in seawater, to low-molecular-weight molecules including pyruvate, formaldehyde, and CO (Conrad et al., 1982; Conrad and Seiler, 1988;

Table 5. Estimates of aerobic methane oxidation rates in various aerobic marine waters

Environment	CH_4 concn	Oxidation rate	Turnover time of CH_4 pool (days)	Reference
Cape Lookout Bight (coastal marine)	0.05–1.5 μM	0.01 μM · day^{-1} (low DIN)	6.3–88	Sansone and Martens, 1978
		0.21 μM · day^{-1} (high DIN)	7.4–325	
Bering Sea	4–43 nM	0.018–0.71 nM · day^{-1}	200–3,000	Griffiths et al., 1982
Big Lagoon, Calif. (stratified coastal salt pond, hypersaline lake)	50–2340 nM	0–110 nM · day^{-1}	60–50,000	Butler et al., 1988
Framvaren Fjord (surface waters)	3 μM	2 μM · day^{-1}	1.5	Lidstrom, 1983
Cariaco Trench (surface waters)	2–12 nM	0.001–0.06 nM · day^{-1}	73–5,800	Ward et al., 1987
North Atlantic				
Near surface	2.4 nM	0.15 nM · year^{-1}	5,800	Scranton and Brewer, 1978
Deep water (> 500 m)	0.4 nM	0.00022 nM · day^{-1}	657,000	

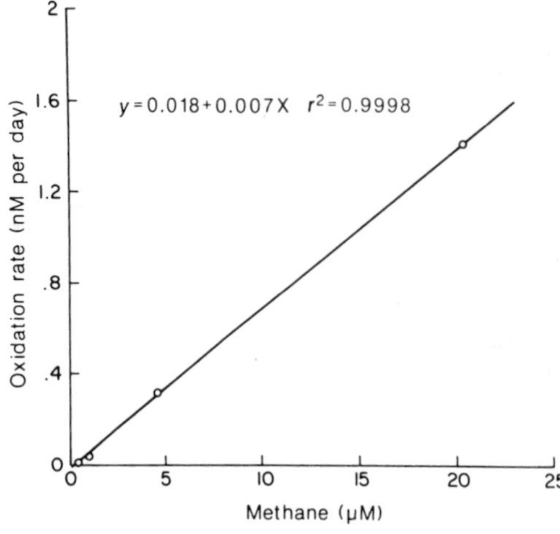

Figure 9. Relationship between methane oxidation rate and methane concentration for water samples collected from 50-m depth over the Cariaco Trench. Water samples were amended with CH_4 to give the concentrations indicated. Endogenous CH_4 concentrations were about 0.003 μM. (From Ward et al., 1987.)

Kieber et al., 1989). These compounds could fuel chemotrophic production which would be uncoupled from algal primary production. The magnitude of this chemotrophic production remains to be determined.

GLOBAL CLIMATE CHANGE AND METHANE CYCLING

Predicting the effects of climate change on complex biogeochemical cycles is a daunting task. This is especially true for the CH_4 cycle because CH_4 itself is a

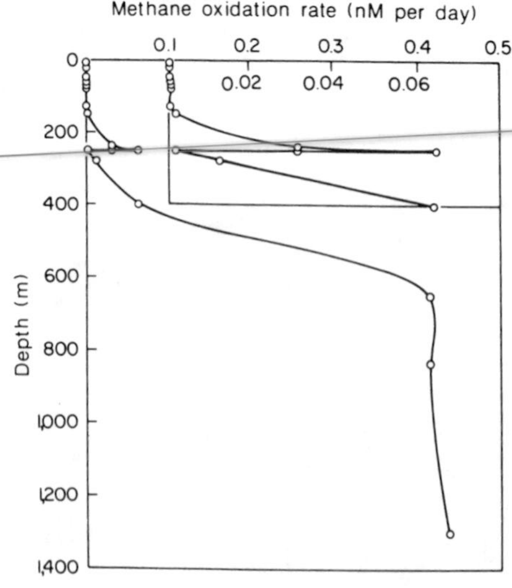

Figure 10. Depth profile of total methane oxidation rate (uptake into particulates + CO_2 production) for the Cariaco Trench. (From Ward et al., 1987.)

greenhouse gas and therefore feedbacks must be considered. As summarized above, we have made great strides in understanding the fundamental factors involved in CH_4 cycling in many aquatic habitats. Each aquatic ecosystem has a different net CH_4 production because of the interactions of factors such as microbial ecology, organic matter supply and lability, electron acceptor supply, inorganic nutrients, temperature, light, pH, and physical mixing. Our ability to understand the nature of CH_4 emissions from these systems depends on our firm understanding of how these factors interact. Given this, we can begin to guess at how climate change may affect the CH_4 cycle, how each type of ecosystem will behave under a given set of conditions, and perhaps how it will evolve into a different type of system. This is important because climate change will not simply involve a temperature change but rather a reshaping of the Earth's biomes, along with hydrologic and mineral cycles. In addition, it will be important to know the relative distribution of various methane-emitting ecosystems and to have reasonable predictions as to how these relative distributions will be altered in the future.

Acknowledgments. I thank the National Science Foundation (grant OCE-8817442) for financial support. The help of Lorene Townsend in obtaining library materials and Eileen Hedick in preparation of the figures is gratefully acknowledged. This is contribution no. 669 of the University of Georgia Marine Institute.

REFERENCES

Abram, J. W., and D. B. Nedwell. 1978. Hydrogen as a substrate for methanogenesis and sulphate reduction in anaerobic salt marsh sediment. *Arch. Microbiol.* **117:**93–97.

Alldredge, A. L., and Y. Cohen. 1987. Oxygen depleted microzones in marine particulate matter. *Science* **235:**689–691.

Alperin, M. J., and W. S. Reeburgh. 1984. Geochemical evidence supporting anaerobic methane oxidation, p. 282–289. *In* R. Crawford and R. Hanson (ed.), *Microbial Growth on C_1 Compounds.* American Society for Microbiology, Washington, D.C.

Alperin, M. J., and W. S. Reeburgh. 1985. Inhibition experiments on anaerobic methane oxidation. *Appl. Environ. Microbiol.* **50:**940–945.

Anthony, C. 1975. The biochemistry of methylotrophic organisms. *Sci. Progr.* **62:**167–206.

Atkinson, L. P., and F. A. Richards. 1967. The occurrence and distribution of methane in the marine environment. *Deep-Sea Res.* **14:**673–684.

Balch, W. E., G. E. Fox, L. J. Magrum, C. R. Woese, and R. S. Wolfe. 1979. Methanogens: reevaluation of a unique biological group. *Microbiol. Rev.* **43:**260–296.

Banat, I. M., D. B. Nedwell, and M. T. Balba. 1983. Stimulation of methanogenesis by slurries of saltmarsh sediment after the addition of molybdate to inhibit sulphate-reducing bacteria. *J. Gen. Microbiol.* **129:**123–129.

Bartlett, K. B., D. S. Bartlett, R. C. Harriss, and D. I. Sebacher. 1987. Methane emissions along a salt marsh salinity gradient. *Biogeochemistry* **4:**183–202.

Blake, D. R., and F. S. Rowland. 1988. Continuing worldwide increase in tropospheric methane, 1978 to 1987. *Science* **239:**1129–1131.

Blotevogel, K. H., U. Fischer, and K. H. Lupkes. 1986. *Methanococcus frisius* sp. nov., a new methylotrophic marine methanogen. *Can. J. Microbiol.* **32:**127–131.

Brooks, J. M., D. F. Reid, and B. B. Bernard. 1981. Methane in the upper water column of the Northwestern Gulf of Mexico. *J. Geophys. Res.* **86:**11029–11040.

Burke, R. A., Jr., D. F. Reid, J. M. Brooks, and D. M. Lavoie. 1983. Upper water column methane geochemistry in the eastern tropical North Pacific. *Limnol. Oceanogr.* **28:**19–32.

Butler, J. H., J. E. Pequegnat, L. I. Gordon, and R. D. Jones. 1988. Cycling of methane, carbon monoxide, nitrous oxide, and hydroxylamine in a meromictic, coastal lagoon. *Est. Coast. Shelf Sci.* **27:**181–203.

Capone, D. G., and R. P. Kiene. 1988. Comparison of microbial dynamics in marine and freshwater sediments: contrasts in anaerobic carbon catabolism. *Limnol. Oceanogr.* **4:**725–749.

Capone, D. G., D. D. Reese, and R. P. Kiene. 1983. Effects of metals on methanogenesis, sulfate reduction, carbon dioxide evolution, and microbial biomass in anoxic salt marsh sediments. *Appl. Environ. Microbiol.* **45:**1586–1591.

Cicerone, R. J., and R. S. Oremland. 1988. Biogeochemical aspects of atmospheric methane. *Global Biogeochem. Cycles* **2:**299–327.

Cicerone, R. J., and J. D. Shetter. 1981. Sources of atmospheric methane: measurements in rice paddies and a discussion. *J. Geophys. Res.* **86:**7203–7209.

Conrad, R., S. Goodwin, and J. G. Zeikus. 1987a. Hydrogen metabolism in mildly acidic lake sediment (Knaack Lake). *FEMS Microbiol. Ecol.* **45:**243–249.

Conrad, R., and H. Schütz. 1988. Methods of studying methanogenic bacteria and methanogenic activities in aquatic environments, p. 301–343. *In* B. Austin (ed.), *Methods in Aquatic Bacteriology.* John Wiley & Sons, Chichester, U.K.

Conrad, R., H. Schütz, and M. Babbel. 1987b. Temperature limitation of hydrogen turnover and methanogenesis in anoxic paddy soil. *FEMS Microbiol. Ecol.* **45:**281–289.

Conrad, R., and W. Seiler. 1988. Influence of the surface microlayer on the flux of nonconservative trace gases (CO, H_2, CH_4, N_2O) across the ocean-atmosphere interface. *J. Atmos. Chem.* **6:**83–94.

Conrad, R., W. Seiler, G. Bunse, and H. Gielh. 1982. Carbon monoxide in seawater (Atlantic Ocean). *J. Geophys. Res.* **87:**8839–8852.

Crill, P. M., and C. S. Martens. 1983. Spatial and temporal fluctuations of methane production in anoxic coastal marine sediments. *Limnol. Oceanogr.* **28:**1117–1130.

Dacey, J. W. H. 1980. Internal winds in water lilies: an adaptation for life in anaerobic sediments. *Science* **210:**1019.

Dacey, J. W. H., and M. J. Klug. 1979. Methane efflux from lake sediments through water lilies. *Science* **203:**1253–1255.

Davis, J. B., and H. E. Yarbrough. 1966. Anaerobic oxidation of hydrocarbons by *Desulfovibrio desulfuricans. Chem. Geol.* **1:**137–146.

De Angelis, M. A., and M. D. Lilley. 1987. Methane in surface waters of Oregon estuaries and rivers. *Limnol. Oceanogr.* **32:**716–722.

DeBont, J. A. M. 1976. Nitrogen fixation by methane-utilizing bacteria. *Antonie van Leeuwenhoek* **42:**245–253.

DeLaune, R. D., C. J. Smith, and J. R. Patrick. 1983. Methane release from Gulf Coast wetlands. *Tellus* **35B:**8–15.

Devol, A. H. 1983. Methane oxidation rates in the anaerobic sediments of Saanich Inlet. *Limnol. Oceanogr.* **28:**738–742.

Devol, A. H., J. J. Anderson, K. Kuivila, and J. W. Murray. 1984. A model for coupled sulfate reduction and methane oxidation in the sediments of Saanich Inlet. *Geochim. Cosmochim. Acta* **48:**993–1004.

Ehhalt, D. H. 1974. The atmospheric cycle of methane. *Tellus* **26:**58–70.

Ehhalt, D. H., and U. Schmidt. 1978. Sources and sinks of atmospheric methane. *Pure Appl. Geophys.* **116:**452–464.

Elliot, S., E. Lu, and F. S. Rowland. 1989. Hydrogen sulfides in oxic seawater, p. 314–327. *In* E. Saltzman and W. Cooper (ed.), *Biogenic Sulfur in the Environment.* American Chemical Society, Washington, D.C.

Fallon, R. D., S. Harrits, R. S. Hanson, and T. D. Brock. 1980. The role of methane in internal carbon cycling in Lake Mendota during summer stratification. *Limnol. Oceanogr.* **25:**357–360.

Franklin, M. J., W. J. Wiebe, and W. B. Whitman. 1988. Populations of methanogenic bacteria in a Georgia salt marsh. *Appl. Environ. Microbiol.* **54:**1151–1157.

Frenzel, P., B. Thebrath, and R. Conrad. 1990. Oxidation of methane in the oxic surface layer of a deep lake sediment (Lake Constance). *FEMS Microbiol. Ecol.* **73:**149–158.

Griffiths, R. P., B. A. Caldwell, J. D. Cline, W. A. Broich, and R. Y. Morita. 1982. Field observations of methane concentrations and oxidation rates in the southeastern Bering Sea. *Appl. Environ. Microbiol.* **44:**435–446.

Hammond, D. E., H. J. Simpson, and G. Mathieu. 1975. Methane and radon-222 as tracers for mechanisms of exchange across the sediment-water interface in the Hudson River estuary, p. 119–132. *In* T. M. Church (ed.), *Marine Chemistry in the Coastal Environment*. ACS Symposium Series vol. 18. American Chemical Society, Washington, D.C.

Hanson, R. S. 1980. Ecology and diversity of methylotrophic organisms. *Adv. Appl. Microbiol.* **26:**3–39.

Harriss, R. C., E. Gorham, D. I. Sebacher, K. B. Bartlett, and P. A. Flebbe. 1985. Methane flux from northern peatlands. *Nature* (London) **315:**652–654.

Harrits, S. M., and R. S. Hanson. 1980. Stratification of aerobic methane-oxidizing organisms in Lake Mendota, Madison, Wisconsin. *Limnol. Oceanogr.* **25:**412–421.

Harvey, H. R., R. D. Fallon, and J. S. Patton. 1989. Methanogenesis and microbial lipid synthesis in anoxic salt marsh sediments. *Biogeochemistry* **7:**111–129.

Henrichs, S. M., and W. S. Reeburgh. 1987. Anaerobic mineralization of marine sediment organic matter: rates and role of anaerobic processes in the ocean carbon economy. *Geomicrobiol. J.* **5:**191–237.

Holzapfel-Pschorn, A., R. Conrad, and W. Seiler. 1986. Effects of vegetation on the emission of methane from submerged paddy soil. *Plant Soil* **92:**223–233.

Idso, S. B., S. G. Allen, and B. A. Kimball. 1990. Growth response of water lily to atmospheric CO_2 enrichment. *Aquat. Bot.* **37:**87–92.

Iversen, N., and T. H. Blackburn. 1981. Seasonal rates of methane oxidation in anoxic marine sediments. *Appl. Environ. Microbiol.* **41:**1295–1300.

Iversen, N., and B. B. Jørgensen. 1985. Anaerobic methane oxidation rates at the sulfate-methane transition in marine sediments from the Kattegat and Skagerrak (Denmark). *Limnol. Oceanogr.* **30:**944–955.

Iversen, N., R. S. Oremland, and M. J. Klug. 1987. Big Soda Lake, Nevada. 3. Pelagic methanogenesis and anaerobic methane oxidation. *Limnol. Oceanogr.* **32:**804–814.

Johnson, K. M., P. G. Davis, and J. McN. Sieburth. 1983. Diel variation of TCO_2 in the upper layer of oceanic waters reflects microbial composition, variation and possibly methane cycling. *Mar. Biol.* **77:**1–10.

Jones, J. G., S. Gardener, and B. M. Simon. 1983. Bacterial reduction of ferric iron in a stratified eutrophic lake. *J. Gen. Microbiol.* **129:**131–139.

Jones, R. D., and R. Y. Morita. 1983. Methane oxidation by *Nitrosococcus oceanus* and *Nitrosomonas europaea*. *Appl. Environ. Microbiol.* **45:**401–410.

Jones, W. J., D. P. Nagle, Jr., and W. B. Whitman. 1987. Methanogens and the diversity of archaebacteria. *Microbiol. Rev.* **51:**135–177.

Kelly, C. A., and D. P. Chynoweth. 1981. The contributions of temperature and of the input of organic matter in controlling rates of sediment methanogenesis. *Limnol. Oceanogr.* **26:**891–897.

Khalil, M. A. K., and R. A. Rasmussen. 1987. Atmospheric methane: trends over the last 10,000 years. *Atmos. Environ.* **21:**2445–2452.

Khalil, M. A. K., and R. A. Rasmussen. 1990. Atmospheric methane: recent global trends. *Environ. Sci. Technol.* **24:**549–553.

Kieber, D. J., J. McDaniel, and K. Mopper. 1989. Photochemical source of biological substrates in sea water: implications for carbon cycling. *Nature* (London) **341:**637–639.

Kiene, R. P. 1988. Dimethyl sulfide metabolism in salt marsh sediments. *FEMS Microbiol. Ecol.* **53:**71–78.

Kiene, R. P., and T. S. Bates. 1990. Biological removal of dimethyl sulfide from seawater. *Nature* (London) **345:**702–705.

Kiene, R. P., and D. G. Capone. 1988. Microbial transformations of methylated sulfur compounds in anoxic salt marsh sediments. *Microb. Ecol.* **15:**275–291.

Kiene, R. P., R. S. Oremland, A. Catena, L. G. Miller, and D. G. Capone. 1986. Metabolism of reduced methylated sulfur compounds in anaerobic sediments and by a pure culture of an estuarine methanogen. *Appl. Environ. Microbiol.* **52:**1037–1045.

Kiene, R. P., and P. T. Visscher. 1987. Production and fate of methylated sulfur compounds from methionine and dimethylsulfoniopropionate in anoxic salt marsh sediments. *Appl. Environ. Microbiol.* **53:**2426–2434.

King, G. M. 1984a. Metabolism of trimethylamine, choline, and glycine betaine by sulfate-reducing and methanogenic bacteria in marine sediments. *Appl. Environ. Microbiol.* **48:**719–725.

King, G. M. 1984b. Utilization of hydrogen, acetate, and "noncompetitive" substrates by methanogenic bacteria in marine sediments. *Geomicrobiol. J.* **3:**275–306.

King, G. M. 1988a. Methanogenesis from methylated amines in a hypersaline algal mat. *Appl. Environ. Microbiol.* **54:**130–136.

King, G. M. 1988b. Distribution and metabolism of quarternary amines in marine sediments, p. 143–173. *In* T. Blackburn and J. Sorensen (ed.), *Nitrogen Cycling in Coastal Marine Environments.* John Wiley & Sons, Inc., New York.

King, G. M. 1990. Regulation by light of methane emissions from a wetland. *Nature* (London) **345:**513–515.

King, G. M., T. Berman, and W. J. Wiebe. 1981. Methane formation in the acidic peats of Okefenokee Swamp, Georgia. *Am. Midland Nat.* **105:**386–389.

King, G. M., M. J. Klug, and D. R. Lovley. 1983. Metabolism of acetate, methanol, and methylated amines in intertidal sediments of Lowes Cove, Maine. *Appl. Environ. Microbiol.* **45:**1848–1853.

King, G. M., P. Roslev, and H. Skovgaard. 1990. Distribution and rate of methane oxidation in sediments of the Florida Everglades. *Appl. Environ. Microbiol.* **56:**2902–2911.

King, G. M., and W. J. Wiebe. 1978. Methane release from soils of a Georgia salt marsh. *Geochim. Cosmochim. Acta* **42:**343–348.

King, G. M., and W. J. Wiebe. 1980. Regulation of sulfate concentrations and methanogenesis in salt marsh soils. *Est. Coast. Mar. Sci.* **10:**215–223.

Kristjansson, J. K., P. Schonheit, and R. K. Thauer. 1982. Different K_s values for hydrogen of methanogenic bacteria and sulfate reducing bacteria: an explanation for the apparent inhibition of methanogenesis by sulfate. *Arch. Microbiol.* **131:**278–282.

Kuivila, K. M., J. W. Murray, A. H. Devol, M. E. Lidstrom, and C. E. Reimers. 1988. Methane cycling in the sediments of Lake Washington. *Limnol. Oceanogr.* **33:**571–581.

Lamontagne, R. A., J. W. Swinnerton, V. J. Linnenbom, and W. D. Smith. 1973. Methane concentrations in various marine environments. *J. Geophys. Res.* **78:**5317–5324.

Lashof, D. A., and D. R. Ahuja. 1990. Relative contributions of greenhouse gas emissions to global warming. *Nature* (London) **344:**529–531.

Lidstrom, M. E. 1983. Methane consumption in Framvaren, an anoxic marine fjord. *Limnol. Oceanogr.* **28:**1247–1251.

Lovley, D. R. 1985. Minimum threshold for hydrogen metabolism in methanogenic bacteria. *Appl. Environ. Microbiol.* **49:**1530–1531.

Lovley, D. R., D. F. Dwyer, and M. J. Klug. 1982. Kinetic analysis of competition between sulfate reducers and methanogens for hydrogen in sediments. *Appl. Environ. Microbiol.* **43:**1373–1379.

Lovley, D. R., and S. Goodwin. 1988. Hydrogen concentrations as an indicator of the predominant terminal electron-accepting reactions in aquatic sediments. *Geochim. Cosmochim. Acta* **52:**2993–3003.

Lovley, D. R., and M. J. Klug. 1982. Intermediary metabolism of organic matter in the sediments of a eutrophic lake. *Appl. Environ. Microbiol.* **43:**552–560.

Lovley, D. R., and M. J. Klug. 1983a. Sulfate reducers can outcompete methanogens at freshwater sulfate concentrations. *Appl. Environ. Microbiol.* **45:**187–192.

Lovley, D. R., and M. J. Klug. 1983b. Methanogenesis from methanol and methylamines and acetogenesis from hydrogen and carbon dioxide in the sediments of a eutrophic lake. *Appl. Environ. Microbiol.* **45:**1310–1315.

Lovley, D. R., and M. J. Klug. 1986. Model for the distribution of sulfate reduction and methanogenesis in freshwater sediments. *Geochim. Cosmochim. Acta* **50:**11–18.

Lovley, D. R., and E. J. Phillips. 1986a. Organic matter mineralization with reduction of ferric iron in anaerobic sediments. *Appl. Environ. Microbiol.* **51:**683–689.

Lovley, D. R., and E. J. Phillips. 1986b. Availability of ferric iron for microbial reduction in bottom sediments of the freshwater tidal Potomac River. *Appl. Environ. Microbiol.* **52:**751–757.

Lovley, D. R., and E. J. Phillips. 1987. Competitive mechanisms for inhibition of sulfate reduction and methane production in the zone of ferric iron reduction in sediments. *Appl. Environ. Microbiol.* **53:**2636–2641.

Lovley, D. R., and E. J. Phillips. 1989. Requirement for a microbial consortium to completely oxidize glucose in Fe(III)-reducing sediments. *Appl. Environ. Microbiol.* **55:**3234–3236.

Luther, G. W., III, and E. Tsamakis. 1989. Concentration and form of dissolved sulfide in the oxic water column of the ocean. *Mar. Chem.* **27:**165–177.

Martens, C. S., and R. A. Berner. 1977. Interstitial water chemistry of anoxic Long Island Sound sediments. 1. Dissolved gases. *Limnol Oceanogr.* **22:**10–25.

Martens, C. S., and J. V. Klump. 1980. Biogeochemical cycling in an organic-rich coastal marine basin. I. Methane sediment-water exchange processes. *Geochim. Cosmochim. Acta* **44:**471–490.

Martens, C. S., and J. V. Klump. 1984. Biogeochemical cycling in an organic-rich coastal marine basin. 4. An organic carbon budget for sediments dominated by sulfate reduction and methanogenesis. *Geochim. Cosmochim. Acta* **48:**1987–2004.

Mathews, E., and I. Fung. 1987. Methane emission from natural wetlands: global distribution, area and environmental characteristics of sources. *Global Biogeochem. Cycles* **1:**61–86.

Mathrani, I. M., and D. R. Boone. 1985. Isolation and characterization of a moderately halophilic methanogen from a solar saltern. *Appl. Environ. Microbiol.* **50:**140–143.

McInerney, M. J., and P. S. Beaty. 1988. Anaerobic community structûre from a nonequilibrium thermodynamic perspective. *Can. J. Microbiol.* **34:**487–493.

Michener, R. H., M. I. Scranton, and P. Novelli. 1988. Hydrogen (H_2) distributions in the Carmans River estuary. *Est. Coast. Shelf Sci.* **27:**223–235.

Mopper, K., and D. J. Kieber. Distribution and turnover of dissolved organic compounds in the water column of the Black Sea. *Deep-Sea Res.*, in press.

Mopper, K., and W. L. Stahovec. 1986. Sources and sinks of low molecular weight organic carbonyl compounds in seawater. *Mar. Chem.* **19:**305–321.

Mountfort, D. O., and R. A. Asher. 1981. Role of sulfate reduction versus methanogenesis in terminal carbon flow in polluted intertidal sediment of Waimea Inlet, Nelson, New Zealand. *Appl. Environ. Microbiol.* **42:**252–258.

Novelli, P. C., A. R. Michelson, M. I. Scranton, G. T. Banta, J. E. Hobbie, and R. W. Howarth. 1988. Hydrogen and acetate cycling in two sulfate-reducing sediments: Buzzards Bay and Town Cove, Mass. *Geochim. Cosmochim. Acta* **52:**2477–2486.

Oremland, R. S. 1979. Methanogenic activity in plankton samples and fish intestines: a mechanism for in situ methanogenesis in oceanic surface waters. *Limnol. Oceanogr.* **24:**1136–1141.

Oremland, R. S. 1988. Biogeochemistry of methanogenic bacteria, p. 641–705. *In* A. Zehnder (ed.), *Biology of Anaerobic Microorganisms.* John Wiley & Sons, Inc., New York.

Oremland, R. S., R. P. Kiene, I. Mathrani, M. J. Whiticar, and D. R. Boone. 1989. Description of an estuarine methylotrophic methanogen which grows on dimethyl sulfide. *Appl. Environ. Microbiol.* **55:**994–1002.

Oremland, R. S., and G. M. King. 1989. Methanogenesis in hypersaline environments, p. 180–190. *In* Y. Cohen and E. Rosenberg (ed.), *Microbial Mats: Physiological Ecology of Benthic Microbial Communities.* American Society for Microbiology, Washington, D.C.

Oremland, R. S., L. Marsh, and D. J. DesMarais. 1982a. Methanogenesis in Big Soda Lake, Nevada: an alkaline, moderately hypersaline desert lake. *Appl. Environ. Microbiol.* **43:**462–468.

Oremland, R. S., L. M. Marsh, and S. Polcin. 1982b. Methane production and simultaneous sulphate reduction in anoxic, salt marsh sediments. *Nature* (London) **296:**143–145.

Oremland, R. S., and S. Polcin. 1982. Methanogenesis and sulfate reduction: competitive and noncompetitive substrates in estuarine sediments. *Appl. Environ. Microbiol.* **44:**1270–1276.

Oremland, R. S., and B. F. Taylor. 1977. Diurnal fluctuations of O_2, N_2 and CH_4 in the rhizosphere of Thalassia testudinum. *Limnol. Oceanogr.* **22:**566–570.

Oremland, R. S., and B. F. Taylor. 1978. Sulfate reduction and methanogenesis in marine sediments. *Geochim. Cosmochim. Acta* **42:**209–214.

Paerl, H. W., and R. G. Carlton. 1988. Control of nitrogen fixation by oxygen depletion in surface-associated microzones. *Nature* (London) **332:**260–262.

Paerl, H. W., and L. E. Prufert. 1987. Oxygen-poor microzones as potential sites of microbial nitrogen fixation in nitrogen-depleted aerobic marine waters. *Appl. Environ. Microbiol.* **53:**1078–1087.

Panganiban, A. T., T. E. Patt, W. Hart, and R. S. Hanson. 1979. Oxidation of methane in the absence of oxygen in lake water samples. *Appl. Environ. Microbiol.* **37:**303–309.

Paterek, J. R., and P. H. Smith. 1985. Isolation and characterization of a halophilic methanogen from Great Salt Lake. *Appl. Environ. Microbiol.* **50:**877–881.

Phelps, T. J., and J. G. Zeikus. 1984. Influence of pH on terminal carbon metabolism in anoxic sediments from a mildly acidic lake. *Appl. Environ. Microbiol.* **48:**1088–1095.

Rasmussen, R. A., and M. A. K. Khalil. 1981. Atmospheric methane (CH_4): trends and seasonal cycles. *J. Geophys. Res.* **86:**9826–9832.

Reeburgh, W. S., and D. T. Heggie. 1977. Microbial methane consumption reactions and their effect on methane distributions in freshwater and marine environments. *Limnol. Oceanogr.* **22:**1–9.

Rimbault, A., P. Niel, H. Virelizier, J. C. Darboard, and G. Leluan. 1988. L-Methionine, a precursor of trace methane in some proteolytic clostridia. *Appl. Environ. Microbiol.* **54:**1581–1586.

Robinson, J. A., and J. M. Tiedje. 1984. Competition between sulfate-reducing and methanogenic bacteria for H_2 under resting and growing conditions. *Arch. Microbiol.* **137:**26–32.

Rodhe, H. 1990. A comparison of the contribution of various gases to the greenhouse effect. *Science* **248:**1217–1219.

Rudd, J. W. M. 1980. Methane oxidation in Lake Tanganyika (East Africa). *Limnol. Oceanogr.* **25:**958–963.

Rudd, J. W. M., A. Furutani, R. J. Flett, and R. D. Hamilton. 1976. Factors controlling methane oxidation in shield lakes: the roles of nitrogen fixation and oxygen concentration. *Limnol. Oceanogr.* **21:**357–364.

Rudd, J. W. M., and R. D. Hamilton. 1975. Factors controlling rates of methane oxidation and the distribution of methane oxidizers in a small stratified lake. *Arch. Hydrobiol.* **4:**522–538.

Rudd, J. W. M., and R. D. Hamilton. 1978. Methane cycling in a eutrophic shield lake and its effects on whole lake metabolism. *Limnol. Oceanogr.* **23:**337–348.

Rudd, J. W. M., and C. D. Taylor. 1980. Methane cycling in aquatic environments. *Adv. Aquat. Microbiol.* **2:**77–150.

Sansone, F. J., and C. S. Martens. 1978. Methane oxidation in Cape Lookout Bight, North Carolina. *Limnol. Oceanogr.* **23:**349–355.

Schink, B., and J. G. Zeikus. 1982. Microbial ecology of pectin decomposition in anoxic lake sediments. *J. Gen. Microbiol.* **128:**393–404.

Schonheit, P., J. K. Kristjansson, and R. K. Thauer. 1982. Kinetic mechanism for the ability of sulfate reducers to out-compete methanogens for acetate. *Arch. Microbiol.* **132:**285–288.

Schütz, H., W. Seiler, and R. Conrad. 1989. Processes involved in formation and emission of methane in rice paddies. *Biogeochemistry* **7:**33–53.

Scranton, M. I., and P. G. Brewer. 1977. Occurrence of methane in the near-surface waters of the western subtropical North Atlantic. *Deep-Sea Res.* **24:**127–138.

Scranton, M. I., and P. G. Brewer. 1978. Consumption of dissolved methane in the deep ocean. *Limnol. Oceanogr.* **23:**1207–1213.

Sebacher, D. I., R. C. Harriss, and K. B. Bartlett. 1985. Methane emissions to the atmosphere through aquatic plants. *J. Environ. Qual.* **14:**40–46.

Sebacher, D. I., R. C. Harriss, K. B. Bartlett, S. M. Sebacher, and S. S. Grice. 1986. Atmospheric methane sources: Alaskan tundra bogs, an alpine fen, and a subarctic boreal marsh. *Tellus* **38B:**1–10.

Seiler, W. 1984. Contribution of biological processes to the global budget of CH_4 in the atmosphere, p. 468–477. *In* M. Klug and C. Reddy (ed.), *Current Perspectives in Microbial Ecology*. American Society for Microbiology, Washington, D.C.

Senior, E., E. B. Lindstrom, I. M. Banat, and D. B. Nedwell. 1982. Sulfate reduction and methanogenesis in the sediment of a salt marsh on the east coast of the United Kingdom. *Appl. Environ. Microbiol.* **43:**987–996.

Sieburth, J. McN. 1988. The nanoalgal peak in the dim oceanic pycnocline: is photosynthesis augmented by microparticulates and their bacterial consortia?, p. 101–130. *In* C. R. Agegian (ed.), *Biogeochemical Cycling and Fluxes between the Deep Euphotic Zone and Other Oceanic Realms*. National Undersea Research Program Report 88–1. U.S. Department of Commerce, Washington, D.C.

Sieburth, J. McN., P. W. Johnson, M. A. Eberhardt, M. E. Sieracki, M. Lidstrom, and D. Laux. 1987. The first methane-oxidizing bacterium from the upper mixing layer of the deep ocean: *Methylomonas pelagica* sp. nov. *Curr. Microbiol.* **14:**285–293.

Sieburth, J. McN., and M. D. Keller. 1991. Methylaminotrophic bacteria in xenic nanoalgal cultures: incidence, significance, and role of methylated algal osmoprotectants. *Biol. Oceanogr.* **6:**443–455.

Sorensen, J. 1988. Dimethylsulfide and methane thiol in sediment pore-water of a Danish estuary. *Biogeochemistry* **6:**201–210.

Sowers, K. R., and J. G. Ferry. 1983. Isolation and characterization of a methylotrophic marine methanogen, *Methanococcoides methylutens* gen. nov., sp. nov. *Appl. Environ. Microbiol.* **45:**684–690.

Strayer, R. F., and J. M. Tiedje. 1978. In situ methane production in a small, hypereutrophic, hard-water lake: loss of methane from sediments by vertical diffusion and ebullition. *Limnol. Oceanogr.* **23:**1201–1206.

Svensson, B. H. 1984. Different temperature optima for methane formation when enrichments from acid peat are supplemented with acetate or hydrogen. *Appl. Environ. Microbiol.* **48:**389–394.

Svensson, B. H., and T. Rosswall. 1984. In situ methane production from acid peat in plant communities with different moisture regimes in a subarctic mire. *Oikos* **43:**341–350.

Thauer, R. K., K. Jungermann, and K. Dekker. 1977. Energy conservation in chemotrophic anaerobic bacteria. *Bacteriol. Rev.* **41:**100–180.

Traganza, E. D., J. W. Swinnerton, and C. H. Cheek. 1979. Methane supersaturation and ATP-zooplankton blooms in near-surface waters of the Western Mediterranean and the subtropical North Atlantic Ocean. *Deep-Sea Res.* **26A:**1237–1245.

Van Neste, A., R. A. Duce, and C. Lee. 1987. Methylamines in the marine atmosphere. *Geophys. Res. Lett.* **14:**711–714.

Ward, B. B., K. A. Kilpatrick, P. C. Novelli, and M. I. Scranton. 1987. Methane oxidation and methane fluxes in the ocean surface layer and deep anoxic waters. *Nature* (London) **327:**226–229.

Ward, D. M., and M. R. Winfrey. 1985. Interactions between methanogenic and sulfate-reducing bacteria in sediments, p. 141–179. *In* H. Jannasch and P. Williams (ed.), *Advances in Aquatic Microbiology*. Harcourt Brace Jovanovich, London.

Westermann, P., and B. K. Ahring. 1987. Dynamics of methane production, sulfate reduction, and denitrification in a permanently waterlogged alder swamp. *Appl. Environ. Microbiol.* **53:**2554–2559.

Westrich, J. T., and R. A. Berner. 1988. The effect of temperature on rates of sulfate reduction in marine sediments. *Geomicrobiol. J.* **6:**99–117.

Whalen, S. C., and W. S. Reeburgh. 1988. A methane flux time series for tundra environments. *Global Biogeochem. Cycles* **2:**399–409.

Whalen, S. C., and W. S. Reeburgh. 1990. Consumption of atmospheric methane by tundra soils. *Nature* (London) **346:**160–162.

Whittenbury, R., K. C. Phillips, and J. F. Wilkinson. 1970. Enrichment, isolation, and some properties of methane-utilizing bacteria. *J. Gen. Microbiol.* **61:**205–218.

Widdel, F. 1986. Growth of methanogenic bacteria in pure culture with 2-propanol and other alcohols as hydrogen donors. *Appl. Environ. Microbiol.* **51:**1056–1062.

Winfrey, M. R. 1984. Microbial production of methane, p. 153–219. *In* R. Atlas (ed.), *Petroleum Microbiology*. Macmillan, New York.

Winfrey, M. R., D. R. Nelson, S. C. Klevickis, and J. G. Zeikus. 1977. Association of hydrogen metabolism with methanogenesis in Lake Mendota sediments. *Appl. Environ. Microbiol.* **33:**312–318.

Winfrey, M. R., and D. M. Ward. 1983. Substrates for sulfate reduction and methane production in intertidal sediments. *Appl. Environ. Microbiol.* **45:**193–199.

Winfrey, M. R., and J. G. Zeikus. 1977. Effect of sulfate on carbon and electron flow during microbial methanogenesis in freshwater sediments. *Appl. Environ. Microbiol.* **33:**275–281.

Yancey, P. H., M. E. Clark, S. C. Hanf, R. D. Bowlus, and G. N. Somero. 1982. Living with water stress: evolution of osmolyte systems. *Science* **217:**1214–1222.

Zehnder, A. J. B. 1978. Ecology of methane formation, p. 349–376. *In* R. Mitchell (ed.), *Water Pollution Microbiology*. Wiley Interscience, New York.

Zehnder, A. J. B., and T. D. Brock. 1980. Anaerobic methane oxidation: occurrence and ecology. *Appl. Environ. Microbiol.* **39:**194–204.

Zeikus, J. G. 1977. The biology of methanogenic bacteria. *Bacteriol. Rev.* **41:**514–541.

Zeikus, J. G., and M. R. Winfrey. 1976. Temperature limitation of methanogenesis in aquatic sediments. *Appl. Environ. Microbiol.* **31:**99–107.

Zhilina, T. N. 1983. New obligate halophilic methane-producing bacterium. *Mikrobiologiya* **52:**375–382.

Zinder, S. H., and T. D. Brock. 1978a. Methane, carbon dioxide, and hydrogen sulfide production from the terminal methiol group of methionine by anaerobic lake sediments. *Appl. Environ. Microbiol.* **35:**344–352.

Zinder, S. H., and T. D. Brock. 1978b. Production of methane and carbon dioxide from methane thiol and dimethyl sulfide by anaerobic lake sediments. *Nature* (London) **273:**226–228.

Chapter 8

Methane and Hydrogen Sulfide in the Pycnocline: a Result of Tight Coupling of Photosynthetic and "Benthic" Processes in Stratified Waters

John McNeill Sieburth

PROLOGUE

It is 25 years since Richards et al. (1965) published their seminal paper on the consequences of organic matter decomposition on the distribution of methane and hydrogen sulfide in the water column of an anoxic fjord. I included this topic (Atkinson and Richards, 1967) in my text *Sea Microbes* (Sieburth, 1979), but from the perspective of a nonparticipant. A year later, Rudd and Taylor (1980) treated this topic comprehensively in their broad review on methane cycling in aquatic environments from their vantage point as participants. Shortly afterwards, my laboratory published our enigmatic results from direct observations on microbial populations and their metabolism in the stratified upper ocean (Burney et al., 1981, 1982; Johnson et al., 1981), which indicated an underestimation of both primary productivity and methane cycling in such waters (Johnson et al., 1983b; Sieburth, 1983). This has made me a participant, and since then I have been exploring the coupling between these two processes (Sieburth, 1987b, 1988a, 1988b; Sieburth and Keller, 1991). Four years of research with my colleagues on a permanently anoxic estuarine lake (Gaines and Pilson, 1972) similar to that studied by Richards et al. (1965), although largely unpublished (Sieburth, 1991), also gives me a perspective on the processes in stratified lakes. Methanogenesis in the water column has received scant attention in the recent reviews by Oremland (1988) and Cicerone and Oremland (1988). The purpose of this chapter is to review the pertinent literature on both methane and hydrogen sulfide production in the water column, to reinterpret this literature, and to propose a new paradigm that explains how "benthic" processes could be largely transferred to the photic pycnocline during stratification.

During my oceanic cruises over a decade ago, observations on bulk water

John McNeill Sieburth • Graduate School of Oceanography, University of Rhode Island Bay Campus, Narragansett, Rhode Island 02882-1197.

indicated diurnal accumulations of dissolved polysaccharides with static populations of bacteria in contrast to noctural decreases in polysaccharides with increasing bacterial populations (Burney et al., 1981, 1982). The diel variations in these processes, plus concurrent observations on the diel cycling of TCO_2 (total CO_2) and O_2 (Johnson et al., 1981; Johnson et al., 1983b), indicated that oceanic production based on bottle studies could be underestimating actual rates by at least 1 order of magnitude. While convening an informal session on productivity at the International Helgoland Symposium "Ecosystems Research" (Sieburth, 1977), Karl Banse stated that there was not enough chlorophyll to account for my claimed underestimation. Subsequently we helped provide the answer by first defining the smaller size fractions in the plankton: the virus-sized particles in the femtoplankton, 0.02 to 0.2 μm; the bacterial-sized particles in the picoplankton, 0.2 to 2.0 μm; and the algal and protozoan-sized particles in the nanoplankton, 2.0 to 20 μm (Sieburth et al., 1978). This was the basis for characterizing natural populations of phototrophic bacteria (Johnson and Sieburth, 1979) and bacterial-sized algae (Johnson and Sieburth, 1982) in the picoplankton and the larger algae in the nanoplankton (Estep et al., 1984), to show their distributions (Davis et al., 1985) and to estimate their predators, the bacterivorous nanoflagellates (Caron et al., 1989). Virus populations in the femtoplankton are now recognized to be everpresent in picoalgae (Johnson and Sieburth, 1982; Sieburth et al., 1988), as prevalent as bacterial populations (Sieracki et al., 1985; Sieburth et al., 1988), and to possibly control in part both bacterial and algal populations (Bergh et al., 1989; Suttle et al., 1990). The smaller phototrophs in the picoplankton and nanoplankton (Johnson and Sieburth, 1979, 1982) are now recognized to account for 50 to 80% of oceanic productivity (Platt and Li, 1986; Furnas and Mitchell, 1988). It is now accepted that the rates of gross production obtained by conventional ^{14}C bottle assays underestimate actual production at least 10-fold due to inaccuracies in determinations (Gieskes et al., 1979; Fitzwater et al., 1982) as well as to conceptual errors (Krupatkina, 1990). Conventional procedures used to study methane production, consumption, and evasion may have some of the same shortcomings. They may not take the intense microbial activities of sporadic microbial aggregations that are excluded by small-volume bottles into consideration. Most also ignore diel differences and the great impact of storm fronts. If primary productivity by bottle assays can be underestimated by at least 10-fold, so then can the production, consumption, and evasion of greenhouse gases.

OUTMODED NOTIONS OF THE FOOD CHAIN AND ITS PROCESSES

It was once believed that there was a rain of detritus to the sea floor where mineralization occurs (Kriss, 1963). We now know that 87 to 95% of primary productivity is mineralized in situ when the upper ocean is stratified (Thurman and Webber, 1984). This is not portrayed in the usual diagrams of the food chain such as that from Florek and Rowe (1983), shown in Fig. 1.

There is obviously a rain of detritus from intact fecal pellets and the more resistant parts of larger microorganisms to the sea floor (Kriss, 1963). But Kriss also

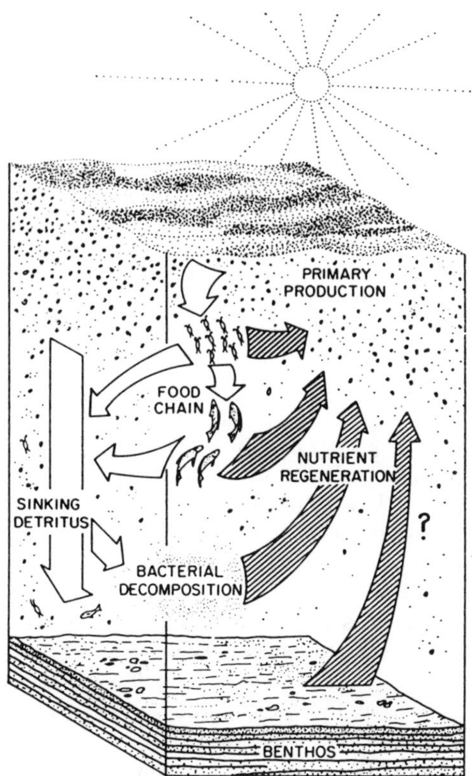

Figure 1. A representative food chain model showing divergent sites for primary production, grazing through the food chain, and bacterial decomposition of sedimented detritus in the benthos. Note that all this is occurring in a water column in which stratification and the physical factors causing it are ignored. (From Florek and Rowe, 1983.)

described how the soft-bodied nanoplankton that dominates primary production in the Black Sea is regenerated in situ. He documented that organic matter and its associated bacteria accumulate in the pycnocline, a rediscovered phenomenon (Garfield et al., 1983). As a newcomer to marine microbiology, I studied algal-bacterial interactions of the Antarctic food chain (Sieburth, 1960a) and reviewed a decade of Soviet aquatic microbiology as an entrée into the field (Sieburth, 1960b). The direct observation and characterization of natural populations by A. E. Kriss and co-workers at the Institute of Microbiology in Moscow have made a lasting impression on my work. Accordingly, the emphasis of my three decades of research on algal-bacterial interactions has been on characterizing the dominant sea microbes in natural populations and describing new species, so that the major players in microbiological processes can be recognized and identified. Without knowing the trophic nature of the microorganisms involved, how can one have a rational basis for conducting meaningful productivity and rate measurements (Sieburth, 1988b)?

The significant nutrient regeneration in the upper water column depends upon the nature of phototrophs and the physical properties of the pycnocline. The smaller cells in the picoplankton and nanoplankton thrive best in the pyconocline (Davis et al., 1985; Murphy and Haugen, 1985; Goldman, 1986; Glover et al., 1987; Sieburth, 1988a). This is due to their negligible sinking rate in the denser water of the pycnocline and their rapid in situ growth that allows them to accumulate there (Glover et al., 1986). Oceanic productivity is maximal in the pycnocline because the microbes in this thin band of stagnant water, such as *Pycnococcus provasolii* (Guillard et al., 1991), are adapted to low-intensity filtered light and would be very susceptible to bleaching and phototoxicity at the higher light intensities and wavelengths in the mixing layer. It is probably not a coincidence that the deep chlorophyll maximum where the smaller phototrophs live is also the pycnocline where methane and hydrogen sulfide are also maximal (see New Paradigms for the Food Chain and Microbial Regeneration, below).

This phenomenon of the accumulation of primary producers in the pycnocline of oxygenated waters can apparently also occur in suboxic pycnoclines. Large, motile, oxygenic phototrophs such as the dinoflagellate *Prorocentrum mariae-lebouriae* accumulate in the pycnocline of Chesapeake Bay when it is seasonally stratified and goes anoxic (Tyler and Seliger, 1978, 1981; Harding and Coates, 1988), while a large biomass of the euglenoid *Euglena proxima* is always present in the pycnocline of the permanently anoxic basins of the Pettaquamscutt Estuary (Miller, 1972; Sieburth, 1987b), except during rare catastrophic physical perturbations. The coexistence of accumulations of oxygenic phototrophs along with anoxygenic phototrophs in the pycnoclines of meromictic lakes (Lindholm and Weppling, 1987) usually goes undetected because of the arbitrary depths selected for large-interval sampling. Close-interval sampling facilitates such observations (Lindholm, 1979). Looking at the data of Brock (1985), algal populations increased as the lower pycnocline in Lake Mendota was approached, indicating the possibility of a shoulder of an unrecognized algal maximum in the pycnocline.

There are many studies on the flux of materials from the sea surface. The consensus of these studies is that during stratification, only 5 to 10% of primary production settles below the pycnocline, indicating that there is a 90 to 95% regeneration or mineralization occurring in situ in the upper layer of the sea (Eppley and Peterson, 1979; Karl and Knauer, 1984; Baker et al., 1985; Gardiner et al., 1985). The same is also true of the meromictic lake Big Soda Lake, Nevada (Cloern et al., 1987). Despite this, conventional models for methane and hydrogen sulfide production (Hanson, 1980; Large, 1983; Brock, 1985) always indicate that they occur wholly within the sediment. They ignore the possibility that these processes can occur throughout the water column (Richards et al., 1965; Winfrey and Zeikus, 1979).

REPORTS OF THE PARADOXICAL OCCURRENCE OF REDUCED GASES IN OXYGENATED OCEANIC SURFACE WATERS

The presence of a number of reduced gases (including H_2, CH_4, and H_2S) at supersaturation in surface waters despite oxygen saturation presents a biochemical and microbiological paradox. Close to shore, the source of this methane can be due

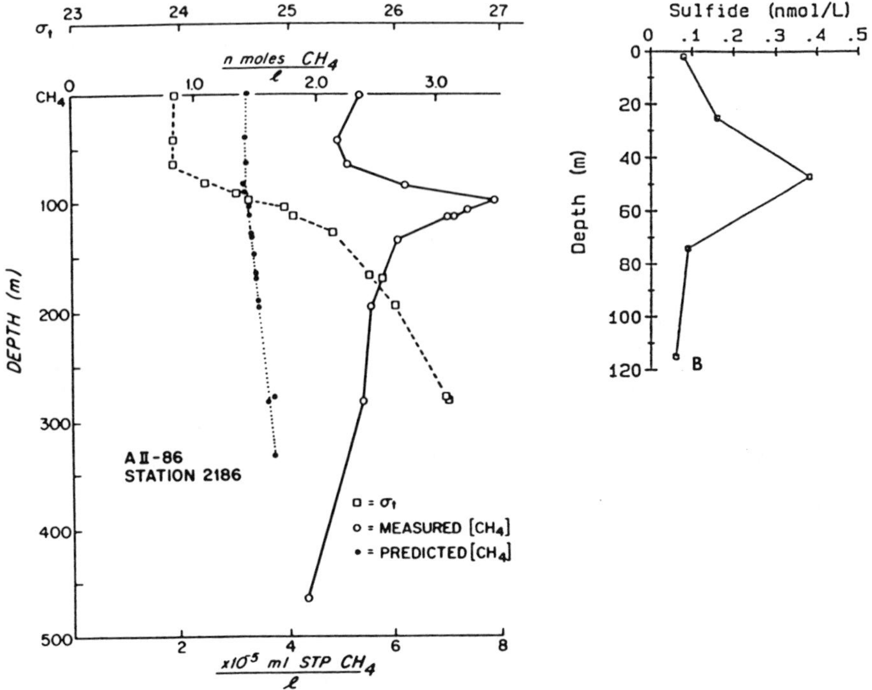

Figure 2. The paradoxical maxima of the anaerobically formed reduced gases methane (A) and hydrogen sulfide (B) in the thermoclines of oxygenated water from the North Atlantic. (A, from Scranton and Brewer, 1977; B, from Cutter and Krahforst, 1988.)

in part to transport from anoxic sediments or to oil seeps. But within the stratified water column, most profiles for methane show a pronounced maximum associated with the pycnocline at depths between 30 and 100 m, where it is 30 to 70% supersaturated relative to the atmospheric equilibrium concentration (Swinnerton et al., 1969; Brooks and Sackett, 1973; Brooks et al., 1973; Lamontagne et al., 1973; Williams and Bainbridge, 1973; Seiler and Schmidt, 1974). This is apparently due to in situ production in the oxygenated water column (Lamontagne et al., 1973; Scranton and Brewer, 1977; Scranton and Farrington, 1977; Brooks et al., 1973). A typical methane profile is shown in Fig. 2A. Traganza et al. (1979) observed a correlation of methane supersaturation with zooplankton ATP. Oremland (1979) demonstrated that particles or plankton concentrated by plankton nets are methanogenic, an observation he deemed highly significant as a mechanism for in situ methanogenesis in oceanic surface waters. Brooks et al. (1981) reported that chlorophyll and ATP maxima in the water column correspond with the methane maxima in the water column of the northwestern Gulf of Mexico. An elaborate study to pinpoint the site of water column methanogenesis in the eastern tropical north Pacific Ocean compared a variety of chemical, biological, and physical data

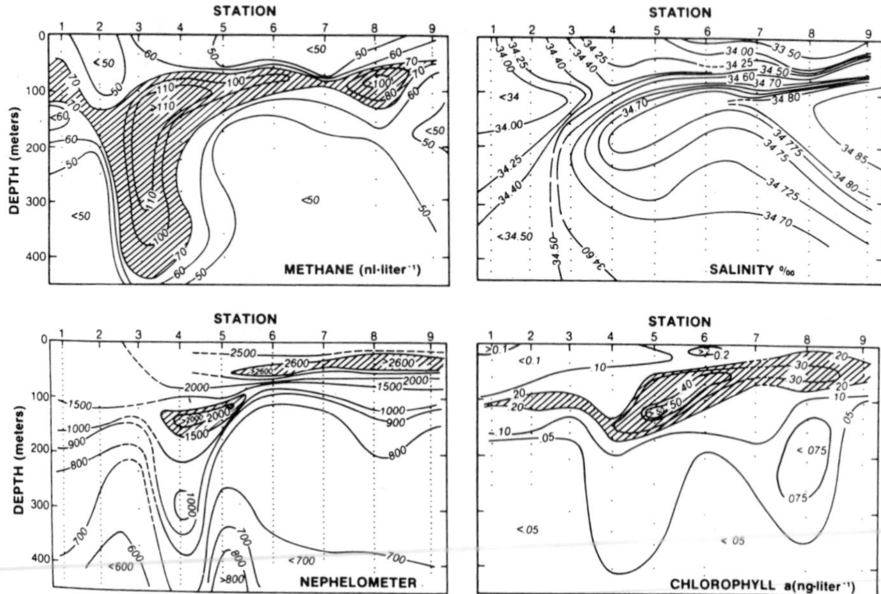

Figure 3. The distribution of methane compared with the isopleths for salinity, turbidity (nephelometer), and chlorophyll *a*, showing the relationships between the concentrations of methane, the physical effects of the halocline, primary production, and total particulates in the eastern tropical North Pacific Ocean. (From Burke et al., 1983.)

and concluded that the methane maximum was due to in situ biological production controlled by physical processes and that methanogenesis must occur within reducing microenvironments (Burke et al., 1983). The relation of methane distribution to particle turbidity (nephelometer), salinity (and thus density), and ultimately to primary production (chlorophyll) for this data set is shown in Fig. 3. There is little doubt that an upper water column methane production exists that is tied to primary production and its decay in situ, but as Rudd and Taylor (1980) conclude, the nature and mechanism of the anoxic microsites required for this production require clarification.

A number of other reduced chemical species are also present in the upper mixed layer and the upper pycnocline. The first vertical profile of hydrogen sulfide showing tenths of nanomoles per liter in the oxygenated upper waters of the North Atlantic was obtained by Cutter and Krahforst (1988) with an indirect gas chromatographic method (Fig. 2B). The same general vertical distribution has been shown with direct measurements using a voltammetric method (Luther and Tsamakis, 1989). The hydrogen sulfide maximum obtained by Cutter and Krahforst (1988) correlated with the top of the major thermocline and a maximum in chlorophyll *a*, and there was a diel variation with a noctural increase and a diurnal decrease. This is similar to the diurnal bacterial inactivity and nocturnal activity shown for the bacterial population as a whole that correlated with the accumulation and utilization of soluble polysaccharides (Burney et al., 1981, 1982; Burney, 1986). This phenomenon is ascribed to inhibition by sunlight (Sieracki and Sieburth, 1985, 1986). A nocturnal production of formaldehyde (indicative of methane utilization) has also been observed and linked to diel methane cycling (Johnson et al., 1983b) and is probably due to the UV sensitivity of methane-oxidizing bacteria (Sieburth et al., 1987). Methane production and sulfate reduction to produce hydrogen sulfide may co-occur in sediments when the methanogens are utilizing methylated substrates (Oremland et al., 1982; King et al., 1983; King, 1984). Why should we not assume that common anoxic microsites are also important for both of these anaerobic processes in the upper ocean? A persistent source of sulfide in the stratified upper ocean would explain why sulfide-utilizing thiobacilli (Tilton et al., 1967a, 1967b) and thiobacilli-like bacteria (Tuttle and Jannasch, 1972) are present in seawater. Sulfide would also supply the reduced sulfur required by methanogens, since most lack assimilatory sulfate reductase and cannot utilize sulfate (Daniels et al., 1986). These processes could be the chemolithotrophy of Karl et al. (1984).

THE PARADOXICAL ENRICHMENT OF O_2-TOLERANT METHANOGENS AND SULFATE-REDUCING BACTERIA FROM OXYGENATED WATERS

Each trophic group of bacteria involved in methanogenesis, methanotrophy, and methylotrophy is conventionally cultured in process-specific growth media. For example, when Oremland (1979) wanted to enrich for methanogens using plankton net concentrates and fish intestinal contents as inocula, he used prereduced media and added either H_2:CO_2, formate, or acetate as a substrate. When I wanted to see if there was sufficient methanogenesis in the stratified Sargasso Sea to support detectable methanotrophs, the seawater samples were incubated in a methane-enriched atmosphere (Sieburth et al., 1987). Such classic enrichment and isolation procedures are illustrated in Fig. 4. Until 1984, I was a firm believer in the use of such trophic-specific culture procedures for initial enrichments. Having been a member of Professor Robert E. Hungate's lab, 1949 to 1950, I was indoctrinated into the strict requirements for anaerobes (Hungate, 1950). These requirements as they apply to methanogens (Oremland, 1988) include (i) the requirement for a strong reducing agent such as sulfide or cysteine to lower the E_h; (ii) a requirement for a lack of any trace of oxygen; and (iii) the related requirement that only other anaerobes provide their growth substrates and habitat. Such anaerobic enrichments, however, exclude aerobes as potentially essential syntrophs of the methanogens.

I never intended to grow methanogens present in oxygenated waters; my previous experiences had made me anaerobe shy. I therefore concentrated on the aerobic methane oxidizers. My purpose in enriching for oceanic methanotrophs was an attempt to identify the trophic nature of a dominant group of bacteria in the pyconocline (the type III bacteria of Johnson and Sieburth [1979]), which have intracytoplasmic membranes suggestive of methane and ammonia oxidizers. When methane was used as a substrate, instead of obtaining these bacteria, we obtained cells with the more elaborate intracytoplasmic membrane stacks of type II metha-

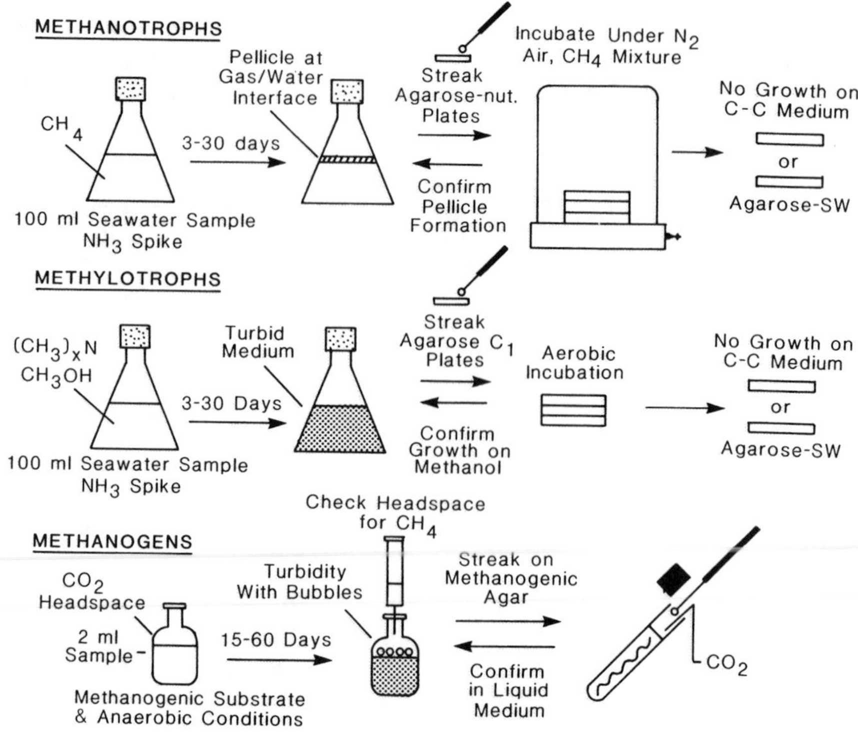

Figure 4. Trophic-specific procedures used to enrich and isolate the component bacteria involved in the metabolism of C_1 compounds. The aerobic bacteria oxidizing methane and the methylated substrates methanol and methylated amines are enriched separately under aerobic conditions. The very oxygen-sensitive methanogens are only enriched in media prereduced with cystine and sulfide.

notrophs (Sieburth et al., 1987). When the ammonia enrichments also failed to enrich our target bacterium, I used monomethylamine (MMA) since it contains both molecules. At the 1983 C_1 conference (Crawford and Hanson, 1984), Leonard Zatman had impressed me with the potential importance of the methylated amines that are the decay products of the quaternary ammonium osmoprotectant glycine betaine.

To my surprise, instead of the usual sporadic occurrence of culturable bacteria (Sieburth, 1971), every 100-ml portion of oxygenated oceanic seawater supplemented with MMA, regardless of ocean or depth, turned turbid with methylamine-oxidizing bacteria similar to the *Methylophaga* species of Janvier et al. (1985). These long motile rods are devoid of intracytoplasmic membranes and unlike the bacterial cell I was after. Initially I was disappointed, but while making transfers one day in 1984, I noted that some of the MMA enrichment cultures had a pellicle (skin of bacteria) at the gas/water interface suggestive of methane-oxidizing bacteria. Could

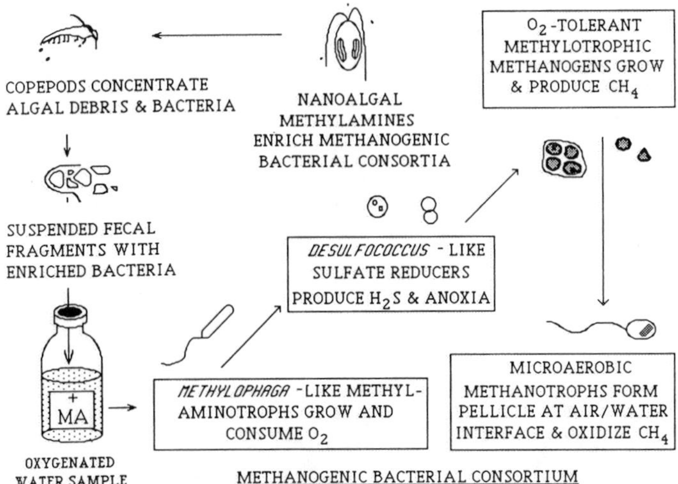

Figure 5. The paradoxical enrichment of a four-compartment methanogenic consortium, two aerobes and two anaerobes, from oxygenated seawater samples supplemented with MMA. Seawater particulates, xenic nanoalgal clones, and copepod fecal pellets are all good inocula for these methanogenic consortia. The assumed sequential enrichment is that methylaminotrophs deplete oxygen and produce formate for the sulfate-reducing bacteria, whose hydrogen sulfide creates the strictly anaerobic conditions required for the growth of the O_2-tolerant methanogens. (Adapted from Sieburth and Keller, 1991.)

the dissolved oxygen initially present in the seawater used to make these enrichments have been consumed sufficiently to allow the growth of obligately anaerobic methanogens? Resazurin added as a redox indicator confirmed oxygen depletion by being reduced to its colorless leuco form. The presence of hydrogen sulfide indicated that the enrichments had become anaerobic and anoxic. The use of serum bottles sealed with rubber septa (to permit sampling of the headspace for methane) showed order-of-magnitude increases in methane concentration. Methanogens, sulfate-reducing bacteria, and methane-oxidizing bacteria which were all being coenriched with the MMA oxidizers were all subcultured on trophic-specific media to verify their presence. The O_2-tolerant obligately anaerobic methanogens were inhibited by prereduced media, but were enhanced by their aerobic bacterial consorts. The proposed scenario for these enrichments has been discussed (Sieburth, 1987b, 1988a, 1988b) and is shown in Fig. 5.

OTHER ANAEROBIC PROCESSES REQUIRING ANOXIC MICROSITES

Besides methanogenesis, which would require anoxic microsites (Scranton and Brewer, 1977; Burke et al., 1983; Sieburth, 1983), the recent demonstration of sulfides in the upper layer of the Atlantic Ocean (Cutter and Krahforst, 1988; Luther and Tsamakis, 1989) would also require a mechanism for the localized depletion of oxygen. Alldrege and Cohen (1987) used microprobes to elegantly demonstrate the microzonal depletion of oxygen within particulates, such as marine snow and fecal pellets, that are suspended in oxygen-saturated waters. Nitrogen fixation has also been shown to occur in oxygenated surface waters and to be associated with oxygen depletion (Guernot and Colwell, 1985; Paerl and Prufert, 1987; Paerl and Bebout, 1987). Another process requiring similar anoxic microsites is fermentation that would produce hydrogen gas, which is also reported in the upper ocean (Schropp et al., 1987a; Schropp et al., 1987b). This process may also be associated with nitrogen fixation (Scranton, 1984). The evidence is starting to accumulate that microaerophilic and anaerobic processes usually associated with the benthos are not rare in the oxygenated upper ocean. The concept of anaerobic microsites accumulating in the thermocline to form a false benthos (Sieburth, 1983) was further elaborated on with the concept that microorganisms form three types of associations in the upper ocean (Sieburth, 1987a). These are the loosely associated "planktonic constellations" that occur throughout the photic zone (Sieburth and Davis, 1982), tighter aggregations of microorganisms forming "planktonic oases" (including anoxic oases in the oxygenated desert of the mixed layer), and "sestonic aggregations" of the smaller debris from zooplankton feeding that occur throughout the water column but accumulate in the oxygen-depleted thermocline (Fig. 6).

THE ALGAL-C_1 BACTERIAL CONNECTION

The four-compartment methanogenic consortium in Fig. 5 is contrary to conventional wisdom. It questions much of the dogma of the cultural requirements of anaerobes such as methanogens discussed above. The strictly anaerobic but O_2-tolerant methanogens appear to be protected and enriched by their natural syntrophic consorts, both aerobic and anaerobic. The C_1-based methanogenic consortium allows obligate aerobes and strict anaerobes to coexist.

Horstmann and Hoppe (1981) reported that a high proportion of bacteria in offshore waters utilized ^{14}C-labeled MMA. The primary producers may be a major source of these amines. A survey of the nanoalgae in the Provasoli-Guillard Culture Collection of Marine Phytoplankton, at the Bigelow Laboratory (West Boothbay Harbor, Maine), confirms this suspicion (Sieburth and Keller, 1991). Of 166 xenic clones tested, 42% have sustained MMA-oxidizing bacteria through years of transfers. The results from nanoalgae isolated from environments ranging from well mixed to well stratified show a trend towards an increased incidence of positive cultures from oceanic realms assumed to have well-developed stratification (see Table 1).

Clones containing the oxidative methylotrophs also contain methylotrophic methanogens (Sieburth, 1988a). This observation explains the paradoxical production of small amounts of methane in algal cultures reported by Scranton and Brewer (1977). I theorize that the methylamines released by algae (Sieburth and Keller, 1991) sustain methylamine-utilizing bacteria (Horstmann and Hoppe, 1981; Janvier

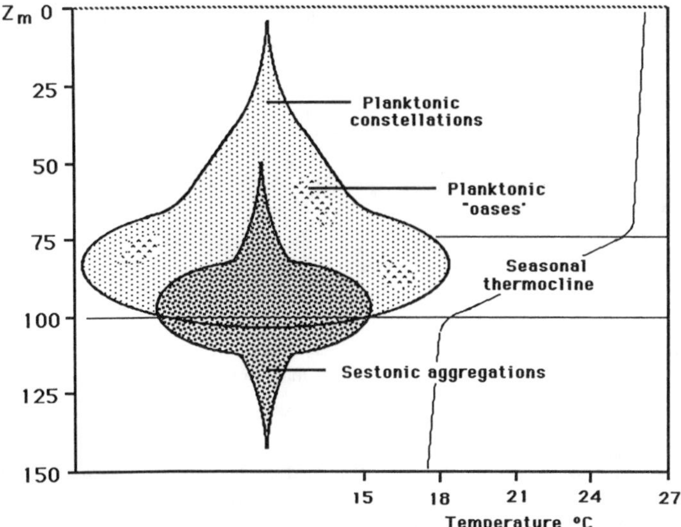

Figure 6. Distribution of three major hypothetical microbial assemblages in the upper ocean. The planktonic constellations are the trophic-dependent microbial repeating units found in just a microliter or two of seawater (Sieburth and Davis, 1982). The planktonic oases are microbial associations consisting of anaerobes, microaerophiles, and aerobes. The sestonic aggregations are composed of algal debris resulting from zooplankton feeding and the fragmentation of fecal pellets which accumulate in the pycnocline and also serve as anoxic microsites. (Adapted from Sieburth, 1987a.)

et al., 1985) whose penultimate organic degradation product is formate (Large, 1983). Sulfate-reducing bacteria that can use formate to produce hydrogen sulfide (Widdel, 1980; Pfennig et al., 1981) then provide both the sulfur and anoxia required by the methanogens that are more fastidious anaerobes. The production of methane would then enrich oxidizers like *Methylomonas pelagica* which can consume methane even at undetectable concentrations of oxygen (Sieburth, 1987b). These methanogenic consortia can persist for weeks and months under either oxic or anoxic conditions. Glucose and ammonia work equally well in enriching similar O_2-tolerant methanogenic consortia from the oxygenated water column of the freshwater pond that fronts my home (unpublished observation).

It has been suggested that zooplankton could play a major role in water column methanogenesis (Oremland, 1979; Traganza et al., 1979). Copepod fecal pellets or their fragments are an excellent inoculum for the enrichment of methanogenic consortia from oxygenated seawater. Copepod fecal pellets caught with sediment traps in a stratified mesocosm readily yielded methanogenic consortia when enriched by methylamine (J. McN. Sieburth and N. J. P. Owens, unpublished data). Since the gut microflora of copepods is greatly reduced after the

Table 1. Incidence of methylamine-oxidizing bacteria in xenic clones of nanoalgae from the Provasoli-Guillard Collection[a]

Water mass	No. of clones positive	Total tested	% Positive
Poorly stratified			
Western North Atlantic slope	19	61	31
Gulf of Mexico, Gulf Stream	2	6	33
Gulf of Mexico, near shore	5	13	38
Better stratified			
Gulf of Mexico	6	13	46
Sargasso Sea	24	45	53
Gulf of Maine	7	9	78

[a] Grouped according to the oceanic realm from which they were obtained. Mean = 63/147 = 43%. (Modified from Sieburth and Keller, 1991.)

passage of fecal pellets (Sochard et al., 1979), they appear to lack a stable intestinal bacterial flora per se (Boyle and Mitchell, 1978; Gowing and Wishner, 1986), and the role of copepods may be as concentrators of algae, whose debris would be enriched with methylated amines and methanogenic bacterial consortia, rather than as a site for methanogen colonization as envisioned by Oremland (1979). Studies on the methane maximum in the upper ocean have pointed to algae, to zooplankton, and to micro environments as the source of methane. They all seem to be correct. Upper water column methanogenesis appears to be dependent upon in situ multitrophic photic-zone processes involving the production, consumption, and decay of organic matter in situ in the pycnocline, so that productivity can be maintained throughout seasonal stratification.

DEPENDENCE OF WATER COLUMN METHANOGENESIS UPON SEASONAL STRATIFICATION

Seasonal studies in oceanic waters are difficult to interpret, as the water masses are continually moving and one never knows if the same water mass is being sampled (Burney et al., 1981, 1982). Exceptions are captive seawater in estuarine basins with high sills and the warm-core rings spun off from the Gulf Stream, which would make excellent macrocosms for studying greenhouse gases. Macrocosms like deep lakes are equally amenable to such studies. A beautiful example of how the benthic process of methanogenesis could largely transfer to the pycnocline during seasonal stratification is shown for Lake Constance in Fig. 7. Note how the methane concentrations in the upper water column exceed those in the water above the sediment. If one assumes that lateral processes are small, then in situ regeneration all the way to methane could be occurring to a significant degree in the pycnocline.

The seasonal transfer of "benthic processes" to the pycnocline does affect oxygen concentrations in the pycnocline. An oxygen minimum occurs in the deep chlorophyll maximum of the open ocean (Schulenberger and Reid, 1981), which is

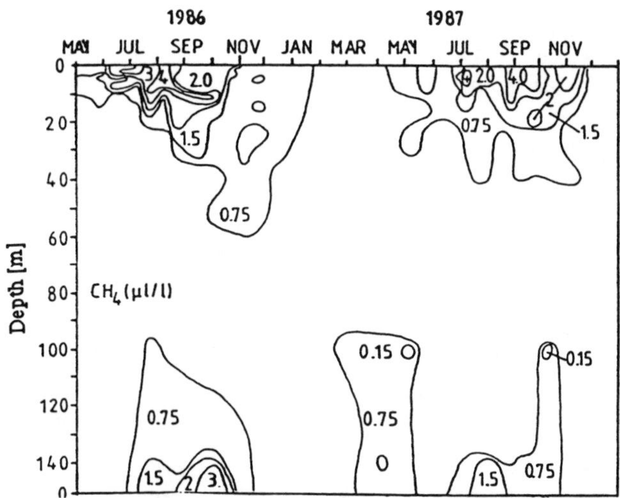

Figure 7. The seasonal distribution of methane in the water column of Lake Constance (Bodensee) indicates that this benthic process occurs to a significant extent in the upper water column when seasonal productivity and stratification would occur. (Adapted from Schuler et al., 1990.)

also the zone of the methane maximum (Burke et al., 1983). Under conditions of severe eutrophication as in Chesapeake Bay, suboxia can lead to anoxia (Officer et al., 1984; Seliger et al., 1985; Jonas and Tuttle, 1990). Here again algae accumulate in the pycnocline (Tyler and Seliger, 1978, 1981; Harding and Coates, 1988), and although their presence has been interpreted as being dormant rather than active, they could be a source of high O_2 flux as explained below.

Active grazing in the algal-rich pycnocline may fuel methanogenesis and sulfate reduction. In habitats with poor ventilation, where the conditions of stratification and anoxia are not seasonal but permanent the year round (Richards et al., 1965; Trüper and Genovese, 1968; Gaines and Pilson, 1972), the production and consumption of methane and hydrogen sulfide must be major factors in creating and maintaining the highly productive oxic-anoxic transition zone.

As anyone who reads the literature on seasonally or permanently stratified lakes knows, the oxic-anoxic interface in the pycnocline is discolored red with the pigments of anoxygenic photobacteria. What is not usually recognized in these studies is that both maximal production and consumption of both methane and hydrogen sulfide can apparently occur in the same zone. A fine-scale profile for a permanently stratified estuarine basin of the Pettaquamscutt Estuary, Rhode Island, is shown in Fig. 8. Turbidity and fluorescence are maximal in the pycnocline, with one pigment-poor peak followed by two superimposed fluorescent peaks. Note that these features, which occur in less than a 2-m interval, would be missed by sampling at 2, 4, 6, 8, and 10 m in this 20-m water column. The upper

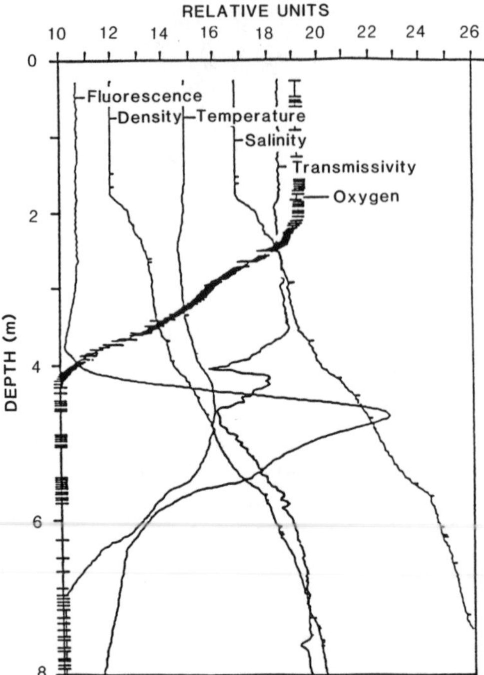

Figure 8. A continuous profile of physical-chemical parameters with a Sea Bird profiler through the microbial assemblages that have accumulated in the inapparent high O_2-flux zone (4 to 6 m) below the detectable oxygen transition zone (2 to 4 m). The fluorescence peak (4 to 6 m) consists of two overlapping populations of phototrophs, the green-colored oxygenic *Euglena proxima* and the red-colored brown-green anoxygenic photobacteria. Routine sampling at 2-m intervals in this 20-m water column would have missed most of the microbial assemblages including those producing and consuming reduced gases in the inapparent high O_2-flux zone dominated by *Euglena proxima*. (From Sieburth, 1988b.)

turbidity peak (low in pigments) is rich in grazers including flagellated and ciliated protozoa. The pigment-rich peak contains the green-colored oxygenic eucaryotic phototroph *Euglena proxima* and a spiral-shaped, red-pigmented, anoxygenic photobacterium whose chlorosomes would place it in the family *Chlorobiaceae*. This co-occurrence of microorganisms producing oxygen but also requiring hydrogen sulfide is apparently typical in this zone of low concentrations but high fluxes of oxygen and reduced gases. I think of this area as the "inapparent high O_2-flux zone." Indrebø et al. (1979a), studying the microbial processes in the anoxic basin of a stratified Norwegian fjord, found maximal rates of sulfate reduction just below the oxic-anoxic interface, while maximal rates of sulfide oxidation and dark CO_2 fixation also occurred in this zone (Indrebø et al., 1979b).

IS THE UPPER OCEAN A SIGNIFICANT SOURCE OF METHANE?

Stan Tyler's global models of methane sources showed the oceans as blank spaces. This is understandable, since the amounts of greenhouse gases evading from the sea surface are very small, and there are major problems in extrapolating limited data sets to a very large area. However, at least 40% of global productivity occurs in the surface layer of the ocean (Woodwell et al., 1978; Bender et al., 1985), mainly during seasonal stratification when in situ regeneration recycles at least 90%

of primary productivity (Eppley and Peterson, 1979; Karl and Knauer, 1984; Baker et al., 1985; Gardiner et al., 1985). The small net flux may tend to minimize the importance of a much larger gross flux in the upper ocean, a flux that might be perturbed by storm events, eutrophication, and global warming.

Those who study and model stratified lakes with an anoxic hypolimnion (Hanson, 1980; Harrits and Hanson, 1980; Rudd and Taylor, 1980; Large, 1983; Brock, 1985; Smith and Oremland, 1987; Iverson et al., 1987; and Cloern et al., 1987, among others) show methane and hydrogen sulfide being produced only in anoxic sediments. This troubles me, and I wonder why it is so. Part of the problem may be that their coarse-scale sampling at intervals of several meters may have totally missed some of the processes that require fine-interval sampling in order to be observed. Another part of the problem may be due to an insufficient characterization of the natural populations of microorganisms, especially those that one would not expect to find in oxygenated/reduced-specific habitats (Sieburth, 1987b).

Both Richards et al. (1965) and Winfrey and Zeikus (1979) point out that the shape of methane distribution in the water column indicates methanogenesis in the water column. The distribution curves for methane (Iverson et al., 1987) and hydrogen sulfide (Smith and Oremland, 1987) in Big Soda Lake indicate to me that a marked "oxidation" must occur between their chemocline and oxycline, the inapparent high O_2-flux zone (see Fig. 9A and B). Oxygen in the pycnocline could be coming from oxygenic phototrophs below 20 m whose pigments may have been obscured by those of the anoxygenic photobacteria (Cloern et al., 1987) (see Fig. 9C) and whose presence was not anticipated or looked for. Smith and Oremland (1987) and Iverson et al. (1987) assume that mineralization occurs only in the benthos, while a companion paper in this four-paper series on the vertical fluxes of particulate matter (Cloern et al., 1987) indicated that some 90% of productivity was regenerated in the "mixolimnion." Regeneration (including methanogenesis and sulfate reduction) cannot be occurring maximally in both the pycnocline and the benthos!

NEW PARADIGMS FOR THE FOOD CHAIN AND MICROBIAL REGENERATION

Two principal osmoprotectants occur in planktonic algae, the sulfonium chemical dimethylsulfoniopropionate (Keller et al., 1989) and the quaternary ammonium chemical glycine betaine (Yancey et al., 1982). Both hydrolyze to yield volatile substances that evade to the atmosphere, dimethyl sulfide from dimethylsulfoniopropionate and the methylated amines (trimethylamine, dimethylamine, and MMA) from glycine betaine. Both dimethyl sulfide (Kiene and Vischer, 1987; Oremland et al., 1989) and the methylamines (Sowers et al., 1984) serve as substrates for methylotrophic methanogens.

The nanoalgae that dominate the pycnocline produce these osmoprotectants. The algae in some classes have a tendancy to favor one osmolyte over another. The algae producing dimethylsulfoniopropionate are dominated by the chromophytes such as the Prymnesiophytes and Dinophytes, while the major classes containing

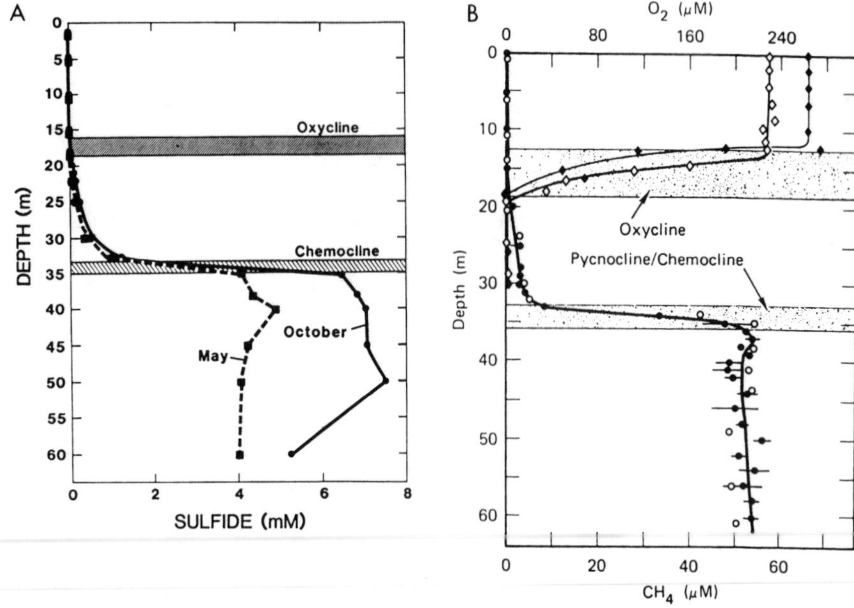

Figure 9. The inapparent high O_2-flux zone may be occurring between the oxycline and chemocline, with the simultaneous production and consumption of the reduced gases (A) hydrogen sulfide and (B) methane, which can accumulate below this zone (●, ○), and dissolved oxygen (◆, ◇). The presence of oxygenic phototrophs responsible for this inapparent high O_2-flux zone may be masked by the simultaneous occurrence of (C) (next page) anoxygenic phototrophs producing bacteriochlorophyll that masks chlorophyll *a*. (A, from Smith and Oremland, 1987; B, from Iverson et al., 1987; C, from Cloern et al., 1987.)

glycine betaine may be the green algae containing the Prasinophytes and Chlorophytes (Sieburth and Keller, 1991). Studies on oceanic greenhouse gases cannot ignore the systematics of the dominant algae in the pycnocline.

The dominant algae in the pycnocline, belonging to the picoplankton and nanoplankton, must be the raw material for methane and hydrogen sulfide production in the upper stratified ocean. The question is whether the decay of this organic matter produces a small, smoky fire, or the ashes of a large one that is not observed? Algal products other than osmoprotectants affect the production of methane. The osmoprotectants may be thought of as the kindling, since they constitute just 1 to 10% of the solutes in the cytoplasm. The logs for the fire must be the structural and reserve biopolymers that dominate the contents of the algal cell. Algae such as diatoms produce galactans as their storage product (Myklestad, 1974; Paulsen and Myklestad, 1978) which, when fermented by the bacteria in algal debris, should yield methanol (Large, 1983). Prasinophytes, whose pigment signature is characteristic of the pycnocline, include such species as *Pycnococcus provasolii* (Guillard et al., 1991). The theca of this and other prasinophytes appear

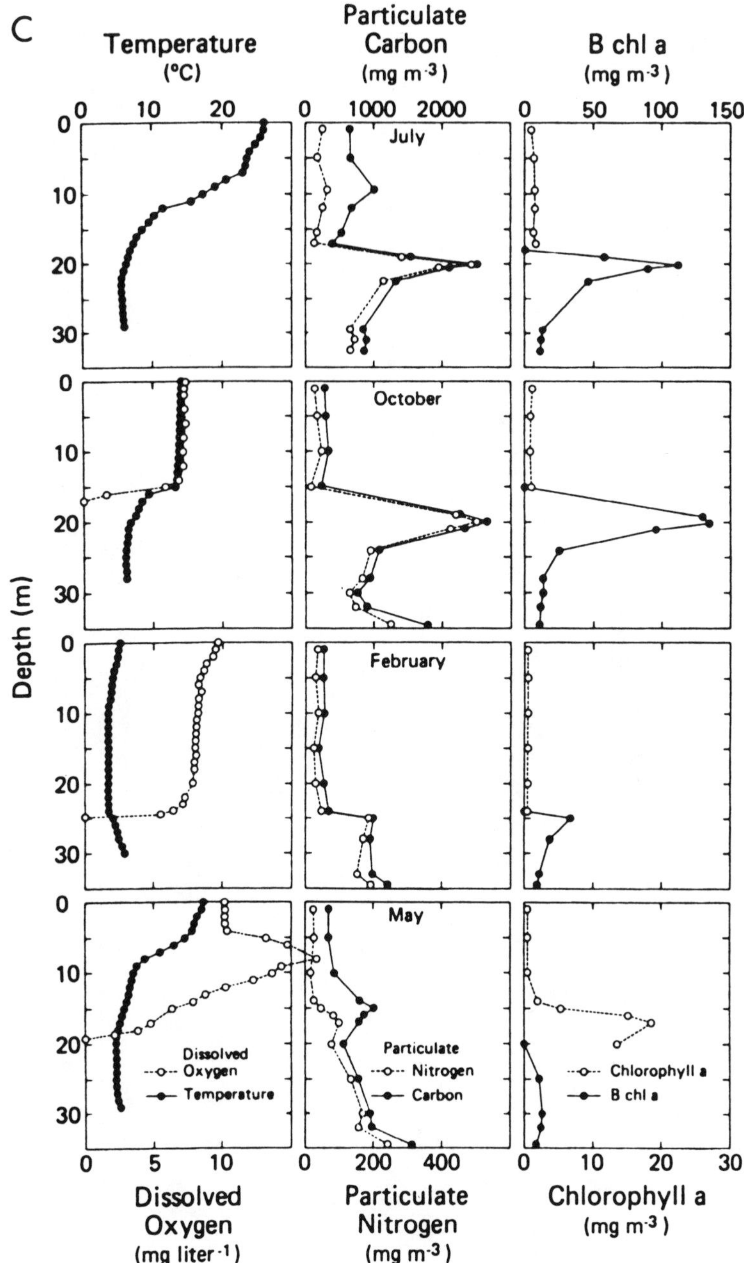

Figure 9. *Continued.*

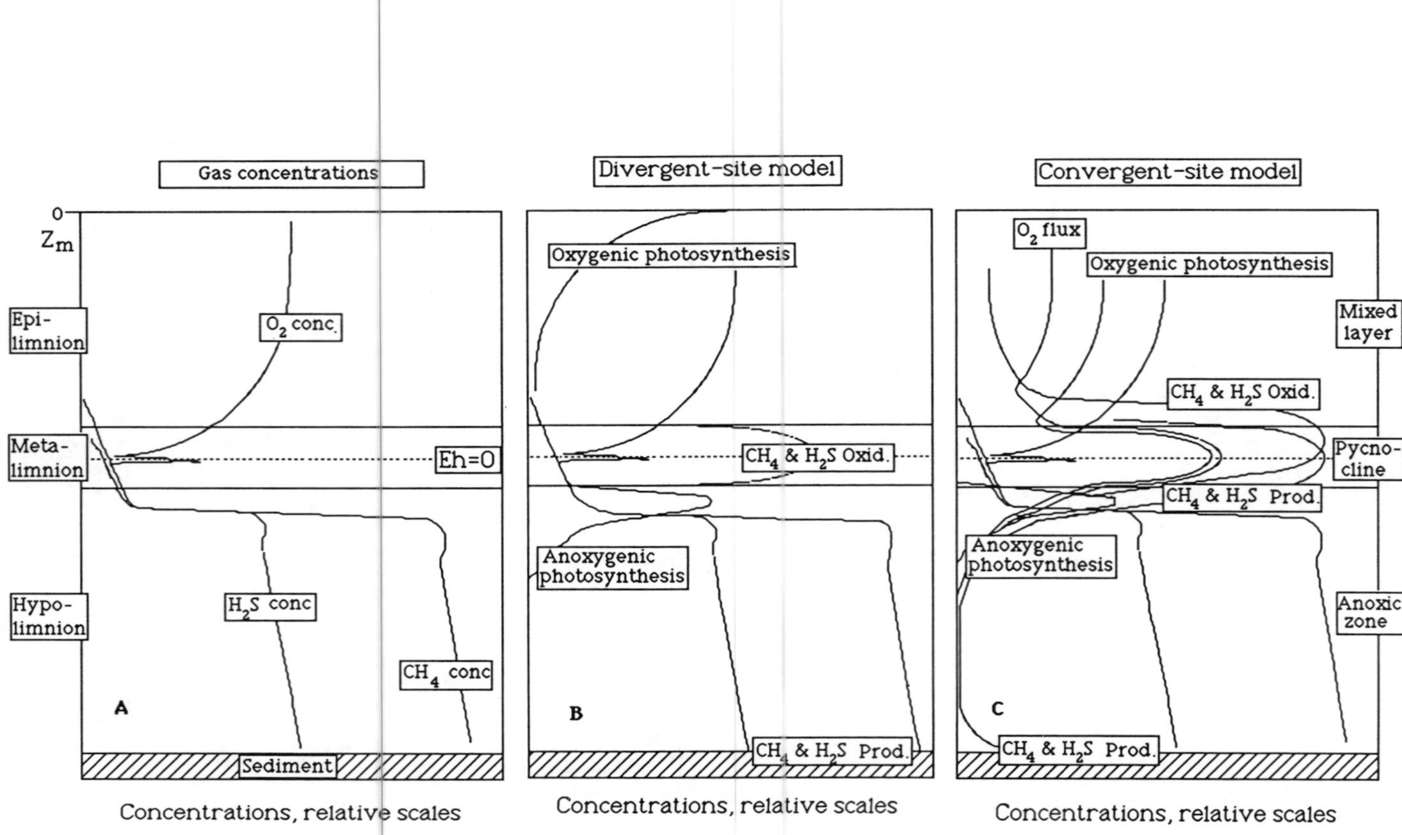
Gas concentrations
Divergent-site model
Convergent-site model
0
Zm
Epi-
limnion
Meta-
limnion
Hypo-
limnion
O_2 conc.
Eh=0
H_2S conc
CH_4 conc
A
Sediment
Concentrations, relative scales
Oxygenic photosynthesis
CH_4 & H_2S Oxid.
Anoxygenic
photosynthesis
B
CH_4 & H_2S Prod.
Concentrations, relative scales
O_2 flux
Oxygenic photosynthesis
Mixed
layer
CH_4 & H_2S Oxid.
Pycno-
cline
CH_4 & H_2S Prod.
Anoxygenic
photosynthesis
Anoxic
zone
C
CH_4 & H_2S Prod.
Concentrations, relative scales

pectinlike (Gooday, 1971) and are probably good methanogenic substrates (Ollivier and Garcia, 1990). Other polysaccharides, such as cellulose and the chitins that are also structural components of the plankton, are also substrates for methane and hydrogen sulfide production (Boyer, 1986). The copepods in the zooplankton encase their fecal pellets in a peritrophic membrane consisting of chitin. Not all fecal pellets raft to the sea floor, as many are stripped of their peritrophic membrane by the copepods, apparently to recycle and conserve this nitrogen-rich biopolymer through the process of coprorhexy (Noji et al., in preparation), which allows the fecal pellet to fragment and become suspended organic matter.

Physical/chemical processes that induce stratification appear to effectively transfer "benthic processes" to the pycnocline. The distribution of reduced substances in the stratified water column is very similar to those across the redox potential discontinuity layer of sandy beaches (as illustrated in Fig. 3 of Fenchel and Reidel, 1970). The essential step of seasonal stratification required for the phytoplankton blooms and their bacterial consorts, zooplankton enhancement of this process through their fecal fragments, and the development of bacterial consortia have been described above. A paradigm to explain this, shown in Fig. 10, contrasts the conventional sites of the processes of regeneration with those for a site in the seasonal or permanent pycnocline where processes are apparently concentrated in the inapparent high O_2-flux zone. The latter includes such sites as the meromictic waters of the Black Sea as well as smaller lakes such as Faro (Trüper and Genovese, 1968), Pettaquamscutt (Gaines and Pilson, 1972; Sieburth, 1987b), and those on Åland (Lindholm and Weppling, 1987).

But the principles must also apply to stratified oxygenated waters. A general paradigm for stratified waters, both oxygenated and those that become seasonally anoxic, is shown in Fig. 11. This diagram shows how an oxygenated ecosystem can become anoxic by eutrophication, poor ventilation, or both. Note that the methanogenic and sulfidogenic consortia on particulates, although concentrated in the pycnocline, are distributed throughout the water column.

DISCREPANCIES BETWEEN HIGH RATES OF IN SITU REGENERATION AND LOW FLUX RATES FOR METHANE

If "benthic"-like processes are large in the seasonal pycnocline, why does this not show up in the flux rates of methane to the atmosphere? The sea/air fluxes

Figure 10. The distribution of gas concentrations in stratified and anoxic basins (A) can be interpreted two ways. The conventional divergent site model (B) shows primary production occurring above and below the pycnocline (metalimnion), oxidation restricted to the suboxic pycnocline (metalimnion), and the production of methane and hydrogen sulfide restricted to the sediment. The proposed "convergent site model" (C) suggests that the pycnocline (metalimnion) is the major site for both primary production and bacterial decay (including the production and oxidation of the reduced gases methane and hydrogen sulfide). Oxygenic photosynthesis in the suboxic pycnocline (metalimnion) is hypothesized to provide the high oxygen flux required for the oxidation of these gases while they are simultaneously produced by anaerobic processes occurring in anoxic microzones created by the oxidative bacteria.

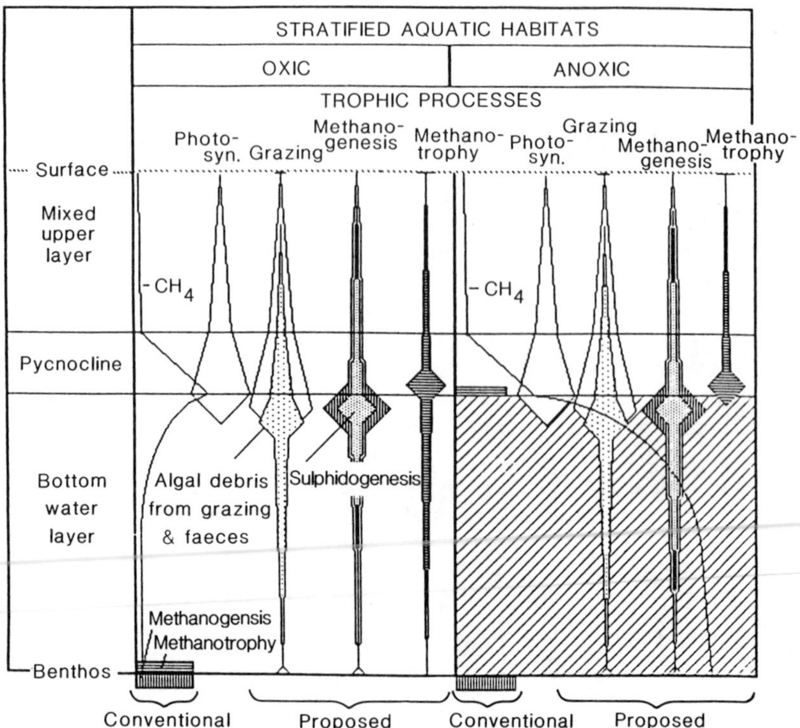

Figure 11. A paradigm to explain the benthic processes in the oxygenated pycnocline and how, with eutrophication and poor ventilation, they can turn the waters below the pycnocline anoxic. The kite diagrams show the relative activity on a vertical scale, in which the algae responsible for photosynthesis are grazed by the zooplankton to form algal and fecal debris. The fermentation of this debris forms methylotrophic substrates and sites where the aerobes and anaerobes of the methanogenic consortia can coexist in an oxygenated water column. Methanogenesis and sulfidogenesis are hypothesized to be maximal in the pycnocline, but to occur throughout the water column rather than being restricted to anoxic sediments as in conventional schemes. This scheme should be applicable to aquatic habitats in general, including oxic and suboxic coastal waters, the stratified open ocean, and anoxic basins, both freshwater and marine.

calculated by the careful workers Conrad and Seiler (1988a) account for only some 0.4% of the annual atmospheric source of methane. Methodology prevented a large oceanic productivity from being measured for decades. Perhaps that is also our problem here.

The role of the surface film and its methanotrophic activity may not be fully appreciated. The development of bacteria in the surface microlayer has been studied in the open sea (Sieburth et al., 1976) and reviewed by Sieburth (1983). The everpresent organic film that expands with wave action like a latex sheet may be an effective damper to gaseous flux and permit the accumulation of gases such as methane at the film, where methanotrophs could develop. Just such a microbiota appears to explain the sink rather than the source of methane in chamber studies (Conrad and Seiler, 1988b). Such measurements are usually made when the sea is calm, however. Turbulent sea states with whitecaps that would preclude such observations occur over 13% of the sea's surface at any one time (E. Monahan, personal communication). Such storm fronts effectively mix the upper water layer and could deplete the concentrations of reduced gases. Low-air-pressure events usually associated with storm systems have been reported to cause the sporadic ebullition of methane bubble from the sediments of Mirror Lake, an 18% increase per millibar drop in air pressure (Mattson and Likens, 1990). Why not also from the pycnocline, the site of the seasonal "false benthos"?

EPILOGUE

During the workshop on which this volume is based, I attempted to outline the tenets of this manuscript and was exposed to the arguments of those who disagree with my interpretations of the literature and my own experiences. This goaded me into a thorough rereading of the literature and focusing on why investigators with common interests can come to such divergent conclusions. The results are not just this manuscript, but a response to the Athens Environmental Research Laboratory's Request for Proposal concerning the bacterial production, consumption, and fluxes of radiatively important trace gases from estuarine and coastal waters. Our proposal was accepted, and an in-depth study is presently being conducted in the permanently stratified basins of the Pettaquamscutt Estuary (Gaines and Pilson, 1972; Sieburth, 1987b). The first year's study has just been completed (Sieburth, 1991). The work included (i) a diel study that showed the radical changes in phototroph distribution and activity with light, which affects the diel distribution of reduced gases; (ii) a transect of the basin which showed how the morphology of the basin and water circulation affect the reduced gases; and most importantly (iii) the mechanism of how storm events and perigean spring tides can not only ventilate anoxic waters, but cause episodic, massive outgassings of radiatively important trace gases that are probably missed in flux measurements. (These data will be reported to the annual meeting of the American Society for Limnology and Oceanography in Halifax, June, 1991.) Next we come back to the Pettaquamscutt basin to make the much-needed flux measurements on radiatively important trace gases including methane, hydrogen sulfide, and carbon dioxide. This last, a major greenhouse gas beyond the scope of this paper, has been a parameter of much interest in my laboratory (Johnson et al., 1981, 1983a; Johnson et al., 1983b). The need for improved TCO_2 determinations led to the application of coulometry (Johnson et al., 1985; Johnson et al., 1987; Johnson and Sieburth, patent pending) in a method which has become the standard for ongoing global ocean flux studies.

What may be required to extend our observations to the pycnocline of coastal waters is a study within the captive waters of a warm core ring (Schink et al., 1982), so that this parcel of confined water could be followed and profiled through diel cycles and storms for as long as it takes to determine fluxes from natural concentrations. Such profiles for greenhouse gases around the clock, to determine diel variation and storm mixing, should give us realistic flux estimations. These determinations are required before we dismiss the slight maxima of methane and hydrogen sulfide as smoke. An open-minded questioning of the ponderables in this essay could lead to work that would determine the true extent of upper water column methanogenesis and sulfate reduction and their contribution to upper ocean processes and to greenhouse gases: a small, smoky fire, or the ashes of a large one that is not observed?

Acknowledgments. This paper is dedicated to the memory of the late Francis Asbury Richards, who got it so right the first time so long ago. I would like to acknowledge my former labmates, Paul W. Johnson and Kenneth M. Johnson, for their technical services. I would also like to thank Henry J. (Hal) Walker (Narragansett Environmental Research Laboratory, U.S. Environmental Protection Agency) and William B. (Barny) Whitman (Department of Microbiology, University of Georgia, Athens) for arranging my attendance at this workshop; Percy L. Donaghay, Mary I. Scranton, and Dr. Whitman for valuable critiques of drafts of this manuscript; and Kenneth W. Estep for Fig. 4 and an early introduction to Macintosh. This work was supported by National Science Foundation grants OCE-8710085 and OCE-8911367, by the University of Rhode Island Sea Grant College Program through grant M/PD 8703, and by the U.S. Environmental Protection Agency through AERL-9005.

REFERENCES

Alldredge, A. L., and Y. Cohen. 1987. Can microscale chemical patches persist in the sea? Microelectrode study of marine snow, fecal pellets. *Science* **235:**689–691.

Atkinson, L. P., and F. A. Richards. 1967. The occurrence and distribution of methane in the marine environment. *Deep-Sea Res.* **14:**673–684.

Baker, E. T., R. A. Feeley, M. R. Landry, and B. Lamb. 1985. Temporal variations in the concentration and settling flux of carbon and phytoplankton pigments in a deep fjordlike estuary. *Estuar. Coastal Shelf Sci.* **21:**859–877.

Bender, M., L. D. Labeyrie, D. Raynaud, and C. Lorius. 1985. Isotopic composition of atmospheric O_2 in ice linked with deglaciation and global primary productivity. *Nature* (London) **318:**349–352.

Bergh, Ø., K. Y. Børsheim, G. Bratbak, and M. Heldal. 1989. High abundance of viruses found in aquatic environments. *Nature* (London) **340:**467–468.

Boyer, J. N. 1986. End products of anaerobic chitin degradation by salt marsh bacteria as substrates for dissimilatory sulfate reduction and methanogenesis. *Appl. Environ. Microbiol.* **52:**1415–1418.

Boyle, P. J., and R. Mitchell. 1978. Absence of microorganisms in crustacean digestive tracts. *Science* **200:**1157–1159.

Brock, T. D. 1985. *A Eutrophic Lake, Lake Mendota, Wisconsin. Ecological Studies,* vol. 55. Springer Verlag, New York.

Brooks, J. M., A. D. Fredericks, W. M. Sackett, and J. W. Swinnerton. 1973. Baseline concentrations of light hydrocarbons in Gulf of Mexico. *Environ. Sci. Technol.* **7:**639–642.

Brooks, J. M., D. F. Reid, and B. B. Bernard. 1981. Methane in the upper water column of the northwestern Gulf of Mexico. *J. Geophys. Res.* **86:**11029–11040.

Brooks, J. M., and W. M. Sackett. 1973. Sources, sinks, and concentrations of light hydrocarbons in the Gulf of Mexico. *J. Geophys. Res.* **78:**5248–5258.

Burke, R. A., Jr., D. F. Reid, J. M. Brooks, and D. M. Lavoie. 1983. Upper water column methane geochemistry in the eastern tropical North Pacific. *Limnol. Oceanogr.* **28:**19–32.

Burney, C. M. 1986. Bacterial utilization of total in situ dissolved carbohydrate in offshore waters. *Limnol. Oceanogr.* **31:**427–431.

Burney, C. M., P. G. Davis, K. M. Johnson, and J. McN. Sieburth. 1981. Dependence of dissolved carbohydrate concentrations upon small scale nanoplankton and bacterioplankton distributions in the western Sargasso Sea. *Mar. Biol.* **65**:289–296.

Burney, C. M., P. G. Davis, K. M. Johnson, and J. McN. Sieburth. 1982. Diel relationships of microbial trophic groups and *in-situ* dissolved carbohydrate dynamics in the Caribbean Sea. *Mar. Biol.* **67**:311–322.

Caron, D. A., P. G. Davis, and J. McN. Sieburth. 1989. Factors responsible for the differences in cultural estimates and direct microscopical counts of populations of bacterivorous nanoflagellates. *Microb. Ecol.* **18**:89–104.

Cicerone, R. J., and R. S. Oremland. 1988. Biogeochemical aspects of atmospheric methane. *Global Geochem. Cycles* **2**:299–327.

Cloern, J. E., B. E. Cole, and S. M. Wienke. 1987. Big Soda Lake (Nevada). 4. Vertical fluxes of particulate matter: seasonality and variations across the chemocline. *Limnol. Oceanogr.* **32**:815–824.

Conrad, R., and W. Seiler. 1988a. Methane and hydrogen in seawater (Atlantic Ocean). *Deep-Sea Res.* **35**:1903–1917.

Conrad, R., and W. Seiler. 1988b. Influence of the surface microlayer on the flux of non-conservative trace gases (CO, H_2, CH_4, H_2) across the ocean-atmosphere interface. *J. Atmos. Chem.* **6**:83–94.

Crawford, R. L., and R. S. Hanson (ed.). 1984. *Microbial Growth on C_1 Compounds.* American Society for Microbiology, Washington, D.C.

Cutter, G. A., and C. F. Krahforst. 1988. Dissolved hydrogen sulfide in surface waters of the western Atlantic. *Geophys. Res. Lett.* **15**:1393–1396.

Daniels, L., N. Belay, and B. S. Rajagopal. 1986. Assimilatory reduction of sulfate and sulfite by methanogenic bacteria. *Appl. Environ. Microbiol.* **51**:703–709.

Davis, P. G., D. A. Caron, P. W. Johnson, and J. McN. Sieburth. 1985. Phototrophic and apochlorotic components of the picoplankton and nanoplankton in the North Atlantic: geographic, vertical, seasonal and diel distribution. *Mar. Ecol. Prog. Ser.* **21**:15–26.

Eppley, R. W., and B. J. Peterson. 1979. Particulate organic matter flux and planktonic new production in the deep ocean. *Nature* (London) **382**:677–680.

Estep, K. W., P. G. Davis, P. E. Hargraves, and J. McN. Sieburth. 1984. Chloroplast containing microflagellates in natural populations of North Atlantic nanoplankton, their identification and distribution; including a description of five new species of *Chrysochromulina* (Prymnesiophycae). *Protistologica* **20**:613–634.

Fenchel, T. M., and R. J. Reidel. 1970. The sulfide system, a new biotic community underneath the oxidized layer of marine sand bottoms. *Mar. Biol.* **7**:255–268.

Fitzwater, S. E., G. A. Knauer, and J. H. Martin. 1982. Metal contamination and its effect on primary production measurements. *Limnol. Oceanogr.* **27**:544–551.

Florek, R. J., and G. T. Rowe. 1983. Oxygen consumption and dissolved inorganic nutrient production in marine coastal and shelf sediments of the Middle Atlantic Bight. *Int. Rev. Gesamte Hydrobiol.* **68**:73–112.

Furnas, M. J., and A. W. Mitchell. 1988. Photosynthetic characteristics of Coral Sea picoplankton (<2 μm size fraction). *Biol. Oceanogr.* **5**:163–182.

Gaines, A. G., Jr., and M. E. Q. Pilson. 1972. Anoxic water in the Pettaquamscutt River. *Limnol. Oceanogr.* **17**:42–49.

Gardiner, W. D., J. B. Southward, and C. D. Hollister. 1985. Sedimentation, resuspension and chemistry of particles in the northwest Atlantic. *Mar. Geol.* **65**:199–242.

Garfield, P. C., T. T. Packard, G. E. Friederich, and L. A. Codispoti. 1983. A subsurface particle maximum layer and enhanced microbial activity in the secondary nitrite maximum of the northeastern tropical Pacific Ocean. *J. Mar. Res.* **41**:747–768.

Gieskes, W. W. C., G. W. Kray, and M. A. Baars. 1979. Current ^{14}C methods for measuring primary production: gross underestimates in oceanic waters. *Neth. J. Sea Res.* **13**:58–78.

Glover, H. E., M. D. Keller, and R. R. L. Guillard. 1986. Light quality and oceanic ultraphytoplankton. *Nature* (London) **319**:142–143.

Glover, H. E., M. D. Keller, and R. W. Spinnard. 1987. The effects of light quality and intensity on photosynthesis and growth of marine eukaryotic and prokaryotic phytoplankton clones. *J. Exp. Mar. Biol. Ecol.* **105**:137–159.

Goldman, J. C. 1986. On phytoplankton growth rates and particulate C:N:P ratios at low light. *Limnol. Oceanogr.* **31**:1358–1363.

Gooday, G. W. 1971. A biochemical and autoradiographic study of the role of the golgi bodies in thecal formation in *Platymonas tetrahele*. *J. Exp. Bot.* **22**:959–971.

Gowing, M. M., and K. F. Wishner. 1986. Trophic relationships of deep-sea calanoid copepods from the benthic boundary layer of the Santa Catalina Basin, California. *Deep-Sea Res.* **33**:939–961.

Guernot, M. L., and R. R. Colwell. 1985. Enumeration, isolation and cultivation of N_2-fixing bacteria from seawater. *Appl. Environ. Microbiol.* **50**:350–355.

Guillard, R. R. L., and M. D. Keller, C. J. O'Kelly, and G. L. Floyd. 1991. *Pycnococcus provasolii* gen. et sp. nov., a coccoid prasinoxanthin-containing phytoplankter from the western North Atlantic and Gulf of Mexico. *J. Phycol.* **27**:39–47.

Hanson, R. S. 1980. Ecology and diversity of methylotrophic organisms. *Adv. Appl. Microbiol.* **26**:3–39.

Harding, L. W., and D. W. Coates. 1988. Photosynthetic physiology of *Prorocentrum mariae-lebouriae* (Dinophyceae) during its subpycnocline transport in Chesapeake Bay. *J. Phycol.* **24**:77–89.

Harrits, S. M., and R. S. Hanson. 1980. Stratification of aerobic methane-oxidizing organisms in Lake Mendota, Madison, Wisconsin. *Limnol. Oceanogr.* **25**:412–421.

Horstmann, V., and H. G. Hoppe. 1981. Competition in the uptake of methylamine/ammonium by phytoplankton and bacteria. *Kieler Meeresforsch. Sonderh.* **5**:110–116.

Hungate, R. E. 1950. The anaerobic mesophilic cellulolytic bacteria. *Bacteriol. Rev.* **14**:1–49.

Indrebø, G., B. Pengerud, and I. Dundas. 1979a. Microbial activities in a permanently stratified estuary. I. Primary productivity and sulfate reduction. *Mar. Biol.* **51**:295–304.

Indrebø, G., B. Pengerud, and I. Dundas. 1979b. Microbial activities in a permanently stratified estuary. II. Microbial activities at the oxic-anoxic interface. *Mar. Biol.* **51**:305–309.

Iverson, N., R. S. Oremland, and M. J. Klug. 1987. Big Soda Lake (Nevada). 2. Pelagic methanogenesis and anaerobic methane oxidation. *Limnol. Oceanogr.* **32**:804–814.

Janvier, M. C., C. Frehl, F. Gremont, and F. Gasser. 1985. *Methylophaga marina* gen. et sp. nov. and *Methylophaga thallasica* sp. nov., marine methylotrophs. *Int. J. Syst. Bacteriol.* **35**:131–139.

Johnston, K. M., C. M. Burney, and J. McN. Sieburth. 1981. Enigmatic ecosystem metabolism measured by direct diel ΣCO_2 and O_2 flux in conjunction with DOC release and uptake. *Mar. Biol.* **65**:49–60.

Johnson, K. M., C. M. Burney, and J. McN. Sieburth. 1983a. Precise and accurate determination by infrared photometry of CO_2 dynamics in marine ecosystems. *Mar. Ecol. Prog. Ser.* **10**:251–256.

Johnson, K. M., P. G. Davis, and J. McN. Sieburth. 1983b. Diel variation of the upper layer of oceanic waters reflects microbial composition, variation, and possibly methane cycling. *Mar. Biol.* **77**:1–10.

Johnson, K. M., A. E. King, and J. McN. Sieburth. 1985. Coulometric TCO_2 analyses for marine studies; an introduction. *Mar. Chem.* **16**:61–82.

Johnson, K. M., and J. McN. Sieburth. Method and apparatus for the coulometric detection of dissolved gases, particularly TCO_2, in seawater. U.S. patent pending, serial no. 591,787.

Johnson, K. M., P. J. leB. Williams, L. Brändström, and J. McN. Sieburth. 1987. Coulometric TCO_2 analysis for marine studies: automation and calibration. *Mar. Chem.* **21**:117–133.

Johnson, P. W., and J. McN. Seiburth. 1979. Chroococcoid cyanobacteria in the sea: a ubiquitous and diverse phototrophic biomass. *Limnol. Oceanogr.* **24**:928–935.

Johnson, P. W., and J. McN. Sieburth. 1982. *In-situ* morphology and occurrence of eucaryotic phototrophs of bacterial size in the picoplankton of estuarine and oceanic waters. *J. Phycol.* **18**:318–327.

Jonas, R. B., and J. H. Tuttle. 1990. Bacterioplankton and organic carbon dynamics in the lower mesohaline Chesapeake Bay. *Appl. Environ. Microbiol.* **56**:747–757.

Karl, D. M., and G. A. Knauer. 1984. Detritus-microbe interactions in the marine pelagic environment: selected results from the Vertex experiment. *Bull. Mar. Sci.* **35:**550–565.

Karl, D. M., G. A. Knauer, J. H. Martin, and B. B. Ward. 1984. Bacterial chemolithotrophy in the ocean is associated with sinking particles. *Nature* (London) **309:**54–56.

Keller, M. D., W. K. Bellows, and R. R. L. Guillard. 1989. Dimethyl sulfide production in marine phytoplankton, p. 167–182. *In* E. S. Salzmann and W. J. Cooper (ed.), *Biogenic Sulfur in the Environment.* ACS Symposium Series no. 393. American Chemical Society, Washington, D.C.

Kiene, R. P., and P. T. Visscher. 1987. Production and fate of methylated sulfur compounds from methionine and dimethylsulfoniopropionate in anoxic salt marsh sediments. *Appl. Environ. Microbiol.* **53:**2426–2434.

King, G. M. 1984. Metabolism of trimethylamine, choline, and glycine betaine by sulfate-reducing and methanogenic bacteria in marine sediments. *Appl. Environ. Microbiol.* **48:**719–725.

King, G. M., M. J. Klug, and D. R. Lovley. 1983. Metabolism of acetate and methylated amines in intertidal sediments of Lowes Cove, Maine. *Appl. Environ. Microbiol.* **45:**1848–1853.

Kriss, A. E. 1963. *Marine Microbiology [Deep Sea]* (English Translation from the Russian edition, 1959, by J. M. Shewan and Z. Kabata). Oliver and Boyd, Edinburg.

Krupatkina, D. K. 1990. Estimates of primary production in oligotrophic waters and metabolism of picoplankton: a review. *Mar. Microb. Food Webs* **4:**87–101.

Lamontagne, R. A., J. W. Swinnerton, V. J. Linnebom, and W. D. Smith. 1973. Methane concentrations in various marine environments. *J. Geophys. Res.* **78:**5317–5324.

Large, P. J. 1983. *Methylotrophy and Methanogenesis. Aspects of Microbiology 8.* American Society for Microbiology, Washington, D.C.

Lindholm, T. 1979. Siphon sampling in meromictic lakes. *Acta Bot. Fenn.* **110:**91–93.

Lindholm, T., and K. Weppling. 1987. Blooms of phototrophic bacteria and phytoplankton in a small brackish lake on Åland, SW Finland. *Acta Acad. Aboens.* **47:**45–53.

Luther, G. W., III, and E. Tsamakis. 1989. Concentration and form of dissolved sulfide in the oxic water column of the ocean. *Mar. Chem.* **27:**165–177.

Mattson, M. D., and G. E. Likens. 1990. Air pressure and methane fluxes. *Nature* (London) **347:**718–719.

Miller, B. T. 1972. The phytoplankton and related hydrography of the Pettaquamscutt River. M.S. Thesis. University of Rhode Island, Kingston.

Murphy, L. S., and E. M. Haugen. 1985. The distribution and abundance of phototrophic ultraplankton in the North Atlantic. *Limnol. Oceanogr.* **30:**47–58.

Myklestad, S. 1974. Production of carbohydrates by marine planktonic diatoms. I. Comparison of nine different species in culture. *J. Exp. Mar. Biol. Ecol.* **15:**261–274.

Noji, T., K. Estep, F. MacIntyre, and F. Norrbin. Manuscript in preparation.

Officer, C. B., R. B. Biggs, J. L. Taft, L. E. Cronin, M. A. Tyler, and W. Boynton. 1984. Chesapeake Bay anoxia: origin, development and significance. *Science* **223:**22–27.

Ollivier, B., and J.-L. Garcia. 1990. Thermophilic methanogenesis from pectin by a mixed defined bacterial culture. *Curr. Microbiol.* **20:**77–81.

Oremland, R. S. 1979. Methanogenic activity in plankton samples and fish intestines: a mechanism for *in-situ* methanogenesis in oceanic surface waters. *Limnol. Oceanogr.* **24:**1136–1141.

Oremland, R. S. 1988. Biogeochemistry of methanogenic bacteria, p. 641–705. *In* A. J. B. Zehnder (ed.), *Biology of Anaerobic Microorganisms.* John Wiley & Sons, Inc., New York.

Oremland, R. S., R. P. Kiene, I. Mathrani, M. J. Whiticar, and D. R. Boone. 1989. Description of an estuarine methylotrophic methanogen which grows on dimethyl sulfide. *Appl. Environ. Microbiol.* **55:**994–1002.

Oremland, R. S., L. M. Marsh, and S. Polcin. 1982. Methane production and simultaneous sulphate reduction in anoxic, salt marsh sediments. *Nature* (London) **296:**143–145.

Paerl, H. W., and B. M. Bebout. 1987. Direct measurements of O_2 depleted microzones in marine *Oscillatoria (Trichodesmium)*: relation to N_2-fixing capabilities. *Science* **241:**441–445.

Paerl, H. W., and L. E. Prufert. 1987. Oxygen-poor microzones as potential sites of microbial N_2 fixation in nitrogen-depleted aerobic marine waters. *Appl. Environ. Microbiol.* **53:**1078–1087.

Paulsen, B. S., and S. Myklestad. 1978. Structural studies of the reserve glucan produced by the marine diatom *Skeletonema costatum* (Grev.) Cleve. *Carbohydr. Res.* **62:**386–388.

Pfennig, N., F. Widdel, and H. G. Trüper. 1981. The dissimilatory sulfate-reducing bacteria, p. 926–940. *In* M. P. Starr, H. Stolp, H. G. Trüper, A. Balows and H. G. Schlegel (ed.), *The Prokaryotes,* vol. 1. Springer-Verlag, Berlin.

Platt, T., and W. K. W. Li (ed.). 1986. Photosynthetic picoplankton. *Can. Bull. Fish. Aquat. Sci.* **214:**1–583.

Richards, F. A., J. D. Cline, W. W. Broenkow, and L. P. Atkinson. 1965. Some consequences of the decomposition of organic matter in Lake Nitinat, an anoxic fjord. *Limnol. Oceanogr.* **10**(R):185–201.

Rudd, J. W. M., and C. D. Taylor. 1980. Methane cycling in the marine environment. *Adv. Aquat. Microbiol.* **2:**77–150.

Schink, D., J. J. McCarthy, T. Joyce, G. Flierl, P. Wiebe, and D. Kester. 1982. Multidisciplinary program to study warm core rings. *EOS Trans. Am. Geophys. Union* **63:**834–835.

Schropp, S. J., J. R. Schwartz, and P. A. LaRock. 1987a. Hydrogen production potential of fermentative microorganisms from the Sargasso Sea. *Geomicrobiol. J.* **5:**149–158.

Schropp, S. J., M. I. Scranton, and J. R. Schwartz. 1987b. Dissolved hydrogen, facultatively anaerobic, hydrogen producing bacteria, and potential hydrogen production rates in the western North Atlantic Ocean and Gulf of Mexico. *Limnol. Oceanogr.* **32:**396–402.

Schulenberger, E., and J. L. Reid. 1981. The Pacific shallow oxygen maximum, deep chlorophyll maximum and primary productivity reconsidered. *Deep-Sea Res.* **28A:**901–919.

Schuler, S., B. Thebrath, and R. Conrad. 1990. Seasonal changes in methane, hydrogen, and carbon monoxide concentrations in a large and a small lake, p. 503–510. *In* M. M. Tilzer and C. Serruya (ed.), *Large Lakes: Ecological Structure and Function.* Springer-Verlag, New York.

Scranton, M. I. 1984. Hydrogen cycling in the waters near Bermuda: the role of the nitrogen fixer *Oscillatoria thiebautii. Deep-Sea Res.* **31:**133–143.

Scranton, M. I., and P. G. Brewer. 1977. Occurrence of methane in the near-surface waters of the western subtropical North Atlantic. *Deep-Sea Res.* **234:**127–138.

Scranton, M. I., and J. W. Farrington. 1977. Methane production in waters off Walvis Bay. *J. Geophys. Res.* **82:**4947–4953.

Seiler, W., and U. Schmidt. 1974. Dissolved nonconservative gases in seawater, p. 219–243. *In* E. D. Goldberg (ed.), *The Sea,* vol. 5. John Wiley & Sons, Inc., New York.

Seliger, H. H., J. A. Boggs, and W. H. Bigglet. 1985. Catastrophic anoxia in the Chesapeake Bay in 1984. *Science* **228:**70–73.

Sieburth, J. McN. 1960a. Acrylic acid, an "antibiotic" principle in *Phaeocystis* blooms in Antarctic waters. *Science* **132:**676–677.

Sieburth, J. McN. 1960b. Soviet aquatic bacteriology: a review of the past decade. *Q. Rev. Biol.* **35:**179–205.

Sieburth, J. McN. 1971. Distribution and activity of oceanic bacteria. *Deep-Sea Res.* **18:**1111–1121.

Sieburth, J. McN. 1977. International Helgoland Symposium: Convenor's report on the informal session on biomass and productivity of microorganisms in planktonic ecosystems. *Helgoländer Wiss. Meerresunters.* **30:**697–704.

Sieburth, J. McN. 1979. *Sea Microbes.* Oxford University Press, New York.

Sieburth, J. McN. 1983. Microbiological and organic-chemical processes in the surface and mixed layers, p. 112–172. *In* P. S. Liss and W. G. N. Slinn (ed.), *Air-Sea Exchange of Gases and Particles.* D. Reidel Publishing Co., Dordrecht, The Netherlands.

Sieburth, J. McN. 1987a. Dominant microorganisms of the upper ocean: form and function, spatial distribution and photoregulation of biochemical processes, p. 173–187. *In* J. D. Burton, P. G. Brewer, and R. Chesselet (ed.), *Dynamic Processes in the Chemistry of the Upper Ocean.* Plenum Press, New York.

Sieburth, J. McN. 1987b. Contrary habitats for redox-specific processes: methanogenesis in oxic waters and oxidation in anoxic waters, p. 11–38. *In* M. A. Sleigh (ed.), *Microbes in the Sea*. Ellis Horwood Limited, Chichester, U.K., and John Wiley & Sons, Inc., New York.

Sieburth, J. McN. 1988a. The nanoalgal peak in the dim oceanic pycnocline: is it sustained by microparticulates and their bacterial consortia?, p. 101–130. *In* C. R. Agegian (ed.), *Biogeochemical Cycling and Fluxes between the Deep Euphotic Zone and Other Oceanic Realms*. National Undersea Research Program report 88-1. National Oceanic and Atmospheric Administration, Washington, D.C.

Sieburth, J. McN. 1988b. The trophic roles of bacteria in marine ecosystems are complicated by synergistic consortia and mixotrophic cometabolism. *Prog. Oceanogr.* **21**:117–128.

Sieburth, J. McN. 1991. The Pettaquamscutt Lakes, a captive estuary for studying global warming. *Maritimes* **35**(2):8–9, 17.

Sieburth, J. McN., and P. G. Davis. 1982. The role of heterotrophic nanoplankton in the grazing and nurturing of planktonic bacteria in the Sargasso and Caribbean Seas. *Ann. Inst. Oceanogr.* **58**(S):285–296.

Sieburth, J. McN., P. W. Johnson, M. E. Eberhardt, M. E. Sieracki, M. Lidstrom, and D. Laux. 1987. The first methane-oxidizing bacterium from the upper mixing layer of the deep ocean: *Methylomonas pelagica* sp. nov. *Curr. Microbiol.* **14**:285–293.

Sieburth, J. McN., P. W. Johnson, and P. E. Hargraves. 1988. Ultrastructure and ecology of *Aureococcus anopgefferens* gen. et sp. nov. (Chrysophyceae), the dominant picoplankter during a bloom in Narragansett Bay, Rhode Island, Summer 1985. *J. Phycol.* **24**:416–425.

Sieburth, J. McN., and M. D. Keller. 1991. Methylaminotrophic bacteria in xenic nanoalgal cultures: incidence, significance, and role of methylated algal osmoprotectants. *Biol. Oceanogr.* **6**:443–455.

Sieburth, J. McN., V. Smetacek, and J. Lenz. 1978. Pelagic ecosystem structure: heterotrophic compartments of the plankton and their relationship to plankton size fractions. *Limnol. Oceanogr.* **23**:1256–1263.

Sieburth, J. McN., P.-J. Willis, K. M. Johnson, C. M. Burnry, D. M. Lavoie, K. R. Hinga, D. A. Caron, F. W. French, III, P. W. Johnson, and P. G. Davis. 1976. Dissolved organic matter and heterotrophic microneuston in the surface microlayers of the North Atlantic. *Science* **194**:1415–1418.

Sieracki, M. E., P. W. Johnson, and J. McN. Sieburth. 1985. Detection, enumeration, and sizing of planktonic bacteria by image-epifluorescence microscopy. *Appl. Environ. Microbiol.* **49**:799–810.

Sieracki, M. E., and J. McN. Sieburth. 1985. Factors controlling the periodic fluctuation in total bacterial populations in the upper ocean: comparison of nutrient, sunlight and predation effects. *Mar. Microb. Food Webs* **1**:35–50.

Sieracki, M. E., and J. McN. Sieburth. 1986. Sunlight-induced growth delay of planktonic marine bacteria in filtered seawater. *Mar. Ecol. Prog. Ser.* **33**:19–27.

Smith, R. L., and R. S. Oremland. 1987. Big Soda Lake (Nevada). 2. Pelagic sulfate reduction. *Limnol. Oceanogr.* **32**:794–803.

Sochard, M. R., D. F. Wilson, B. Austin, and R. R. Colwell. 1979. Bacteria associated with the surface and gut of marine copepods. *Appl. Environ. Microbiol.* **37**:750–759.

Sowers, K. R., J. L. Johnson, and J. G. Ferry. 1984. Phylogenetic relationships among the methylotrophic methane-producing bacteria and emendation of the family *Methanosarcinaceae*. *Int. J. Syst. Bacteriol.* **34**:444–450.

Suttle, C. A., A. M. Chan, and T. Cottrell. 1990. Infection of phytoplankton by viruses and reduction of primary production. *Nature* (London) **347**:467–469.

Swinnerton, J. A., V. J. Linnebom, and C. H. Cheek. 1969. Distribution of CH_4 and CO between the atmosphere and natural waters. *Environ. Sci. Technol.* **3**:836–838.

Thurman, H. V., and H. H. Webber. 1984. *Marine Biology*. Charles E. Merrill Pub. Co., Columbus.

Tilton, R. C., A. B. Cobet, and G. E. Jones. 1967a. Marine thiobacilli. I. Isolation and distribution. *Can. J. Microbiol.* **13**:1521–1528.

Tilton, R. C., A. B. Cobet, and G. E. Jones. 1967b. Marine thiobacilli. II. Culture and ultrastructure. *Can. J. Microbiol.* **13**:1529–1534.

Traganza, E. D., J. W. Swinnerton, and C. H. Cheek. 1979. Methane supersaturation and ATP-zooplankton blooms in near surface waters of the western Mediterranean and the subtropical North Atlantic Ocean. *Deep-Sea Res.* **26:**1237–1245.

Trüper, H.-G., and S. Genovese. 1968. Characterization of photosynthetic sulphur bacteria causing red water in Lake Faro (Messina, Italy). *Limnol. Oceanogr.* **13:**225–232.

Tuttle, J. H., and H. W. Jannasch. 1972. Occurrence and types of thiobacillus-like bacteria in the sea. *Limnol. Oceanogr.* **17:**532–542.

Tyler, M. A., and H. H. Seliger. 1978. Annual subsurface transport of the red tide flagellate to its bloom area: water circulation patterns and organism distributions in the Chesapeake Bay. *Limnol. Oceanogr.* **23:**227–246.

Tyler, M. A., and H. H. Seliger. 1981. Selection for the red tide organism: physiological responses to the physical environment. *Limnol. Oceanogr.* **26:**310–324.

Widdel, F. 1980. Anaerober Abbau von Fettsaüren und Benzoesaüre durch neu isolierte Arten Sulfat-reduziender Bakterien. Doctoral Thesis. University of Göttingen, Göttingen, Germany.

Williams, R. T., and A. E. Bainbridge. 1973. Dissolved CO, CH_4, and H_2 in the southern ocean. *J. Geophys. Res.* **78:**2691–2694.

Winfrey, M. R., and J. G. Zeikus. 1979. Microbial methanogenesis and acetate metabolism in a meromictic lake. *Appl. Environ. Microbiol.* **37:**213–221.

Woodwell, G. M., R. H. Whittaker, W. A. Reiners, G. E. Likens, C. C. Delwiche, and D. B. Botkin. 1978. The biota and the world carbon budget. *Science* **199:**141–146.

Yancey, P. H., M. E. Clark, S. C. Hand, R. D. Bowlus, and G. N. Somero. 1982. Living with water stress: evolution of osmolyte systems. *Science* **217:**1214–1222.

Chapter 9

Biogenic Sources of Methane

Terry L. Miller

Two types of anaerobic microbial ecosystems produce significant amounts of CH_4 (Table 1). It is useful to distinguish between them. Intestinal tract ecosystems (e.g., the complex forestomach [rumen] of ruminants and the large intestine of humans) do not completely convert substrates to CH_4 and CO_2. They accumulate significant quantities of acetate, propionate, and butyrate. The predominant substrates for methanogens in these ecosystems are H_2 and CO_2. Other ecosystems, e.g., swamps, rice paddies, terrestrial and marine aquatic sediments, and anaerobic sewage digestion systems, are complete bioconversion systems. Acetate, propionate, butyrate, and H_2 and CO_2 are formed in initial stages. However, propionate and butyrate are converted to acetate and H_2 at later stages. Acetate, in addition to H_2 and CO_2, is a substrate for methanogens in complete bioconversion systems.

The principal natural substrates for both fermentations are polysaccharides, proteins, and lipids. Production of acetate, propionate, butyrate, H_2, and CO_2 requires the integrated activities of different microbial species of complex anaerobic microbial communities. The overall fermentations are not simple summations of the activities of individual species. The use of H_2 for the production of CH_4 has a pivotal influence on the activities of the nonmethanogenic organisms in these systems. A major difference between the two types of systems is the turnover time. Intestinal bioconversion systems have turnover times of approximately 1 to 2 days. In contrast, the turnover times of complete bioconversion systems are weeks to months. Microbial conversion of acetate, propionate, and butyrate to CH_4 and CO_2 requires organisms that have relatively long generation times. These organisms cannot be sustained in intestinal systems because of their slow rates of multiplication.

This review describes the characteristics of animal intestinal tract ecosystems and how methanogens alter the catabolic pathways of the resident nonmethanogenic microbes. The principles governing incomplete bioconversion that apply to complete bioconversion are presented. Finally, a strategy is proposed for reducing the contribution of rumen CH_4 to global CH_4.

Terry L. Miller • Wadsworth Center for Laboratories and Research, New York State Department of Health, Albany, New York 12201-0509.

Table 1. Incomplete and complete anaerobic bioconversion of hexose to CH_4

Incomplete (rumen):
$57.5\ C_6H_{12}O_6 \longrightarrow 65\ CH_3COOH + 20\ CH_3CH_2COOH + 15\ CH_3CH_2CH_2COOH + 35\ CH_4 + 60\ CO_2 + 25\ H_2O$
Complete (aquatic sediments, waste digestors):
$57.5\ C_6H_{12}O_6 \longrightarrow 172.5\ CH_4 + 172.5\ CO_2$

INTESTINAL FERMENTATIONS

The best example of intestinal fermentations occurs in the rumen (Hungate, 1966). A ruminant can be thought of as a fermentation factory. The animal ingests plant polymers (grasses, corn, silage, etc.) that are the raw material for fermentation in the rumen. The plant material is chewed, swallowed, and transported to the ruminant's complex stomach, called the reticulo-rumen, or more simply the rumen. A massive community of microorganisms in the rumen, including fungi, bacteria, and protozoa, ferment the plant polymers. The fermentation products are short-chain fatty acids—acetate, propionate, and butyrate—and the gases CH_4 and CO_2. The acids are absorbed into the bloodstream. They are the animal's principal sources of energy and carbon. Gases are waste products and are removed by belching. The fermentation provides nutrients and energy for growth of the microbial populations. Microorganisms and undigested food are semicontinuously removed from the rumen and pass to the lower part of the digestive tract. Digestive processes then occur that are similar to those of simple-stomached animals, like humans. Lower-tract digestion of rumen microorganisms provides the animal with its major sources of water-soluble vitamins and amino acids.

The rumen environment is anaerobic. The gas phase is essentially all CO_2 (65%) and CH_4 (27%), although traces of other gases (N_2, H_2, O_2) are present (Hungate, 1966). Large amounts of fermentation gases are produced. A 500-kg bovine eructates ca. 200 liters of CH_4 per day. Rumen liquid contents have a low oxidation/reduction potential of ca. −350 mV and a pH of ca. 6.5. The temperature is 39°C.

It is important to emphasize the semicontinuous nature of the fermentation system. A rumen of a 500-kg bovine has ca. 70 liters of liquid contents. During a single day, 100 liters of liquid enters the rumen in saliva and drinking water. To maintain a 70-liter volume, 100 liters has to leave the rumen each day.

Extracellular enzymes of the predominant microbial species hydrolyze insoluble plant polymers, primarily cellulose, hemicellulose, and starch. The hydrolysis products are soluble sugars. They enter the cells of the hydrolytic species and also enter cells of species that cannot hydrolyze the insoluble substrates. Fermentation of the hydrolysis products produces acetate, propionate, butyrate, H_2, and CO_2. Methanogens simultaneously use the H_2 to reduce CO_2 to CH_4. Only traces of H_2 (10^{-4} atm [ca. 0.01 kPa]) accumulate in the system (Hungate, 1966). The use of H_2 by methanogens influences the production of H_2 and other products by fermentative microorganisms.

Most rumen microorganisms use the Embden-Meyerhof-Parnas pathway to

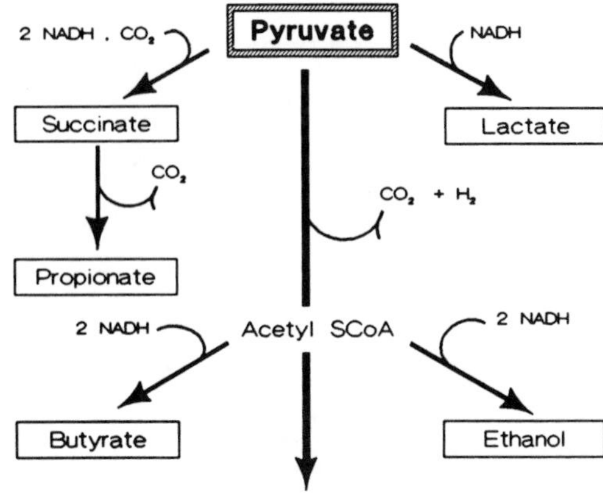

Figure 1. Oxidation of NADH and the formation of electron sink products by rumen microbes. The compounds in the open boxes are electron sink products formed by individual test tube cultures of rumen microbes.

oxidize sugar units to pyruvate (Wolin and Miller, 1988). At the same time, the intracellular electron carrier coenzyme, NAD, is reduced to NADH. NADH has to be reoxidized to NAD to continue the fermentation of sugars. Different species of rumen bacteria, fungi, and protozoa use different strategies for reoxidizing NADH (Fig. 1). These strategies ultimately result in the excretion of the electron sink fermentation products.

Some species have pathways that use pyruvate and NADH to form butyrate. Others use pyruvate and NADH to produce succinate, which is decarboxylated to propionate. Some species use pyruvate and NADH to make lactate, and others make ethanol. Lactate and ethanol are important products of fermentation by individual test tube cultures of major rumen species. However, they are not normally formed in the animal's forestomach. Almost all species oxidize pyruvate to acetyl coenzyme A (acetyl-SCoA), H_2, and CO_2. Acetyl-SCoA is converted to acetate and energy in the form of ATP. Acetate itself is not an electron sink product.

The use of H_2 for CH_4 production by methanogens alters product formation by the fermentative organisms. A predominant rumen cellulolytic species, *Ruminococcus albus*, produces ethanol, acetate, H_2, and CO_2 in monoculture (Fig. 2). Yet, it does not form ethanol in the rumen or when cocultured with a methanogen. *R. albus* produces H_2 by oxidizing pyruvate to acetyl-SCoA, H_2, and CO_2 (Miller and Wolin, 1973). Hydrogen does not inhibit the reaction. The organism also produces H_2 by oxidizing NADH to NAD and H_2. Hydrogen inhibits the reaction (Glass et al., 1977). When H_2 accumulates in monocultures, the organism does not produce H_2 from NADH. NADH is reoxidized by the enzymes used to produce ethanol from acetyl-SCoA. When H_2 is removed by methanogens, H_2 is produced from NADH. Acetyl-SCoA is completely transformed to acetate and ATP and is not reduced to ethanol. The maintenance of a low partial pressure of H_2 by methanogens increases the production of H_2, acetate, and ATP by *R. albus* and eliminates its production of ethanol.

Table 2 shows the products formed by *R. albus* grown on cellulose in

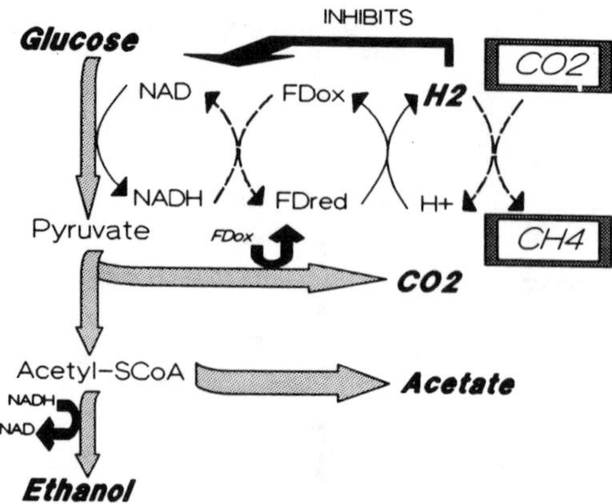

Figure 2. Fermentation products of *R. albus* in monoculture and in coculture with a methanogen. The products listed in bold print are produced when *R. albus* is grown in monoculture. In the presence of an H_2-using methanogen, the reactions indicated with broken arrows proceed and the coculture products are acetate, CO_2, and CH_4.

continuous culture by itself and in the presence of a methanogen (Pavlostathis et al., 1990). In the presence of the methanogen, acetate concentrations increase at the expense of ethanol. The increased amounts of H_2 production are reflected in the amounts of methane produced by the methanogen. Each mole of CH_4 produced requires the use of 4 mol of H_2.

Wolin and his colleagues called the interaction between *R. albus* and a methanogen "interspecies H_2 transfer" (Iannotti et al., 1973; Wolin, 1982). The transfer allows the coupling of oxidative reactions of fermentation pathways to reductive reactions of methanogenesis. Interspecies H_2 transfer eliminates the

Table 2. Cellulose fermentation products by *R. albus* monoculture and by *R. albus* + *Methanobrevibacter smithii* coculture[a]

Products	Amt produced (mol/100 mol of hexose used)	
	R. albus	*R. albus* + *M. smithii*
Acetate	89 ± 6	151 ± 10
Ethanol	81 ± 9	22 ± 2
Formate	14 ± 4	0
CO_2	156 ± 3	98 ± 7
CH_4	0	75 ± 5
H_2	140 ± 13	0[b]
Cells[c]	12 ± 1.6	12 ± 1.9
C-recovery	94.8 ± 2.3	95.8 ± 6.5
Oxidation/reduction	1.02 ± 0.08	0.96 ± 0.03

[a] Data are from Pavlostathis et al. (1990).

[b] Total H_2 produced is calculated from: $4\ H_2 + CO_2 \rightarrow CH_4 + 2\ H_2O$. The amount of H_2 produced and used in the coculture = ($4 \times CH_4$) = 300 mol/100 mol of hexose used.

[c] *R. albus* cells only.

production of electron sink products formed during the reoxidation of NADH in monocultures of the fermenting species. Other fermentative rumen microorganisms that have different reactions for reoxidizing NADH can participate in interspecies H_2 transfer interactions with methanogens. Organisms that produce lactate, propionate, and succinate in monoculture produce acetate when cocultured with methanogens (for more detail see Wolin [1982] and Wolin and Miller [1988]).

The rumen fermentation exemplifies how intestinal microbial ecosystems make CH_4 and how it influences the activities of the fermentative bacteria. Production of CH_4 is ubiquitous in ruminants such as bovines, sheep, and deer (Hungate, 1966). Although not as well studied, similar fermentations occur in the forestomachs of nonruminants such as kangaroos and elephants and in monkeys that have complex forestomachs (Bauchop, 1971; Hungate, 1966). A foregut fermentation was recently described in the hoatzin, a subtropical leaf-eating bird (Grajal et al., 1989).

Although methanogenesis in ruminants has tremendous economic importance, information about the species of methanogens in the rumen is incomplete. *Methanobrevibacter ruminantium* (formerly *Methanobacterium ruminantium*) was the first methanogen isolated from the bovine rumen (Smith and Hungate, 1958). It was present in high numbers and formed CH_4 from H_2 and CO_2. It requires 2-mercaptoethanesulfonic acid (coenzyme M) for growth. Coenzyme M-requiring and nonrequiring strains of *Methanobrevibacter* species were isolated from bovine rumen contents (Lovley et al., 1984; Miller et al., 1986a). They did not react with rabbit antisera raised against *Methanobrevibacter ruminantium*. There is a single report of the isolation of *Methanomicrobium mobile* from high dilutions of rumen contents of a cow (Paynter and Hungate, 1968). *Methanosarcina* species are present in low concentrations (10^4/ml of rumen contents) in the rumen of cows, where they grow on methylamines formed from choline (Patterson and Hespell, 1979). All information is based on studies of a small number of bovines.

Large forestomachs support a microbial fermentation of ingested plants that precedes host digestion in the small intestine. In simple-stomached animals like humans, pigs, horses, and rodents, host digestion precedes microbial fermentation of dietary plant substrates. Microorganisms in the colon or large intestine ferment the undigested and nonabsorbed material that exits the small intestine.

The human colon illustrates the general features of these habitats (Fig. 3). Like all intestinal fermentation systems, it is anaerobic. Residues of host digestion enter the colon from the ileum, the terminal portion of the small intestine. The major dietary substrates available for colonic fermentation are the plant cell wall components cellulose and hemicellulose, starch, and pectin. Some oligosaccharides such as stachyose in beans, trehalose in mushrooms and, in some individuals, lactose are not digested in the small intestine. They are fermented by colonic bacteria. Almost any organic substance in food that is not digested by enzymes in the mouth, stomach, or small intestine or is not absorbed before reaching the colon is fermented in the colon. Gastroenterologists call substrates fermented in the colon "malabsorbed substrates."

The colonic fermentation is accomplished by a complex community of different species of bacteria. In contrast to ruminal fermentations, protozoa and fungi are not

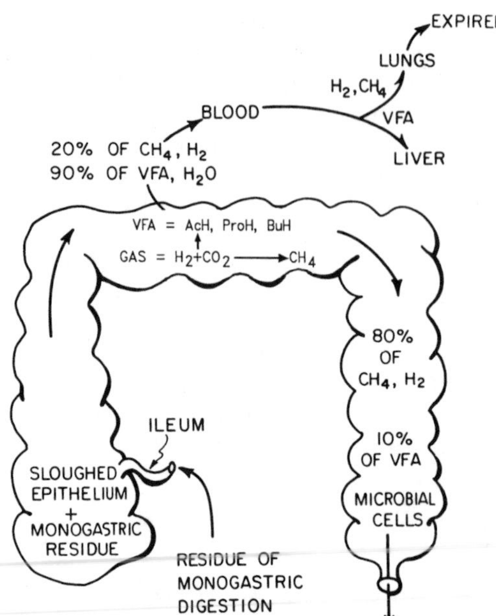

Figure 3. Schematic of the human colonic fermentation. VFA, volatile fatty acids; AcH, acetate; ProH, propionate; BuH, butyrate.

involved. Polysaccharides and oligosaccharides are fermented by pathways similar to those that occur in the rumen. The end products of fermentation are acetate, propionate, and butyrate and the gases H_2 and CO_2 (Wolin and Miller, 1983).

The molar proportions of acetate:propionate:butyrate are similar to those produced in the rumen (Weaver et al., 1988, 1989; Wolin, 1981). Most of the volatile fatty acids are absorbed into the blood and metabolized in tissues. Gastroenterologists call this fermentation a "salvage fermentation"; i.e., the host uses the acid fermentation products as sources of carbon and energy. Humans and pigs salvage about 5 to 10% of diet energy (McNeil, 1984). Rabbits and ponies salvage about 40% of diet energy (McNeil, 1984). When methanogens are present, they use fermentation H_2 to reduce CO_2 to CH_4. Unlike the ruminant, most of the H_2 and CH_4 exit the body as flatus. About 20% is absorbed into the bloodstream, transported to the lungs, and excreted in breath.

Methanogenic fermentations are important in only some humans (Miller and Wolin, 1986; Weaver et al., 1986). The colonic concentrations of methanogens vary over a range of 10 logs depending on the subject. Some individuals have only a few methanogens per gram dry weight (gdw) of feces and some have 10^{10}/gdw. About one-third of the population have 10^8 to 10^{10} methanogens per gdw. The latter concentrations are sufficient to produce between 30 ml and 3 liters of CH_4 per day (Miller and Wolin, 1986). Only about 10% of humans on a Western diet produce liter amounts of CH_4. The equation shown in Table 3 describes the methanogenic fermentation in humans. The methanogens influence the acetate production by the fermentative bacteria.

Table 3. Methanogenic and nonmethanogenic human colonic fermentations

Methanogenic	
230 Glucose ⟶	230 acetate + 77 propionate + 77 butyrate + 133 CH_4 + 248 CO_2 + 116 H_2O
Nonmethanogenic	
185 Glucose ⟶	230 acetate + 76 propionate + 77 butyrate + 115 CO_2 + 115 H_2O

There is scant information on the frequency of significant methanogenic fermentations in the colon of species other than *Homo sapiens* and pigs (Christensen and Thorbek, 1987; Miller and Wolin, 1986). It occurs in some strains of rats but not others and in some species of termites but not others (Breznak and Switzer, 1986; Lajoie et al., 1988).

As in forestomach fermentations, the predominant colonic methanogens are those that use H_2 to reduce CO_2 to CH_4 (Miller and Wolin, 1986; Miller et al., 1986b). Methanogens were isolated from feces of humans and other animals including the rat, horse, cow, pig, monkey, baboon, rhinoceros, hippopotamus, giant panda, goose, turkey, and chicken. *Methanobrevibacter* is a major intestinal genus. Species of this genus are found in a variety of domestic and wild animals (Miller and Wolin, 1986). Although *Methanogenium* species were isolated from a turkey and a chicken, a *Methanobrevibacter* species was isolated from a goose (Miller et al., 1986b). *Methanobrevibacter smithii* is the numerically dominant methanogen in the human large bowel (Weaver et al., 1986; Miller and Wolin, 1986). *Methanosphaera stadtmanae* is present in lower concentrations than *Methanobrevibacter smithii* (Miller and Wolin, 1983, 1985). It is unique, however, because it makes CH_4 only by using H_2 to reduce methanol. Methanol is produced in the colon by nonmethanogenic bacteria that hydrolyze pectin and other methylated plant polysaccharides to methanol.

Table 4 shows the differences in the magnitude of CH_4 production in domestic ruminants, pigs, and humans. Ruminant CH_4 production is universal (Hungate, 1966). Significant CH_4 production occurs only in a few humans who have high

Table 4. Comparison of daily intestinal methane production

Species	Body wt (kg)	Amt produced (liters/day)	
		CH_4	H_2 used[a]
Cow[b]	500	200	800
Sheep[c]	40	30	120
Pig[d]	100	8	32
Humans[e]	50	3	12

[a] Calculated according to $4\ H_2 + CO_2 \rightarrow CH_4 + 2\ H_2O$.
[b] From data in Kleiber et al. (1945).
[c] From data in Hofmeyer et al. (1984) and Murray et al. (1976).
[d] From data in Christensen and Thorbek (1987).
[e] From calculations in Miller and Wolin (1986).

Table 5. Estimate of intestinal CH_4 production from livestock in the United States in 1989[a]

Livestock	No. of animals	Methane production per year		
		Liters[b]	mol	g
Dairy cows	1.03×10^7	7.52×10^{11}	3.35×10^{10}	5.35×10^{11}
Cattle	9.95×10^7	7.26×10^{12}	3.24×10^{11}	5.19×10^{12}
Sheep	1.08×10^7	7.88×10^{10}	3.52×10^9	5.63×10^{10}
Swine	5.25×10^7	9.56×10^{10}	4.27×10^9	6.84×10^{10}
Total		8.19×10^{12}	3.65×10^{11}	5.85×10^{12}

[a] The livestock numbers are from *Information Please Almanac*, 43rd ed., Houghton Mifflin Co., Boston, 1990, p. 67.
[b] To account for possible differences in age and diet variation, the daily CH_4 production values in Table 4 for sheep and pigs were revised downward to 20 and 5 liters of CH_4 per day per animal, respectively.

concentrations of methanogens in their colons. A rough estimate is that 10% of adult humans produce 1 to 3 liters per day. The daily CH_4 production values for cattle, sheep, and pigs given in Table 4 are from direct in vivo experiments on a small number of animals. Very little is known about the variability of livestock CH_4 production as a function of diet or age. Kleiber et al. (1945) reported that lactating cows on a production diet produced 230 liters of CH_4 per day while dry cows on a maintenance diet produced 180 liters of CH_4 per day. Studies with pigs showed that colonic CH_4 production increases with age (Christensen and Thorbek, 1987). Despite the limitations in our knowledge of the daily CH_4 production in individual animals, domestic livestock are clearly a significant biogenic source of global CH_4 (Table 5).

COMPLETE BIOCONVERSION ECOSYSTEMS

In complete bioconversion systems, long turnover times allow the conversion of acetate, propionate, and butyrate to CH_4 and CO_2. Conversion of the acids to CH_4 requires additional microbial processes. Acetate is converted to CH_4 by methanogens that produce CH_4 and CO_2 from the acid (Jones et al., 1987). Propionate and butyrate are converted to acetate and H_2 by organisms that grow only when the H_2 is used by methanogens. Obligate interspecies H_2 transfer is called syntrophy. The reactions that yield energy for the growth of both species require coordination of the production of H_2 from the acids with H_2 utilization by methanogens.

The first recognition of syntrophic methanogenesis occurred when Bryant et al. (1967) discovered that two different species were required for the conversion of ethanol and CO_2 to acetate and CH_4. A bacterium designated as "S-organism" grows poorly and produces small amounts of H_2 and acetate with ethanol as an energy source. When it is cocultured with a methanogen, both organisms grow well and produce large amounts of CH_4 and acetate. Table 6 shows the equations and free energy changes for the reactions.

Table 6. Interspecies H_2 transfer between the S-organism and a methanogen

Culture			$\Delta G_0'$ (kJ/reaction)[a]
S-organism monoculture:			
$C_2H_5OH + NAD^+$	$\longrightarrow$	$CH_3CHO + NADH + H^+$	+ 23.8
$NADH + H^+$	$\longrightarrow$	$NAD^+ + H_2$	+ 18.0
$CH_3CHO + H_2O$	$\longrightarrow$	$CH_3COO^- + H_2 + H^+$	− 32.2
Sum: $C_2H_5OH + H_2O$	$\longrightarrow$	$CH_3COO^- + 2\,H_2 + H^+$	+ 9.6
Methanogen monoculture:			
$4\,H_2 + H^+ + HCO_3^-$	$\longrightarrow$	$CH_4 + 3\,H_2O$	− 135.6
S-organism + methanogen coculture:			
$2\,C_2H_5OH + 2\,H_2O$	$\longrightarrow$	$2\,CH_3COO^- + 4\,H_2 + 2\,H^+$	+ 19.2
$4\,H_2 + H^+ + HCO_3^-$	$\longrightarrow$	$CH_4 + 3\,H_2O$	− 135.6
Sum: $2\,C_2H_5OH + HCO_3^-$	$\longrightarrow$	$2\,CH_3COO^- + CH_4 + H^+ + H_2O$	− 116.4

[a] From Thauer et al. (1977).

The common feature of syntrophic associations is the thermodynamic barrier to the reduction of protons to H_2. It is the only reductive reaction the S-organism uses to oxidize ethanol. The barrier is overcome by coupling the formation of H_2 to the reduction of CO_2 to CH_4. The free energy change for the oxidation of NADH to NAD and H_2 for standard concentrations of reactants and products at pH 7 is positive. The standard concentration of H_2 is 1 atmosphere (ca. 101.3 kPa). Syntrophic methanogenesis reduces the partial pressure of H_2 to a very low value. The free energy of the reaction is lowered to a negative, thermodynamically favorable value in the syntrophic coculture.

Figure 4 shows the influence of the partial pressure of H_2 on the free energy changes for the syntrophic production of acetate and H_2 from ethanol, propionate, and butyrate. The partial pressure of H_2 must be reduced to below 10^{-4} atm (ca.

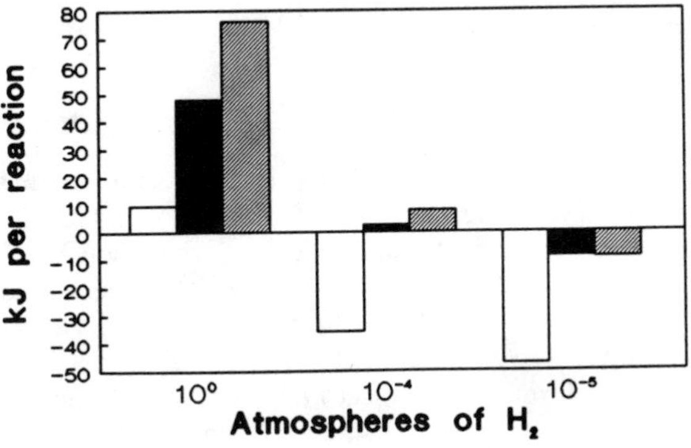

Figure 4. Effect of the partial pressure of H_2 on the free energy ($\Delta G_0'$) for conversion to acetate and H_2. Partial pressure is in atmospheres. The $\Delta G_0'$ is in kilojoules (kJ) per reaction.

0.01 kPa) to obtain a negative free energy for the oxidation of propionate or butyrate to acetate. Besides alcohols and organic acids, syntrophic interspecies H_2 transfer was demonstrated with pure culture systems with organisms that produce H_2 from aromatic compounds, sugars, and amino acids. Syntrophic methanogenesis is important for the conversion of many organic compounds to acetate in complete bioconversion ecosystems (see Wolin and Miller [1986] for a review).

The microorganisms that oxidize butyrate and propionate to acetate and H_2 grow very slowly even in the presence of methanogens. The generation time for *Syntrophomonas wolfei* growing on butyrate is ca. 3 days (McInerney et al., 1981). The generation time for *Syntrophobacter wolinii* growing on propionate is ca. 7 days (Boone and Bryant, 1980). Therefore, the complete bioconversion of these substrates requires ecosystems with long turnover times. The conversion of acetate to CH_4 and CO_2 completes the bioconversion process. Acetate mineralization is accomplished by those special methanogens that can produce CH_4 from the methyl group of acetate (Jones et al., 1987). Acetate is the precursor of about 67% of the CH_4 produced in complete bioconversion ecosystems.

POSSIBLE STRATEGIES FOR REDUCTION OF BIOGENIC CH_4

It is not practical to consider strategies for diminishing the CH_4 produced in complete bioconversion systems. However, current knowledge of colonic fermentations suggests possible strategies for reducing the contribution from agriculturally important livestock.

Most humans do not produce large amounts of CH_4. Recent studies show that the kinds and amounts of acids produced by the methanogenic and nonmethanogenic intestinal fermentations are very similar (Weaver et al., 1989; Wolin, 1981). Table 3 shows the equation that describes the nonmethanogenic colonic fermentation of most humans on a Western mixed diet. Studies of the nonmethanogenic fermentations in humans, rats, and termites suggest that methanogenic and nonmethanogenic fermentations are the same in all general features except one. In the methanogenic fermentation, H_2 produced by nonmethanogenic species is used to reduce CO_2 to CH_4. In the nonmethanogenic fermentation, H_2 is used to reduce CO_2 to acetate. This is accomplished by bacteria that carry out the distinctive reactions of the Wood-Ljungdahl pathway for reducing CO_2 to acetate (Ljungdahl, 1986). The overall reaction is:

$$4\,H_2 + 2\,CO_2 \rightarrow CH_3COOH + 2\,H_2O$$

Lajoie et al. (1988) first provided evidence in support of the hypothesis by Wolin and Miller (1983) that reduction of CO_2 is a major pathway of acetate production in nonmethanogenic human colons. Table 7 shows that the fecal suspensions from humans who do not harbor significant concentrations of methanogens have a fecal flora that uses H_2 to reduce CO_2 to acetate. Subject B harbors high concentrations of methanogens (10^{10}/gdw of feces), and most of the H_2 is used for CH_4 formation. Subject J does not harbor methanogens, and all the H_2 is used for acetate formation. Subject D has moderate concentrations of methanogens

Table 7. Formation of ^{13}C-labeled acetate from $^{13}CO_2$ and H_2 by human fecal microbes[a]

Subject	Total (μmol)					Actual/ theoretical (%)
	CH_4	[^{13}C]acetate		H_2 used		
		Single + double	Double	Actual	Theoretical[b]	
B	782	0	0	3,252	3,128	104
D	24	71	45	376	380	99
J	0	56	37	248	224	111

[a] Data are from Lajoie et al. (1988). Fecal suspensions (5 ml, 10% wt/vol) were incubated with 80% H_2–20% CO_2 for 24 h. Quantitative nuclear magnetic resonance spectroscopy was used to determine the labeled carbon positions in acetate.

[b] Theoretical H_2 used for CH_4 and acetate formation is calculated from: $4\ H_2 + CO_2 \rightarrow CH_4 + 2\ H_2O$ and $4\ H_2 + 2\ CO_2 \rightarrow CH_3COOH + 2\ H_2O$.

(10^8/gdw of feces), and H_2 is used for CH_4 formation as well as CO_2 reduction to acetate. Lajoie et al. (1988) used ^{13}C-labeled CO_2 to confirm that both carbons of acetate originated from CO_2 in subjects J and D (Table 7). Similar incubations with plant polysaccharides and labeled CO_2 show that CO_2 reduction accounts for 35% of the total acetate produced during fermentation when methanogens are present in low concentrations ($< 10^8$/gdw of feces; unpublished data).

CO_2 reduction to acetate occurs in nonmethanogenic rats, guinea pigs, and rabbits (Lajoie et al., 1988; Prins and Lankhorst, 1977). Breznak and Switzer (1986) showed that significant amounts of acetate, rather than CH_4, are formed from CO_2 in lower and higher termite guts. It is likely that CO_2 reduction to acetate is a common pathway for acetate formation in many other colonic ecosystems.

It may be useful to classify intestinal tract fermentations as CO_2-CH_4 fermentations and CO_2-acetate fermentations. Reduction of CO_2 to CH_4 is a major feature of the CO_2-CH_4 fermentation of ruminants, other forestomached animals, and some colonic ecosystems. CO_2 reduction to CH_4 is negligible in the colonic CO_2-acetate fermentation.

Conversion of the CO_2-CH_4 fermentation of ruminants to a CO_2-acetate fermentation would simultaneously benefit agricultural production and significantly diminish the contribution of ruminants to global CH_4 formation. CH_4 production by cattle and sheep represents a loss of carbon and energy to the animal. The fermentation product enters the atmosphere by eructation. Elimination of CH_4 production has been a goal of major animal nutrition research programs. Conversion of CO_2 to acetate instead of CH_4 would increase the microbial production of a rumen fermentation product that is absorbed into the blood and used as a major source of carbon and energy by cattle and sheep.

Biotechnology could be applied to replacing the methanogenic microbial community of the rumen with a community that produces the CO_2-acetate fermentation. A new CO_2-using acetogen, *Acetitomaculum ruminis*, was isolated from 10^{-7} to 10^{-8} ml of bovine rumen contents (Greening and Leedle, 1989). Earlier studies showed that concentrations of CO_2-reducing acetogenic bacteria in the bovine rumen were similar to those of methanogens (Leedle and Greening, 1988). Genthner et al. (1981) isolated *Eubacterium limosum*, a CO_2-reducing acetogen, from the rumen of a sheep fed a high-molasses diet. These results suggest that the microbial potential for a CO_2-acetate fermentation already exists in the rumen. Research directed toward devising methods for effecting the shift to a CO_2-acetate fermentation could produce practical methods for increasing the efficiency of animal production and decreasing the contribution of animals to the production of global CH_4.

REFERENCES

Bauchop, T. 1971. Stomach microbiology of primates. *Annu. Rev. Microbiol.* **25:**429–436.

Boone, D. R., and M. P. Bryant. 1980. Propionate-degrading bacterium, *Syntrophobacter wolinii* sp. nov. gen. nov., from methanogenic ecosystems. *Appl. Environ. Microbiol.* **40:**626–632.

Breznak, J. A., and J. M. Switzer. 1986. Acetate synthesis from H_2 and CO_2 by termite microbes. *Appl. Environ. Microbiol.* **52:**623–630.

Bryant, M. P., E. A. Wolin, M. J. Wolin, and R. S. Wolfe. 1967. *Methanobacillus omelianskii*, a symbiotic association of two species of bacteria. *Arch. Mikrobiol.* **59:**20–31.

Christensen, K., and G. Thorbek. 1987. Methane excretion in the growing pig. *Br. J. Nutr.* **57:**355–361.

Genthner, B. R. S., C. L. Davis, and M. P. Bryant. 1981. Features of rumen and sewage sludge strains of *Eubacterium limosum*, a methanol- and H_2-CO_2-utilizing species. *Appl. Environ. Microbiol.* **42:**12–19.

Glass, T. L., M. P. Bryant, and M. J. Wolin. 1977. Partial purification of ferredoxin from *Ruminococcus albus* and its role in pyruvate metabolism and reduction of nicotinamide adenine dinucleotide by H_2. *J. Bacteriol.* **131:**463–472.

Grajal, A., S. D. Strahl, R. Parra, M. G. Dominguez, and A. Neher. 1989. Foregut fermentation in the Hoatzin, a neotropical leaf-eating bird. *Science* **245:**1236–1238.

Greening, R. C., and J. A. Z. Leedle. 1989. Enrichment and isolation of *Acetitomaculum ruminis*, gen. nov., sp. nov.: acetogenic bacteria from the bovine rumen. *Arch. Microbiol.* **151:**399–406.

Hofmeyer, H. S., N. Slabbert, and J. P. Pienaar. 1984. Partitioning of methane production between ruminal and hindgut fermentation in sheep. *Can. J. Animal Sci.* **64:**171–172.

Hungate, R. E. 1966. *The Rumen and Its Microbes.* Academic Press, Inc., New York.

Iannotti, E. L., D. Kafkewitz, M. J. Wolin, and M. P. Bryant. 1973. Glucose fermentation products of *Ruminococcus albus* grown in continuous culture with *Vibrio succinogenes*: changes caused by interspecies transfer of H_2. *J. Bacteriol.* **114:**1231–1240.

Jones, W. L., D. P. Nagle, Jr., and W. B. Whitman. 1987. Methanogens and the diversity of archaebacteria. *Microbiol. Rev.* **51:**135–177.

Kleiber, M., W. M. Regan, and S. W. Mead. 1945. Measuring food values with dairy cows. *Hilgardia* **16:**511–571.

Lajoie, S. F., S. Bank, T. L. Miller, and M. J. Wolin. 1988. Acetate production from hydrogen and [^{13}C]carbon dioxide by the microflora of human feces. *Appl. Environ. Microbiol.* **54:**2723–2727.

Leedle, J. A. Z., and R. C. Greening. 1988. Methanogenic and acidogenic bacteria in the bovine rumen: postprandial changes after feeding high- or low-forage diets once daily. *Appl. Environ. Microbiol.* **54:**502–506.

Ljungdahl, L. G. 1986. The autotrophic pathway of acetate synthesis in acetogenic bacteria. *Annu. Rev. Microbiol.* **40:**415–450.

Lovley, D. R., R. C. Greening, and J. G. Ferry. 1984. Rapidly growing rumen methanogenic organism that synthesizes coenzyme M and has a high affinity for formate. *Appl. Environ. Microbiol.* **48:**81–87.

McInerney, M. J., M. P. Bryant, R. B. Hespell, and J. W. Costerton. 1981. *Syntrophomonas* *wolfei* gen. nov., sp. nov., an anaerobic, syntrophic, fatty acid-oxidizing bacterium. *Appl. Environ. Microbiol.* **41:**1029–1039.

McNeil, N. I. 1984. The contribution of the large intestine to energy supplies in man. *Am. J. Clin. Nutr.* **39:**338–342.

Miller, T. L., and M. J. Wolin. 1973. Formation of hydrogen and formate by *Ruminococcus albus. J. Bacteriol.* **116:**836–846.

Miller, T. L., and M. J. Wolin. 1983. Oxidation of hydrogen and reduction of methanol as the sole energy source for a methanogen isolated from human feces. *J. Bacteriol.* **153:**1051–1055.

Miller, T. L., and M. J. Wolin. 1985. *Methanosphaera stadtmaniae* gen. nov., sp. nov.: a species that forms methane by reducing methanol with hydrogen. *Arch. Microbiol.* **141:**116–122.

Miller, T. L., and M. J. Wolin. 1986. Methanogens in human and animal intestinal tracts. *Syst. Appl. Microbiol.* **7:**223–229.

Miller, T. L., M. J. Wolin, Z. Hongxue, and M. P. Bryant. 1986a. Characteristics of methanogens isolated from bovine rumen. *Appl. Environ. Microbiol.* **51:**201–202.

Miller, T. L., M. J. Wolin, and E. A. Kusel. 1986b. Isolation and characterization of methanogens from animal feces. *Syst. Appl. Microbiol.* **8:**234–238.

Murray, R. M., A. M. Bryant, and R. A. Leng. 1976. Rates of production of methane in the rumen and large intestine of sheep. *Br. J. Nutr.* **36:**1–14.

Patterson, J. A., and R. B. Hespell. 1979. Trimethylamine and methylamine as growth substrates for rumen bacteria and *Methanosarcina barkeri. Curr. Microbiol.* **3:**79–83.

Pavlostathis, S. G., T. L. Miller, and M. J. Wolin. 1990. Cellulose fermentation by continuous cultures of *Ruminococcus albus* and *Methanobrevibacter smithii. Appl. Microbiol. Biotechnol.* **33:**109–116.

Paynter, M. J. B., and R. E. Hungate. 1968. Characterization of *Methanobacterium mobilis*, sp. n., isolated from the bovine rumen. *J. Bacteriol.* **95:**1943–1951.

Prins, R. A., and A. Lankhorst. 1977. Synthesis of acetate from CO_2 in the cecum of some rodents. *FEMS Microbiol. Lett.* **1:**255–258.

Smith, P. H., and R. E. Hungate. 1958. Isolation and characterization of *Methanobacterium ruminantium* n. sp. *J. Bacteriol.* **75:**713–718.

Thauer, R. K., K. Jungermann, and K. Decker. 1977. Energy conservation in chemotrophic anaerobic bacteria. *Bacteriol. Rev.* **41:**100–180.

Weaver, G. A., J. A. Krause, T. L. Miller, and M. J. Wolin. 1986. Incidence of methanogenic bacteria in a sigmoidoscopy population: an association of methanogenic bacteria and diverticulosis. *Gut* **27:**698–704.

Weaver, G. A., J. A. Krause, T. L. Miller, and M. J. Wolin. 1988. Short chain fatty acid distributions of enema samples from a sigmoidoscopy population: an association of high acetate and low butyrate ratios with adenomatous polyps and colon cancer. *Gut* **29:**1539–1543.

Weaver, G. A., J. A. Krause, T. L. Miller, and M. J. Wolin. 1989. Constancy of glucose and starch fermentations by two different human faecal microbial communities. *Gut* **30:**1–19.

Wolin, M. J. 1981. Fermentation in the rumen and human large intestine. *Science* **213:**1463–1468.

Wolin, M. J. 1982. Hydrogen transfer in microbial communities, p. 323–356. *In* A. T. Bull and J. H. Slater (ed.), *Microbial Interactions and Communities*, vol. 1. Academic Press, Inc. (London), Ltd., London.

Wolin, M. J., and T. L. Miller. 1983. Carbohydrate fermentation, p. 147–165. *In* D. J. Hentges (ed.), *Human Intestinal Microflora in Health and Disease*. Academic Press, Inc., New York.

Wolin, M. J., and T. L. Miller. 1986. Bioconversion of organic carbon to CH_4 and CO_2. *Geomicrobiol. J.* **5:**239–259.

Wolin, M. J., and T. L. Miller. 1988. Microbe-microbe interactions, p. 343–359. *In* P. N. Hobson (ed.), *The Rumen Microbial Ecosystem*. Elsevier Applied Science, London.

Chapter 10

Physiology of Nitrifying and Denitrifying Bacteria

Lesley A. Robertson and J. Gijs Kuenen

INTRODUCTION

Nitrification (the oxidation of reduced nitrogen compounds, generally ammonia, to nitrite, nitrate, or other nitrogen oxides) and denitrification (the reduction of oxidized nitrogen compounds to gases, especially but not always N_2) are two of the dominant driving forces in the biological nitrogen cycle. Thus far, there are only a limited number of species known to be able to grow at the expense of energy generated by nitrification, the best studied of which are members of the genera *Nitrosomonas* (which oxidize ammonia to nitrite via hydroxylamine) and *Nitrobacter* (which oxidize nitrite to nitrate). These gram-negative bacteria are obligate chemolithoautotrophs and depend on CO_2 fixation for biosynthetic carbon. In denitrification, the nitrogen oxides serve as terminal electron acceptors rather than donors, and the denitrifiers are considerably more diverse, as is illustrated by the examples shown in Table 1. There are both gram-negative and gram-positive denitrifiers, with representatives from most physiological (e.g., from obligate chemolithoautotrophs to chemoheterotrophs) and morphological (e.g., from spirals through rods to cocci) types. The complete reduction of nitrate proceeds via nitrite, nitric oxide, and nitrous oxide, but not all denitrifiers can carry out the complete reduction from nitrate to N_2; the enzymes most commonly missing are nitrate reductase or nitrous oxide reductase.

Because of their respective natural roles in returning dinitrogen originally fixed in (organic) nitrogenous compounds by nitrogen-fixing bacteria to the atmosphere, and also because of their importance in nitrogen removal in today's wastewater treatment systems, both the nitrifiers and the denitrifiers have been the subject of considerable research effort over the years, both in the laboratory and in the field. This research has been extensively reviewed (Stouthamer, 1988; Payne, 1981; Kuenen and Robertson, 1987; Winkler, 1981; Wood, 1986). An examination of the history of research into the nitrogen cycle, especially where denitrification and nitrification are concerned, reveals a number of difficulties and sometimes even controversies in the interpretation of data, the resolution of which has required the

---

Lesley A. Robertson and J. Gijs Kuenen • Kluyver Laboratory for Biotechnology, Delft University of Technology, Julianalaan 67, 2628 BC Delft, The Netherlands.

Table 1. Examples of denitrifying reactions and some of the bacteria which employ them[a]

Reaction	Denitrifying species
Nitrate to nitrite	*Thiobacillus thioparus, Lysobacter antibioticum*
Nitrate to nitrous oxide	*Aquaspirillum itersonii*, pseudomonads
Nitrate to nitrogen	*Paracoccus denitrificans, Thiobacillus denitrificans, Alcaligenes eutropha, Hyphomicrobium* sp., *Halobacterium* sp.
Nitrite to nitrogen	*Neisseria* and *Flavobacterium* spp., *Alcaligenes faecalis* TUD
Nitrous oxide to nitrogen	*Vibrio succinogenes*

[a] Adapted from Kuenen and Robertson, 1987.

controlled experimentation possible only in the laboratory. Both batch and continuous culture techniques are used during the work. While most people are familiar with some aspects of batch cultures, this may not be the case for continuous cultures (also known as chemostats), so a brief discussion of the merits of the technique is therefore appropriate here.

Culture Techniques

Batch culture methods can provide a great deal of valuable information about the physiological capabilities of different bacteria. However, they are generally uncontrolled, and factors such as pH, O_2, substrate, and product concentrations all change continuously as the cultures grow. It is thus not always clear which particular environmental parameter is effective. Moreover, to obtain reasonable cell densities, all of the nutrients are provided in excess and at relatively high concentrations at the start of the experiment. It is rarely that such a situation will occur in nature. Chemostats, on the other hand, provide a means of maintaining suspended cultures under defined conditions and at specified growth rates (Herbert et al., 1956; Veldkamp and Jannasch, 1972; Kuenen and Harder, 1982). Medium is continuously pumped into the culture vessel, and a constant working volume is maintained by the simultaneous removal of culture, usually by means of an overflow or a pump. After a period of adaption, a steady state is reached during which all of the cells in the chemostat are growing at a growth rate equivalent to the dilution rate. This allows precisely defined and reproducible experimentation with cultures whose growth is limited by a known factor (usually the substrate concentration). It is of paramount importance that in a steady state, the rate of metabolism of the limiting nutrient is sufficiently high to keep the concentration of that nutrient very low in the culture. Low concentrations are, in general, more relevant to the behavior of bacteria in (semi)natural environments. Indeed, it has been shown that the physiological and competitive behavior of bacteria in the presence of very low concentrations of substrate differs from that observed in batch (nutrient excess) cultures. For example, growth on mixed substrates generally gives rise to diauxy in batch cultures, but mixotrophy in the chemostat (see, for example, Kelly and Kuenen, 1984). In addition, as the environmental parameters which may cause problems during batch culture (e.g., pH, or nutrient or product concentration) do not change during steady-state growth, and can even be controlled, the physiological status of the cells is constant.

This chapter will illustrate how data from laboratory experiments have shed new (or different) light on various aspects of the nitrogen cycle. Some of the controversies which may be of importance in our understanding of the generation of NO and N_2O will also be discussed.

SIMULTANEOUS NITRIFICATION AND DENITRIFICATION

Aerobic Denitrification

For over 100 years, the occurrence of "aerobic denitrification" (i.e., active denitrification in the presence of significant amounts of oxygen) has been a matter of dispute. Despite regular reports of aerobic denitrification, the difficulties of making accurate measurements of parameters such as dissolved oxygen and gas production made the production of convincing results difficult (for a review of the historical arguments see Robertson and Kuenen, 1984a, 1990). However, the availability of modern oxygen- and ion-specific electrodes, culture vessels, and analytical techniques has made it possible to establish that a group of denitrifiers are able to simultaneously utilize oxygen and nitrate or nitrite, even when the dissolved oxygen concentration approaches air saturation. Indeed, these bacteria simultaneously use both electron acceptors (Kuenen and Robertson, 1987). One of the best-studied members of this group is *Thiosphaera pantotropha*, a mixotrophic, colorless sulfur bacterium which was isolated from a denitrifying, sulfide-oxidizing wastewater treatment plant (Robertson and Kuenen, 1983).

During comparisons of the newly isolated strain with known denitrifiers (e.g., *Thiobacillus versutus* and *Paracoccus denitrificans*) in anaerobic respirometry experiments, it was observed that aerobically grown *T. pantotropha* began to denitrify immediately when it was supplied with substrate and nitrate. Similarly grown cultures of the other strains required 2 to 4 h to induce their denitrifying enzymes (Robertson and Kuenen, 1984b). Oxygen and nitrate electrodes were used to monitor the activity of these cultures, and simultaneous nitrate and oxygen removal in *T. pantotropha* suspensions was clearly observed (Robertson et al., 1986). When grown in batch cultures (Robertson and Kuenen, 1984b) with acetate as the substrate, *T. pantotropha* cultures provided with both oxygen (at a dissolved oxygen concentration of 80% air saturation) and nitrate grew more rapidly than similar cultures which had only one electron acceptor (Table 2). Moreover, the protein yield at the end of the experiments with the cultures supplied with nitrate and oxygen was lower than that obtained when oxygen was the sole electron acceptor (Table 2) and higher than when nitrate served this function (40 mg liter^{-1}). As denitrification tends to generate less energy, and thus lower biomass yields, than oxygen respiration, this observation indicated that both pathways were in use by the culture. Sufficient nitrate had disappeared from the oxygen plus nitrate cultures to account for half of the acetate dissimilated. Obviously, the other half of the

Table 2. Comparison of the maximum specific growth rates (μ_{max}), final protein concentrations, and nitrate reduced from aerobic batch cultures[a]

Organism	μ_{max} (h^{-1})			Protein (mg $liter^{-1}$)		Δ Nitrate (mM)
	O_2	O_2/NO_3^-	NO_3^-	O_2	O_2/NO_3^-	
Pseudomonas sp.	0.1	0.41	0.15	78	68	5.0
Alcaligenes faecalis	0.17	0.25	0.01	30	14	4.1
Pseudomonas aureofaciens	0.19	0.21	0.07	66	66	5.0
Thiosphaera pantotropha	0.28	0.34	0.25	81	60	5.5
Paracoccus denitrificans	0.28	0.28		92	88	<1

[a] Cultures were grown with acetate as the substrate and ammonia as the nitrogen source. Results obtained with a *Paracoccus denitrificans* strain which does not denitrify aerobically have been added for comparison (data from Robertson et al., 1989a).

acetate was dissimilated with oxygen as the terminal electron acceptor. A possible explanation of the higher growth rates obtained when both electron acceptors were supplied could be that *T. pantotropha* has a rate-limiting step in the transfer of electrons from its substrate to oxygen. The provision of a second electron acceptor, in this case nitrate, would allow it to use an additional branch in the electron transport chain, overcoming the problem.

Since these batch culture experiments indicated that the denitrifying enzymes of *T. pantotropha* were not only constitutive, but indeed active during aerobic growth (Robertson and Kuenen, 1984b; Robertson et al., 1986), a further detailed analysis of this phenomenon was made from both ecological and physiological points of view. To make accurate nitrogen balances, aerobic denitrification was quantified using chemostat cultures and carefully defined environmental conditions. As soon as these studies were initiated, it became apparent that *T. pantotropha* was also able to nitrify, provided that an organic substrate was available, an ability termed heterotrophic nitrification.

Heterotrophic Nitrification

During the batch culture experiments to discover whether *T. pantotropha* was denitrifying aerobically, nitrite was substituted for nitrate in a series of experiments, and it was observed that the nitrite concentration increased before eventually decreasing to 0. This phenomenon only occurred in the presence of an organic substrate, ammonia, and oxygen, indicating that *T. pantotropha* is a heterotrophic nitrifier. In other words, *T. pantotropha* can catalyze the oxidation of ammonia to nitrite provided that an organic electron donor (in this case acetate) is available. Subsequent experiments revealed that the nitrifying enzymes of *T. pantotropha* were remarkably similar to those of autotrophic nitrifiers such as *Nitrosomonas europaea* in that the ammonia monooxygenase required NAD(P)H, the hydroxylamine oxidoreductase was light sensitive (Robertson and Kuenen, 1988), and the nitrification was sensitive to the same chemicals which inhibit autotrophic nitrification (e.g.,

Table 3. Nitrification and denitrification rates obtained with chemostat cultures of *T. pantotropha*[a]

N compound	Dilution rate (h^{-1})	Nitrification rate (nmol ammonia min^{-1} mg protein^{-1})	Denitrification rate (nmol nitrate min^{-1} mg protein^{-1})
Ammonia	0.02	13	13
	0.05	43	43
	0.10	94	94
Ammonia/nitrate	0.02	8	38
	0.04	12	107
	0.18	26	507
Ammonia/nitrite	0.02	10	13
	0.04	21	41
	0.17	44	177

[a] Grown aerobically (dissolved oxygen = 80% air saturation) on acetate with different nitrogen compounds in the medium (adapted from Robertson et al., 1988).

allylthiourea, nitrapyrin, reduced sulfur compounds). Nitrite only accumulated in the presence of nitrite or an inhibitor of nitrite reductase, and it became clear that it was simultaneously reducing all or most of the nitrite to N_2 (Kuenen and Robertson, 1987; Robertson et al., 1988).

As it has proved impossible to separate the two phenomena, aerobic denitrification and heterotrophic nitrification will be considered further together.

Heterotrophic Nitrification and Aerobic Denitrification in the Chemostat

Chemostat experiments carried out with acetate-limited *T. pantotropha* cultures (Robertson et al., 1988) revealed that both the nitrification and the denitrification rates increased as the growth rate increased (Table 3) and as the dissolved oxygen concentration decreased. For example, a culture growing at a dilution rate of 0.04 h^{-1} in the presence of ammonia and nitrate at a dissolved oxygen concentration of 25% air saturation nitrified and denitrified at rates of 12 and 105 nmol min^{-1} mg of protein^{-1}, respectively. In similar cultures maintained at only 5% air saturation, these rates had risen to 33 and 393 nmol min^{-1} mg of protein^{-1}, respectively. Moreover, the presence of nitrate or nitrite in the medium reduced the amount of nitrification taking place (Table 3). If nitrate or nitrite was not provided and ammonia oxidation was inhibited in some way (e.g., by the presence of hydroxylamine in the medium), *T. pantotropha* synthesized large amounts of poly-β-hydroxybutyrate. These observations, considered with the higher growth rates observed during growth in batch on oxygen and nitrate together, led to the hypothesis that both aerobic denitrification and heterotrophic nitrification are mechanisms by which a rate-limiting step in the flow of electrons to oxygen can be

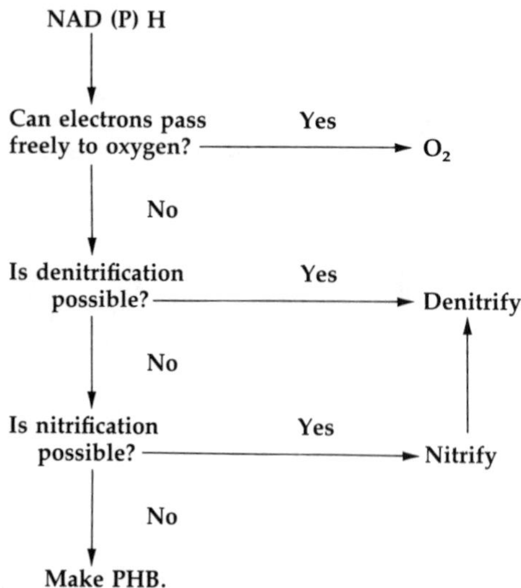

Figure 1. Flow chart to show the hypothesis worked out to explain the physiological controls behind aerobic denitrification and heterotrophic nitrification by *T. pantotropha*. Each of the Yes/No decisions would be controlled by the degree of reduction of the cytochromes involved. (From Robertson et al., 1988.) PHB, poly-β-hydroxybutyrate.

overcome. The model is shown as a flow chart in Fig. 1. It is assumed that denitrification has priority over heterotrophic nitrification because the latter costs, rather than generates, energy.

T. pantotropha Is Not Unique

It was also important to establish whether *T. pantotropha* was unique in its ability to respire oxygen and denitrify simultaneously. A second line of research was therefore to screen other denitrifiers for the ability to use oxygen and nitrate or nitrite simultaneously (Robertson et al., 1989a; Robertson et al., 1989b). The results of the initial experiments with known heterotrophic nitrifiers, where stimulation of growth rate and depression of aerobic protein yields were the parameters tested, are shown in Table 2. It can be seen that, with the exception of the negative control, *Paracoccus denitrificans* (a strain which does not nitrify significantly and which requires anaerobiosis for denitrification), all of the strains tested had used a significant amount of nitrate. *Pseudomonas aureofaciens* did not show the expected lower yields, and its growth rate was only slightly stimulated by the presence of both electron acceptors. This species is atypical in other features, as its denitrification pathway terminates at N_2O, and further work is required to understand its physiology.

Because they were simultaneously denitrifying, these cultures accumulated only insignificant amounts of nitrite, indicating that the reported low activities of

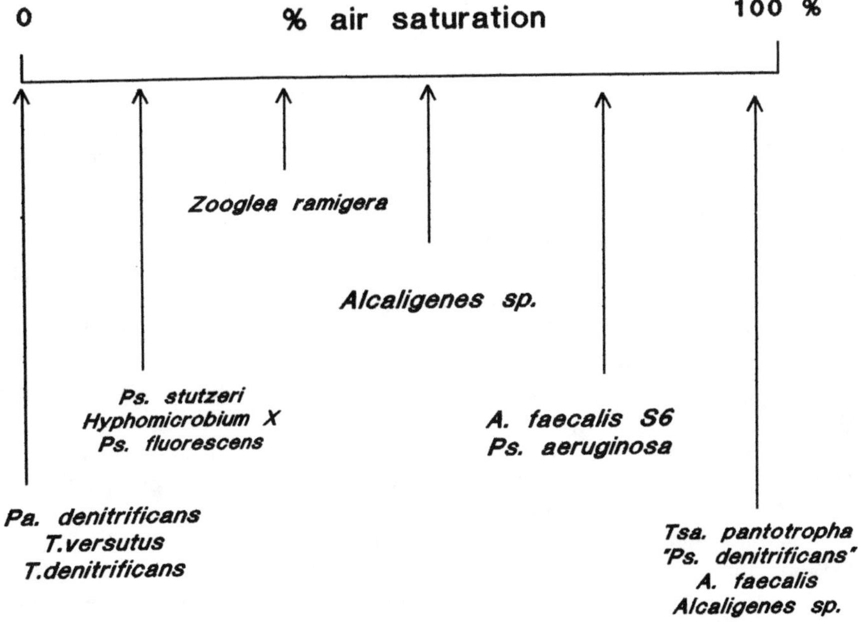

Figure 2. "Spectrum" indicating the thresholds at which denitrification by different bacteria begins to be affected by the dissolved oxygen concentration. Percent air saturation is shown rather than oxygen concentrations to allow for the different growth temperatures of the various bacteria. (Redrawn from Robertson and Kuenen, 1990.) Genera: *Pa.*, *Paracoccus*; *T.*, *Thiobacillus*; *Ps.*, *Pseudomonas*; *A.*, *Alcaligenes*; *Tsa.*, *Thiosphaera*.

heterotrophic nitrifiers might, at least in some cases, be underestimates of in vivo nitrification rates. Nitrification is traditionally judged from the amounts of nitrification products which accumulate, and these results serve to emphasize a need for the making of element balances for cultures in order to measure nitrogen losses and assess both nitrification and denitrification activity. Aerobic denitrification has now been reported for a number of other species (e.g., Bazylinski and Blakemore, 1983; Krul, 1976; Lloyd et al., 1987; Strand et al., 1988; Trevors and Starodub, 1987), and it will be interesting to investigate whether these are also heterotrophic nitrifiers.

Ecological Significance

In view of the wide range of species from very different genera which are now known to be able to nitrify and denitrify simultaneously (see Fig. 2 for a compilation of our and published data), the ecological significance of these bacteria requires reevaluation. Heterotrophic nitrification is slower than autotrophic nitrification when specific activities are compared (by a factor of 10^2 to 10^3 rather than the 10^4 previously believed), but when the vastly superior numbers of heterotrophs present in, for example, soil are considered, it can be seen that these organisms can make a significant contribution to nitrogen turnover rates.

It must not be expected that all of these species will behave in the same way. Recent work (Robertson et al., 1989a; van Niel, 1991) has shown that two other nitrifying/denitrifying species show similarities to each other and to *T. pantotropha*, but that they also differ in some respects. For example, all of the aerobic denitrifiers thus far tested have the copper-based nitrite reductase rather than cytochrome *cd* (Robertson et al., 1989a; Robertson and Kuenen, 1990). *Pseudomonas* sp. was very similar to *T. pantotropha*, except that its denitrifying nitrate reductase was inducible. The remainder of its denitrification pathway was constitutive (Robertson et al., 1989a). Denitrification by *Alcaligenes faecalis* TUD proved to be partially inhibited by oxygen at concentrations of 50% air saturation, and the organism completely lacked a nitrate reductase (van Niel, 1991).

As yet, there are insufficient data available to allow the evaluation of the activity of bacteria capable of heterotrophic nitrification and simultaneous aerobic denitrification in the field. Inhibitors are frequently used in order to estimate microbial activity in a particular situation. The use of inhibitors can prove valuable. For example, acetylene can block N_2O reduction. As N_2O is much easier to measure than N_2, this can make the measurement of denitrification rates in, for example, soil or water samples much more convenient and accurate. However, care should be taken in the interpretation of results, as these inhibitors are seldom completely specific, and all necessary controls must be done. For example, acetylene also inhibits autotrophic nitrification, and therefore it is essential that NO_x be already present in or added to the test system because it will not be generated biologically once acetylene has been added to the experimental system. Another example of the potential pitfalls is that a commonly used inhibitor of nitrification, nitrapyrin, is not soluble in water and is frequently used as a solution in acetone. Hall (1984) showed that in some cases the acetone alone can be as inhibitory as the nitrapyrin solution. This has proved to be the case not only with the obligate autotrophs, but also with heterotrophic nitrifiers such as *T. pantotropha*. A final example of a potential problem associated with inhibitors involves the so-called autotroph-specific inhibitors such as nitrapyrin and allylthiourea. For a long time, it was believed that these compounds only inhibited autotrophic nitrification, and thus only the activity which was not inhibited by them was considered to be due to heterotrophic nitrifiers. Recent research (Robertson et al., 1989b) has shown that nitrification by many heterotrophic bacteria is, in fact, also sensitive to nitrapyrin, allylthiourea, and thiosulfate. It seems likely that at least some heterotrophic nitrifying activity may have been included in estimates of autotrophic activity.

This, of course, leaves the problem of how to distinguish between heterotrophic and autotrophic activity in the field. Most heterotrophic nitrifiers appear to also be aerobic denitrifiers (see above), and it might therefore be profitable to measure the production of ^{15}N-labeled gases from ^{15}N-labeled ammonia in the presence of an organic substrate. Alternatively, ammonia-stimulated CO_2 fixation would give a measure of autotrophic activity. A less direct method might involve the determination of the type of nitrite reductase dominant in a denitrifying community. All of the aerobic denitrifiers thus far tested have possessed a copper-based nitrite reductase rather than the more usual cytochrome *cd* (Kuenen and Robertson, 1987; Robertson et al., 1989a). A larger number of species must be screened before it is certain whether this is universal, but the two nitrite reductases are sensitive to different inhibitors (copper chelators and azide, respectively) and it might be possible to use this fact to detect the presence of bacteria capable of aerobic denitrification and heterotrophic nitrification in a community. Last, but certainly not least, there is the wide range of modern tools which are becoming available to taxonomists, including fluorescent-antibody stains and DNA fingerprinting. The most obvious drawback to the use of such methods in natural systems is the need for extensive data bases (derived from known, pure cultures) in order to interpret experimental results. This information is gradually accumulating, and the drawback is therefore only transient.

"Anaerobic Nitrification"

Following the confirmation that aerobic denitrification takes place, at least under some conditions, it did not, perhaps, come as a total surprise that anoxic nitrification would also be found to occur. Using *Nitrosomonas*, *Nitrosococcus*, and *Nitrosolobus* species, Poth (1986) showed the production of $^{15}N_2O$ and $^{15}N_2$ from $^{15}NO_2^-$ under oxygen stress. It was postulated during this work that the $^{15}NO_2^-$ was serving as an electron acceptor, so that any available oxygen could be used by the ammonia monooxygenase. However, it has since been reported (van de Graaf et al., 1990) that a mixed culture from a wastewater treatment system is capable of nitrification (and, by definition, denitrification) under fully anaerobic conditions, implying that ammonia monooxygenase may not be the sole ammonia-oxidizing enzyme available to these bacteria. Poth suggested that it should be possible to grow autotrophic nitrifiers anaerobically, while denitrifying, provided that they were provided with hydroxylamine rather than ammonia, but this limitation need possibly not apply for the bacteria involved in this reaction.

It should be mentioned here that it has been shown that some nitrite-oxidizing *Nitrobacter* species can grow anaerobically as heterotrophs, with nitrate serving as the terminal electron acceptor (Bock et al., 1986). This finding stresses, once again, that the traditional concepts of nitrification and denitrification need rigorous adjustment.

IMPLICATIONS FOR ECOLOGY AND THE ATMOSPHERE

Nitric oxide and nitrous oxide are both important pollutants of the atmosphere. Ammonia, nitrate, and nitrite can all constitute pollution problems in groundwater, lakes, and rivers, as well as providing the starting materials for denitrification, but are also important for the maintenance of soil fertility. An understanding of the processes underlying the cycling of these and more complex nitrogen compounds is therefore essential if we are to understand, predict, or control leaching or gaseous emissions. If aerobic denitrification is as widespread in the field as it appears to be in the laboratory, this implies that denitrification is not limited to anaerobic zones caused by water saturation or to large soil aggregates resulting in anaerobic microsites. Although only a few denitrifying species are known to terminate their denitrification at N_2O, it has been shown in a number of cases that N_2O reductase is one of the first denitrifying enzymes to be inhibited by increasing amounts of oxygen. It is therefore not unlikely that some aerobic denitrification may result in the emission of N_2O rather than N_2, particularly in poorly controlled wastewater treatment plants. As some nitrifying bacteria have also been shown to produce N_2O under "oxygen stress," sewage and effluent treatment systems are both potential sources of atmospheric NO_x.

If NO_x emission into the atmosphere is to be controlled, a two-pronged attack on the problem is clearly needed. First, accurate measurements of the actual release of N_2O and NO from likely sources such as marshes, sulfide-rich sediments (sulfide also inhibits nitrous oxide reductase), and water treatment plants are necessary. For example, Sorensen (1987) showed that sulfide can inhibit N_2O reduction by marine bacteria, resulting in measurable concentrations of N_2O in seawater. Second, more laboratory studies, with controlled conditions, are needed in order to reveal the mechanisms which determine whether N_2O, NO, or N_2 is released from a particular system. The factors which determine whether NO remains enzyme bound or is released require especially a great deal of further research.

REFERENCES

Bazylinski, D. A., and R. P. Blakemore. 1983. Denitrification and assimilatory nitrate reduction in *Aquaspirillum magnetotacticum*. *Appl. Environ. Microbiol.* **46:**1118–1124.

Bock, E., H.-P. Koops, and H. Harms. 1986. Cell biology of nitrifying bacteria, p. 17–38. *In* J. I. Prosser (ed.), *Nitrification*. IRL Press, Oxford.

Hall, G. 1984. Measurement of nitrification rates in lake sediments: comparison of the nitrification inhibitors nitrapyrin and allylthiourea. *Microb. Ecol.* **10:**25–36.

Herbert, D., R. Elsworth, and R. C. Telling. 1956. The continuous culture of bacteria: a theoretical and experimental study. *J. Gen. Microbiol.* **14:**601–622.

Kelly, D. P., and J. G. Kuenen. 1984. Ecology of the colourless sulphur bacteria, p. 211–240. *In* G. Codd (ed.), *Aspects of Microbial Metabolism and Ecology*. Academic Press, Inc., New York.

Krul, J. M. 1976. Dissimilatory nitrate and nitrite reduction under aerobic conditions by an aerobically and anaerobically grown *Alcaligenes* sp. and by active sludge. *J. Appl. Bacteriol.* **40:**245–260.

Kuenen, J. G., and W. Harder. 1982. Microbial competition in continuous culture, p. 342–367. *In* R. G. Burns and J. H. Slater (ed.), *Experimental Microbial Ecology*. Blackwell Scientific Publishers, Ltd., Oxford.

Kuenen, J. G., and L. A. Robertson. 1987. Ecology of nitrification and denitrification, p. 162–218. *In* J. A. Cole and S. Ferguson (ed.), *The Nitrogen and Sulphur Cycles*. Cambridge University Press, Cambridge.

Lloyd, D., L. Boddy, and K. J. P. Davies. 1987. Persistence of bacterial denitrification capacity under aerobic conditions: the rule rather than the exception. *FEMS Microbiol. Ecol.* **45:**185–190.

Payne, W. J. 1981. *Denitrification*. John Wiley & Sons, Inc., New York.

Poth, M. 1986. Dinitrogen production from nitrite by a *Nitrosomonas* isolate. *Appl. Environ. Microbiol.* **52:**957–959.

Robertson, L. A., R. Cornelisse, P. de Vos, R. Hadioetomo, and J. G. Kuenen. 1989a. Aerobic denitrification in various heterotrophic nitrifiers. *Antonie van Leeuwenhoek* **56:**289–299.

Robertson, L. A., R. Cornelisse, R. Zheng, and J. G. Kuenen. 1989b. The effect of thiosulfate and other inhibitors of autotrophic nitrification on heterotrophic nitrifiers. *Antonie van Leeuwenhoek* **56:**301–309.

Robertson, L. A., and J. G. Kuenen. 1983. *Thiosphaera pantotropha* gen. nov. sp. nov., a facultatively anaerobic, facultatively autotrophic sulphur bacterium. *J. Gen. Microbiol.* **129:**2847–2855.

Robertson, L. A., and J. G. Kuenen. 1984a. Aerobic denitrification—old wine in new bottles? *Antonie van Leeuwenhoek* **50:**525–544.

Robertson, L. A., and K. G. Kuenen. 1984b. Aerobic denitrification: a controversy revived. *Arch. Microbiol.* **139:**351–354.

Robertson, L. A., and J. G. Kuenen. 1988. Heterotrophic nitrification in *Thiosphaera pantotropha*—oxygen uptake and enzyme studies. *J. Gen. Microbiol.* **134:**857–863.

Robertson, L. A., and J. G. Kuenen. 1990. Combined heterotrophic nitrification and aerobic denitrification in *Thiosphaera pantotropha* and other bacteria. *Antonie van Leeuwenhoek* **57:**139–152.

Robertson, L. A., B. H. A. van Kleeff, and J. G. Kuenen. 1986. A microcomputer-based method for semi-continuous monitoring of biological activities. *J. Microbiol. Methods* **5:**237–242.

Robertson, L. A., E. W. J. van Niel, R. A. M. Torremans, and J. G. Kuenen. 1988. Simultaneous nitrification and denitrification in aerobic chemostat cultures of *Thiosphaera pantotropha*. *Appl. Environ. Microbiol.* **54:**2812–2818.

Sorensen, J. 1987. Nitrate reduction in marine sediment: pathways and interactions with iron and sulfur cycling. *Geomicrobiol. J.* **5:**401–421.

Stouthamer, A. H. 1988. Dissimilatory reduction of oxidized nitrogen compounds, p. 245–303. *In* A. J. B. Zehnder (ed.), *Environmental Microbiology of Anaerobes*. John Wiley & Sons, Inc., New York.

Strand, S. E., A. J. McDonnell, and R. F. Unz. 1988. Oxygen and nitrate reduction kinetics of a non-flocculating strain of *Zooglea ramigera*. *Antonie van Leeuwenhoek* **54:**245–255.

Trevors, J. T., and M. E. Starodub. 1987. Effect of oxygen concentration of denitrification in freshwater sediment. *J. Basic Microbiol.* **27:**387–391.

Van de Graaf, A. A., A. Mulder, H. Slijkhuis, L. A. Robertson, and J. G. Kuenen. 1990. Anoxic ammonium oxidation, p. 388–391. *In* C. Christiansen, L. Munck, and J. Villadsen (ed.), *Proceedings of the 5th European Congress on Biotechnology*, vol. I. Munksgaard International Publisher, Copenhagen.

van Niel, E. W. J. 1991. Nitrification by heterotrophic denitrifiers and its relationship to autotrophic nitrification. Ph.D. thesis. Delft University of Technology, Delft, The Netherlands.

Veldkamp, H., and H. W. Jannasch. 1972. Mixed culture studies with the chemostat. *J. Appl. Chem. Biotechnol.* **22:**105–123.

Winkler, M. 1981. *Biological Treatment of Waste-Water*. Ellis Horwood, Publishers, Chichester, U.K.

Wood, P. 1986. Nitrification as a bacterial energy source, p. 39–62. *In* J. I. Prosser (ed.), *Nitrification*. IRL Press, Oxford.

Chapter 11

Ecology of Nitrification and Denitrification in Soil Evaluated at Scales Relevant to Atmospheric Chemistry

Peter M. Groffman

The purpose of this paper is to review the ecological factors controlling nitrification and denitrification in soil, focusing on how these factors are expressed at scales relevant to the chemistry of the regional and global atmosphere. Research at these scales is not common in microbiology, and questions about the contribution of microbial processes to "global change" present distinct and complex conceptual and practical challenges. Understanding ecological controls of microbial processes at large scales is important for identifying environments likely to have significant fluxes of radiatively important trace gases and for understanding how these fluxes may change in response to climatic changes.

A common theme that has emerged in nitrogen (N) cycle research is that, while the physiology and biochemistry of many microbial processes are relatively well understood, it has been difficult to translate this understanding into useful information about how these processes are regulated in the environment (Robertson, 1986). While pure culture and enzyme studies have elucidated the biochemical pathways and physiology of nitrification and denitrification, multiple physical, chemical, and biological factors interact to complicate ecological controls over these processes in nature. The result of this complexity is high spatial and temporal variation in activity, which has made it difficult to establish unambiguous relationships between field-scale parameters and nitrification and denitrification activity. This variability has inhibited attempts to establish relationships between environmental parameters and microbial activity at large scales (e.g., landscape, regional, global).

This chapter will briefly review the well-established physiological controls of nitrification and denitrification at the organismal scale and then discuss conceptually how these controls are expressed at progressively larger spatial and temporal scales. Examples of some studies that have attempted to establish quantitative relationships between large-scale ecological variables and nitrification and denitrification activity will be given.

Peter M. Groffman • Department of Natural Resources Science, University of Rhode Island, Kingston, Rhode Island 02881.

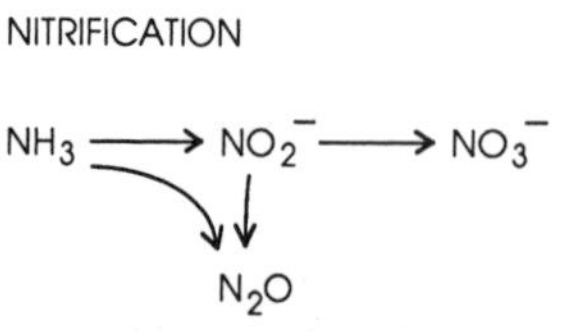

Figure 1. Simplified biochemical autotrophic nitrification pathway.

NITRIFICATION

Nitrification refers to the oxidation of ammonium (NH_4^+, chemoautotrophic nitrification) or organic N compounds (heterotrophic nitrification) to nitrite (NO_2^-) or nitrate (NO_3^-) (Fig. 1). This process is of concern in atmospheric chemistry because it leads to the production of nitrous oxide (N_2O) by at least two mechanisms. First, intermediates between NH_4^+ and NO_2^-, or NO_2^- itself, can chemically decompose to N_2O, especially under acid conditions (Ritchie and Nicholas, 1972). Nitrifying organisms can also produce N_2O during the reduction of NO_2^- under anaerobic or microaerophilic conditions (Poth and Focht, 1985). Nitrification is often considered to be the dominant source of N_2O in "aerobic" soils (Bremner and Blackmer, 1978; Sahrawat and Keeney, 1986).

Nitrifier activity in pure culture is strongly controlled by the availability of NH_4^+ (Table 1). Although nitrification can occur under anaerobic or microaerophilic conditions (Poth and Focht, 1985), high rates of nitrification are generally associated with strongly aerobic conditions (Focht and Verstraete, 1977). At the field scale, nitrification is strongly controlled by NH_4^+ supply either from fertilizer or from mineralization of organic N compounds. Soil water, which is a strong controller of O_2 availability in soil, is also a strong field-scale controller of nitrification (Goodroad and Keeney, 1984).

To address ecological control of nitrification at scales above the field level, we need to identify the factors that control the variability of NH_4^+ and O_2 supply in soil at the scale in question. Soil type (texture and drainage), which is a strong controller of organic matter levels, moisture dynamics, and N mineralization (Jenny, 1980), is an important landscape-scale controller of nitrification. If the landscape is viewed as an aggregate of "field"-scale units consisting of distinct soil types (Hole, 1978), landscape-scale patterns of nitrification can be systematically evaluated. Another approach to evaluate nitrification at the landscape scale is to

Table 1. Factors controlling nitrification activity at different scales of investigation

Scale of investigation	Controlling factors
Organism	Ammonium, oxygen
Field	Ammonium supply, soil water
Landscape	Soil type, plant community type
Regional	Geomorphology, land use
Global	Biome type, climate

DENITRIFICATION

$NO_3^- \longrightarrow NO_2^- \longrightarrow NO \longrightarrow N_2O \longrightarrow N_2$

Figure 2. Simplified biochemical denitrification pathway.

focus on plant community type. Plant community patterns are often strongly related to N availability (Pastor et al., 1984), and different plant communities should thus exhibit distinct patterns of nitrification (Vitousek and Melillo, 1979; Matson and Vitousek, 1987). Consideration of plant community type can be extended to encompass different agricultural production systems, which often differ in the nature and extent of N enrichment and should thus have distinct patterns of nitrification and associated N_2O flux.

At the regional scale we need to identify distinct landscape units that systematically cover the range of conditions affecting nitrification. The distribution of soil types and plant communities across a region is largely controlled by geomorphology and land use patterns. Factors such as the distribution of glacial landforms and alluvial deposits strongly influence soil texture, fertility, microclimate, and other factors that influence soil type and plant community composition (Swanson et al., 1988). Ultimately, global ecological patterns of nitrification control can be studied by analysis of the distribution of regional units by climate and biome types.

Factors such as geomorphology and land use are not usually considered by microbiologists in traditional studies of nitrification in soil. However, we can conceptually establish links between geomorphology, soil type, plant community composition, and NH_4^+ supply, and ultimately to nitrification activity. If we need to address the significance of N_2O production by nitrification to atmospheric chemistry, we must then use these conceptual links as a basis for studies that quantify relationships between these large-scale factors and nitrification activity.

DENITRIFICATION

Denitrification refers to anaerobic respiration and reduction of NO_3^- and NO_2^- to N gases (NO, N_2O, N_2) by organisms that normally use O_2 for respiration (Fig. 2). Although the process has been demonstrated to occur in the presence of O_2 (see Robertson and Kuenen, this volume), vigorous denitrification activity is generally associated with anaerobic conditions in soil. Denitrification is also strongly controlled by the presence of NO_3^- and NO_2^- and by the presence of organic compounds, since most denitrifying organisms are heterotrophic. The partitioning of the gaseous products of denitrification, which is of particular interest as far as atmospheric chemistry is concerned, has been shown to be related to a variety of factors including the level of available NO_3^- and NO_2^-, organic C supply, O_2 level, and pH (Firestone, 1982).

Similar to nitrification, field-scale control of denitrification is affected primarily by N supply (in this case of NO_3^- rather than NH_4^+) and by soil water (which controls O_2) (Table 2). The ecology of denitrification is more complex than that of

Table 2. Factors controlling denitrification activity at different scales of investigation

Scale of investigation	Controlling factors
Organism	Oxygen, nitrate, carbon
Field	Soil water, nitrate supply, carbon supply
Landscape	Soil type, plant community type
Regional	Geomorphology, land use
Global	Biome type, climate

nitrification since organic C supply is a controller of heterotrophic denitrifier activity and of O_2 level as well (through respiratory consumption of O_2). Numerous studies have attempted to quantify rates of denitrification in relation to moisture status, fertilizer additions, cropping systems, and other field-scale variables (Colbourn and Dowdell, 1984; Tiedje, 1988; Nieder et al., 1989). Most of these studies have been hindered by high spatial and temporal variability in activity, with coefficients of variation frequently exceeding 100%, even with intensive sampling of small field plots (Folonoruso and Rolston, 1984; Burton and Beauchamp, 1985; Parkin et al., 1987). This variability is largely due to the complexity of factors which control denitrification. Soil water, NO_3^-, and C supply all exhibit significant variation in time and space in soil. Denitrification variability, which is the product of the variability of these three factors, is thus accordingly high.

At the landscape scale, soil type and plant community type are useful conceptual regulators of denitrification (Groffman et al., 1988). Soil texture and drainage are strong controllers of O_2 availability and indirectly regulate NO_3^- supply through their influence on nitrification. Similarly, plant community type affects NO_3^- supply by controlling the nitrification rate. Both soil type and plant community type have strong effects on the decomposition of plant material and thus influence C supply to denitrifiers.

Regional- and global-scale control of denitrification should be similar to that of nitrification, with a focus on geomorphology, land use patterns, climate, and biome types. This convergence of control at large scales is conceptually logical since both nitrification and denitrification are likely to occur only in ecosystems with high or excess N availability. Only if N availability is high will nitrification occur, and the occurrence of nitrification is necessary for denitrification to occur. This conceptual relationship between N availability and N_2O flux from denitrification and nitrification should be extremely useful for evaluating what types of ecosystems are likely to have high N_2O fluxes and for evaluating how ecosystem changes induced by climate change will affect these fluxes.

APPROACHES TO QUANTIFYING ECOLOGICAL CONTROLS OF NITRIFICATION AND DENITRIFICATION AT LANDSCAPE AND REGIONAL SCALES

These conceptual approaches to ecological control of nitrification and denitrification at large scales can guide landscape- and regional-scale studies of these processes. A variety of studies (only a few of which can be cited here) have addressed ecological control of nitrification and denitrification at landscape and regional scales. In addition to these ecological studies, there have been several fairly comprehensive "surveys" of N gas fluxes (primarily N_2O) from different ecosystem types (Sahrawat and Keeney, 1986; Keller et al., 1986; Schmidt et al., 1988). I will focus only on the ecological studies to illustrate how information from soil science, geology, and ecosystem ecology can provide a framework for studying microbial processes and ecosystem-atmosphere interactions. Examination of widely different areas (from temperate forest to tallgrass prairie to tropical forest) reveals several common mechanisms. I will give two examples in detail (based on my own work) and then more briefly review several studies conducted in other areas.

General Approach

First, it is necessary to establish landscape- and regional-scale experimental designs. In any study at any scale, it is essential to establish a valid experimental design so that the range of conditions present within the given study area is systematically examined, without bias. There are few practical and conceptual tools available for directing experimental design at landscape and regional scales, and many studies fail in this regard (Jeffers, 1988). In field-scale studies, experimental plots are selected, or blocked, to account for variations in soil type or exposure. At scales larger than the field, selection of sampling units becomes considerably more problematic, requiring careful consideration of the factors that control the process in question (O'Neill et al., 1986). To study nitrification and denitrification at the field scale, sampling of small experimental "plots" is based on soil type, slope, vegetation type, or any of a range of field-scale parameters to cover the range of conditions within the field unit. For landscape-scale studies, it is necessary to stratify the landscape into distinct "field" units that cover the range of conditions within the landscape unit. At the regional scale, landscapes are the experimental units that are sampled to cover the range of conditions within the region (Woodmansee, 1988). The globe can be treated as an aggregation of regional units. It is essential to precisely identify the scale of study and the specific study area so as to determine what factor will be used to stratify the area and what sites will be sampled to produce a complete experimental design. As the conceptual outline of nitrification and denitrification controls suggests, and the following examples will show, there can be a variety of stratification factors used, depending on the nature of the study area.

The second essential element of landscape- and regional-scale studies of the ecology of microbial processes is the establishment of quantitative relationships between the microbial process in question and some large-scale ecological variables. These relationships can range from simple regression equations to complex simulation models (Schimel et al., 1988; Groffman and Tiedje, 1989a; Matson et al., 1989), but they are essential for extrapolating data from experimental sites to larger scales for production of landscape- and regional-scale estimates of flux. Establish-

Table 3. Annual N loss to denitrification in nine forest soil classes in lower Michigan, 1985[a]

Soil drainage	N loss (kg ha^{-1} year^{-1}) from profile texture:		
	Clay loam	Loam	Sand
Well drained	18	10	0.6
Somewhat poorly drained	17	11	0.8
Poorly drained	40	24	0.5

[a] Adapted from Groffman and Tiedje (1989b).

ing these relationships involves developing quantitative measures for large-scale ecological variables (not a trivial exercise for variables such as soil type or plant community type) and an extensive program of field measurements to quantify the relationships.

Finally, the production of landscape- and regional-scale estimates of microbial processes is dependent on the availability of accurate, large-area data sets of ecologically relevant variables. These data sets are frequently based on remote sensing and are part of large-area geographic information systems (GISs). These data sources are increasing in sophistication and availability, in many cases more rapidly than our understanding of microbial processes. Linking these data sources with the quantitative relationships discussed above is the final step in the production of large-area estimates of microbial processes.

North Temperate Forest

Experimental design

The conceptual approach discussed here has been applied to landscape- and regional-scale studies of denitrification in forested components of southern Michigan (Groffman and Tiedje, 1989a, 1989b). Soil toposequences (catenas) consisting of well-, somewhat poorly, and poorly drained soils were selected as landscape units for this study. Each soil of different drainage class was a "field" experimental unit within the catena landscape unit. Catenas of different texture (loam, clay loam, sand) associated with different glacial landforms were the landscape experimental units for the regional-scale analysis. The soils of different drainage class covered the range of conditions within the catena landscape units, and the catenas of different texture covered the range of conditions across the region.

Field- and landscape-scale results

The experimental design produced nine sites for intensive sampling: three drainage classes times three textural types. Estimates of annual N gas production by denitrification in each of the nine soils were produced by intensive field sampling over the course of a year (Table 3). Denitrification, since it is an anaerobic process, is strongly controlled by soil water. Therefore, poorly drained soils had higher rates of denitrification than well-drained soils. Fine-textured soils (clays and loams) had higher rates of activity than sandy soils because fine-textured soils have smaller pores that hold water more tightly and more easily become anaerobic than the large pores present in coarse-textured sandy soils. While the annual N loss values for the fine-textured, poorly drained areas appear to be high from an ecosystem N budget perspective, it is important to note that these sites represent relatively small areas in the landscape and that they receive surface and subsurface inputs of water and NO_3^- from much larger upland areas (including agricultural areas). The high losses are reasonable in this perspective and illustrate the usefulness of a true landscape-scale experimental design.

Most of the variability (86%) in annual N loss to denitrification was explained with a multiple regression model using soil texture (percent sand) and soil drainage index (Schaetzl, 1986) as predictor variables. This strong relationship showed that the large-scale factors (soil texture and drainage) were effective predictors of denitrification and could be used to extrapolate results from the experimental sites to produce estimates of annual denitrification N loss for the entire region.

The fact that 86% of the variability in annual denitrification N loss could be explained with texture and drainage was surprising. Previous studies of relationships between hourly or daily denitrification rates and field-scale environmental parameters such as soil moisture or NO_3^- have generally been unable to explain more than 50% of the variability in denitrification activity (Mosier et al., 1983; Robertson and Tiedje, 1984; Burton and Beauchamp, 1985; Davidson and Swank, 1987; Myrold, 1988). Increasing the scale of investigation in both time (annual loss rather than hourly or daily rates) and space (landscape rather than field scale) appeared to be useful for overcoming these variability problems. If variability problems had worsened with increases in scale, the regional-scale analyses would not have been possible.

The positive results at the landscape scale are consistent with the theory of "coherent levels" outlined by O'Neill (1988), who defined coherent levels as "the scale at which predictive power is maximized." While investigators have long suspected that wet, fine-textured soils have more denitrification than dry soils, spatial and temporal variability has made it difficult to establish useful, predictive relationships at small scales. If the coherent level for denitrification (and other biogeochemical processes) is at the landscape or regional scale, this would be a great aid for producing large-area estimates of these processes.

Regional-scale analysis

Estimates of annual denitrification N loss for the region were produced using a GIS (geographic information system) containing soils and land use information. The GIS consisted of 333-m^2 pixels containing soil information digitized from the soil association map of Michigan and land use classes derived from Landsat thematic mapper imagery. Each soil series in the region was assigned one of the nine annual N-loss-to-denitrification values in Table 3, based on its texture and drainage characteristics.

Table 4 shows land use and N gas production by denitrification in different

Table 4. Land use and denitrification for different types of soils in southern lower Michigan[a]

Soil type	% of region	% in agriculture	% in forest	% of regional forest	Annual denitrification (10^6 kg of N ha^{-1} $year^{-1}$)	% of regional denitrification
Clays	15	74	13	9	3.0	22
Loams	63	73	18	47	10.2	73
Sands	22	41	48	44	0.7	5
Total					13.9	

[a] All soils south of approximately 44°N latitude. Total land area = 6.95×10^6 ha; total forest area = 1.71×10^6 ha. Adapted from Groffman et al. (submitted).

types of forest soils in lower Michigan. Fine-textured soils are more heavily cropped in this region than sandy soils since the coarse texture of sandy soils makes them poorly suited for intensive agriculture. Sandy soils accounted for over 40% of the forest area in the region, but only 5% of the denitrification. Clay soils contributed more than twice as much denitrification as the proportion of regional forest area that they represented.

The estimate of N gas production of 1.39×10^7 kg of N $year^{-1}$ is higher (by 50%) on a per unit area basis than estimates of denitrification from a range of rain forest types in West Africa (Robertson and Rosswall, 1986) and lower (by 50%) than estimates of gaseous N losses from the Amazon basin (Salati et al., 1982). Although the tropics have been suggested to be the dominant source of biogenic N_2O production (Matson and Vitousek, 1990a), denitrification rates in the temperate forest soils in this study were as high as tropical rates at both field (Robertson and Tiedje, 1988) and regional scales. In general, tropical soils have more dynamic N cycles than temperate soils, with higher rates of mineralization, nitrification, and N_2O fluxes (Matson and Vitousek, 1990a). In addition, high soil acidity in the tropics may increase N_2O production relative to N_2 during denitrification. Increases in rainfall and soil acidity in temperate zones could thus have significant impacts on the global N_2O budget.

As the regional-scale analysis suggests, the high forest denitrification in southern lower Michigan is caused by the presence of large areas of medium- and fine-textured, fertile soils with relatively high pH and poor natural drainage. Another factor contributing to the high forest denitrification may be the high inputs of N fertilizer to this heavily agricultural region (Lowrance and Groffman, 1987). Excess N may move from agricultural zones to lower-lying forest areas through surface and subsurface hydrologic transport and atmospheric deposition, possibly stimulating denitrification in forested components of the landscape. This analysis suggests that the effects of agricultural activities on trace gas fluxes may be expressed more at the landscape scale than at the field scale.

Table 5. Denitrification in tallgrass prairie[a]

Land use class	Denitrification (g of N ha^{-1} day^{-1}) in soil cores:		
	Unamended	Water amended	Water plus nitrate amended
Unburned	35 (7.2)	25 (8.5)	928 (107)
Burned	10 (3.9)	6 (1.9)	560 (78)
Burned/grazed	13 (4.3)	33 (5.4)	947 (116)
Cultivated	2 (0.98)	51 (16)	131 (51)

[a] Konza Prairie Research Natural Area near Manhattan, Kansas. Values are mean rates from four sample dates in 1987 and early 1988. Values are the mean (standard error in parentheses). Adapted from Groffman et al. (in preparation).

Tallgrass Prairie

Experimental design

The tallgrass prairie landscape at the Konza Prairie Research Natural Area near Manhattan, Kansas, was stratified by topography and land use (Groffman et al., in preparation). Primary land use differences in this region are related to burning and/or grazing of native prairie grasses. Sites for sampling were located on unburned, burned, burned and grazed, and agricultural (row crop) sites. The land use stratification was combined with a topographic stratification, with unburned, burned, and burned and grazed sites located in topslope, midslope, and toeslope positions. There was only one agricultural site (in a toeslope position), since this land use does not commonly occur in the hillier, upslope areas of this region of Kansas. This design produced 10 sites for sampling: the three land use classes times the three topographic positions, plus the agricultural site. This stratification was based on the thesis that burning and grazing strongly affect water and nutrient fluxes in prairie soils (Knapp and Seastedt, 1986; Seastedt and Hayes, 1988) and therefore should control patterns of nitrification and denitrification across the landscape. Topography is a strong controller of soil texture and drainage in prairie landscapes (Schimel et al., 1985b) and should thus also be useful for stratification in this region. Toeslope soils are expected to be deeper, have finer texture, and hold more moisture than topslope and midslope soils.

Results

Results suggest that there are distinct patterns of nitrification and denitrification across the tallgrass prairie landscape. Denitrification was consistently higher in unburned sites than in burned sites and was lowest in the agricultural site (Table 5). This result was as expected since the unburned sites had the highest soil water and nitrate levels of the native prairie sites. Denitrification was quite low in the agricultural site, despite high levels of soil NO_3^-.

Amendment studies suggested that low NO_3^- levels and soil moisture limited denitrification in the native prairie sites (Table 5). There was little increase in denitrification in response to water additions, but very large increases were observed in response to water and NO_3^- additions. It is likely that intense

Table 6. Microbial biomass C and nitrification and denitrification enzyme activity in tallgrass prairie[a]

Land treatment	Microbial biomass C ($\mu g\ g^{-1}$)	Nitrification ($\mu g\ g^{-1}\ day^{-1}$)	Denitrification enzyme activity ($ng\ g^{-1}\ h^{-1}$)
Unburned	1,255	2.51	327
Burned	1,290	3.22	314
Burned/grazed	1,257	4.82	233
Cultivated	327	0.41	30

[a] Microbial biomass C (chloroform fumigation-incubation method [Jenkinson and Powlson, 1976]), nitrification enzyme activity (chlorate inhibition assay [Belser and Mays, 1980]), and denitrification enzyme activity (short-term anaerobic assay [Smith and Tiedje, 1979]) in different land use classes in the tallgrass prairie landscape at the Konza Prairie Research Natural Area near Manhattan, Kansas. Values are mean rates from four sample dates in 1987. Adapted from Groffman et al. (in preparation).

competition among microorganisms and plants keeps levels of available N low in the native prairie soils.

In the agricultural site, denitrification was significantly stimulated by water additions, but the response to water and NO_3^- was much less marked than in the native sites. The water amendment results are consistent with the high levels of NO_3^- present in the agricultural soils. The lack of response to water and NO_3^- suggests that levels of available carbon and/or the denitrifying bacteria populations may be low in these soils relative to the native prairie soils. Levels of microbial biomass C (as measured by the chloroform fumigation-incubation method [Jenkinson and Powlson, 1976]), nitrification enzyme activity (as measured by the chlorate inhibition method [Belser and Mays, 1980]), and denitrification enzyme activity (as measured by the short-term anaerobic assay described by Smith and Tiedje [1979]) were markedly low in the agricultural site relative to the native prairie sites (Table 6). Schimel et al. (1985a) also found low levels of microbial biomass in cultivated prairie soils relative to native soils.

These results suggest that the inherent biological potential for N_2O emission is lower in agricultural soils than in native soils in the tallgrass prairie landscape. This result is somewhat surprising since we expected to observe high N_2O emissions from N-enriched agricultural sites. However, cultivation reduces levels of total, microbial, and available C in soils (Buyanovsky et al., 1987; Aguilar et al., 1988; Burke et al., 1989b) and consequently reduces the level of microbial activity, including production of N_2O. Emissions from agricultural soils directly related to additions of N fertilizer (Breitenbeck et al., 1980) may be a significant source of agriculturally related N_2O in the tallgrass prairie region, but our results suggest that native sites may also be important sources of N_2O. The strong NO_3^- response that we observed in the native soils suggests that native sites that receive inputs of NO_3^- from adjacent agricultural areas may have especially high fluxes.

Our topographic stratification was not very useful, as we did not observe consistent patterns of denitrification with slope position. This stratification was based on the hypothesis that deep, fertile soils, which should have high rates of nitrification and denitrification, would be found at the bottom of the slope. Analysis of soil patterns in this area shows that loess deposition (wind-blown, fine-textured particles) created areas of deep, fine-textured soils in topslope positions (Soil Conservation Service, 1975). These patterns need to be incorporated into our landscape- and regional-scale experimental designs.

Regional-scale analysis

The work at Konza Prairie was part of a larger effort, funded by the National Aeronautics and Space Administration, directed towards the use of remote sensing to quantify land surface variables. This effort, called FIFE (First ISLSCP Field Experiment), is part of the International Satellite Land Surface Climatology Program (ISLSCP) (Sellers et al., 1988). FIFE provided an excellent opportunity to test the usefulness of different types of remote sensing information for study of the ecology of nitrification and denitrification.

Two different approaches were used to estimate denitrification and N_2O flux for the Konza Prairie site on a large scale. The first method was based on establishing simple predictive relationships between denitrification and N_2O flux and soil water. The large-area information on soil moisture produced by a microwave radiometer (an aircraft-mounted sensor [Schmugge, 1983]) or by soil moisture models could then be used to produce large-area estimates of these fluxes. The accuracy and precision of these estimates will be dependent on the strength of the relationships between soil moisture and flux and on the ability of the microwave radiometer and moisture models to depict soil moisture patterns across the landscape.

We also plan to establish relationships between land use and/or plant productivity and N_2O and N_2 flux as tools for producing landscape- and regional-scale estimates of flux. Remote-sensing-based maps of land use and/or plant productivity (Tucker and Sellers, 1986; Hall et al., 1988; Running and Coughlan, 1988) patterns can then be used to extrapolate data from experimental sites to larger areas. This approach to producing large-area estimates is more static than the soil moisture-based approach, since it is not based on a continuous statistical relationship. However, the land use/plant productivity approach may be more appropriate for landscape- and regional-scale applications because land use and plant community dynamics are more appropriate large-scale controllers of microbial processes than soil moisture, which is more likely to be a useful field-scale controller.

Northern Hardwood Forests

Several studies have analyzed the effects of soil texture on plant community composition and N cycling on Blackhawk Island, a small island in the Wisconsin River dominated by old-growth forests. These studies showed that soil texture exerts a strong influence on soil moisture dynamics at a site, which affects plant community composition, which influences plant canopy chemistry and litter quality, which influences soil N cycling (Pastor et al., 1984; McClaugherty et al.,

1985). The distribution of soil textural types is controlled by patterns of glacial geology (Pastor et al., 1982). In this landscape, fine-textured soils hold more water and support vegetation with higher litter quality (lower lignin:N ratio) than coarse-textured soils. Sites with higher litter quality have more dynamic soil N cycles, with higher rates of mineralization and nitrification (Pastor et al., 1984). Denitrification is likely higher on fine-textured sites as well.

The Blackhawk Island work and other studies in northern hardwood forests (Zak et al., 1986) have shown that there are coherent patterns of N cycling among different forest ecosystem types that are constrained by landscape- and regional-scale geologic factors that influence soil texture. Understanding these constraints will be useful for "scaling-up" site-specific information on N gas fluxes to scales relevant to atmospheric chemistry. Data from the Blackhawk Island studies have been synthesized into simulation models capable of depicting forest dynamics and N cycles over large areas (Pastor and Post, 1986). Further, remote sensing has been used to detect differences in forest canopy chemistry (Wessman et al., 1988) at this site. Since canopy chemistry is directly related to litter quality, which is directly related to soil microbial processes, this type of remote sensing may be very useful for landscape- and regional-scale assessments of forest-atmosphere exchange (Aber et al., 1990).

Shortgrass Prairie

Landscape variation in soil properties and N cycling has been extensively studied in shortgrass steppe. Geologic variation along soil toposequences (catenas) strongly influences soil texture, water, plant community, and N cycling dynamics in this landscape (Schimel et al., 1985a, 1985b). Bottom-slope positions routinely have relatively fine-textured soils, with higher soil moisture and plant productivity and more dynamic N cycles than topslope or hillside positions. On a regional scale, geologic variation in soil parent materials controls differences in soil texture among catenas.

Simulation models of plant and N dynamics have been developed that depict N_2O production associated with nitrification and denitrification from shortgrass prairie soils (Parton et al., 1988). These models have been run over regional scales using a GIS to provide climatic and soils information to drive the model (Burke et al., 1990). The strength of these models is their ability to link soil-atmosphere gas exchange to soil and plant dynamics in an ecosystem context. This linkage allows for more integrated and realistic simulation of how processes change in time and space, which is crucial for analysis of "global change" questions.

Sagebrush Steppe

Burke (1989) used multivariate statistical techniques to stratify an experimental design in a sagebrush-dominated landscape in Wyoming. In this landscape, topography controls soil moisture availability through effects on soil texture and winter snow depth. Soil moisture in turn controls plant community composition, N turnover (Burke et al., 1989a), and N_2O fluxes (Matson and Vitousek, 1990b). Remote sensing measurements of differences in plant community composition (Reiners et al., in press) were used as input to vegetation-driven N_2O flux models to produce landscape-scale estimates of N_2O flux (Matson and Vitousek, 1990b). Since plant community composition differences are the integrative product of the same ecological factors that influence microbial trace gas fluxes (water and nutrient availability), remote sensing of plant variables should be useful for large-scale soil-atmosphere gas exchange studies in many settings.

Tropical Forests

As in temperate ecosystems, ecological control of nitrification and denitrification in tropical soils is related to broader-scale ecosystem parameters. Matson and Vitousek (1987) observed strong relationships between potential net N mineralization and nitrification and N_2O flux in a wide range of moist to wet tropical forest sites. In a more detailed study, Matson et al. (in press) stratified an 850-km^2 area of the Amazon basin by forest community type. In this area, plant community type, nitrification, and N_2O flux were controlled by soil texture, which exerts a strong influence on soil fertility.

Using information on relationships between N_2O fluxes associated with nitrification and denitrification with reference to soil fertility, elevation, seasonal climate, and human disturbance, Matson and Vitousek (1990a) produced estimates of N_2O flux from all tropical forests. These estimates were considerably more refined than previous estimates and showed how relatively readily available information on ecosystem type can be used to great advantage in global budgets.

CONCLUSIONS

It appears relatively straightforward to establish conceptually how ecological controls over nitrification and denitrification are expressed at various scales. It also appears to be possible to quantify relationships between these controls and N_2O flux associated with nitrification and denitrification at landscape and regional scales. Variability problems that have made it difficult to establish unambiguous relationships between environmental parameters and N_2O flux appear to decrease with increases in scale. If this variability decrease is consistent, landscape- and regional-scale studies may be useful both for increasing our understanding of the nature of microbial processes in soil and for understanding how these processes contribute to the chemistry of the atmosphere. These studies will be useful for identifying areas likely to have high rates of N_2O flux and for evaluating how ecosystem-scale changes induced by climate change are likely to affect this flux.

There is a strong need for quantifying relationships between ecological variables and N_2O flux in a variety of landscapes and regions and for linking these relationships to large-area data sets of ecological parameters to produce landscape- and regional-scale estimates of flux. While the conceptual basis for these relationships is well established, there is a need for specific data in a wide range of ecosystems and for increased understanding of ecological regulation of microbial processes in the field before they can be generally quantified.

Ultimately it will be necessary to produce estimates of microbial gas fluxes on a scale compatible with general circulation models of the atmosphere so that relationships between these fluxes and atmospheric chemistry can be directly assessed. It will also be necessary to develop techniques for direct measurement of trace gas fluxes over large scales, using micrometeorological (Baldocchi et al., 1988) or aircraft (Matson and Hariss, 1988) approaches. These measurements will be needed to validate large-area trace gas flux predictions derived from models.

REFERENCES

Aber, J. D., C. A. Wessman, D. L. Peterson, J. M. Melillo, and J. H. Fownes. 1990. Remote sensing of litter and soil organic matter decomposition in forest ecosystems, p. 87–103. *In* R. J. Hobbs and H. A. Mooney (ed.), *Remote Sensing of Biosphere Functioning*. Springer-Verlag, New York.

Aguilar, R., E. F. Kelly, and R. D. Heil. 1988. Effects of cultivation on soils in northern Great Plains rangeland. *Soil Sci. Soc. Am. J.* **52:**1081–1085.

Baldocchi, D. D., B. B. Hicks, and T. P. Meyers. 1988. Measuring biosphere-atmosphere exchanges of biologically related gases with micrometeorological methods. *Ecology* **69:**1331–1340.

Belser, L. W., and E. L. Mays. 1980. Specific inhibition of nitrite oxidation by chlorate and its use in assessing nitrification in soils and sediments. *Appl. Environ. Microbiol.* **39:**505–510.

Breitenbeck, G. A., A. M. Blackmer, and J. M. Bremner. 1980. Effects of different nitrogen fertilizers on emission of nitrous oxide from soil. *Geophys. Res. Lett.* **7:**85–88.

Bremner, J. M., and A. M. Blackmer. 1978. Nitrous oxide: emission from soils during nitrification of fertilizer nitrogen. *Science* **199:**295–296.

Burke, I. C. 1989. Control of nitrogen mineralization in a sagebrush steppe landscape. *Ecology* **70:**1115–1126.

Burke, I. C., W. A. Reiners, and D. S. Schimel. 1989a. Organic matter turnover in a sagebrush steppe landscape. *Biogeochemistry* **7:**11–31.

Burke, I. C., D. S. Schimel, C. M. Yonker, W. J. Parton, L. A. Joyce, and W. K. Lauenroth. 1990. Regional modeling of grassland biogeochemistry using GIS. *Landscape Ecol.* **4:**45–54.

Burke, I. C., C. M. Yonker, W. J. Parton, C. V. Cole, K. Flach, and D. S. Schimel. 1989b. Texture, climate, and cultivation effects on soil organic matter content in U.S. grassland soils. *Soil Sci. Soc. Am. J.* **53:**800–805.

Burton, D. L., and E. G. Beauchamp. 1985. Denitrification rate relationships with soil parameters in the field. *Commun. Soil Sci. Plant Anal.* **16:**539–549.

Buyanovsky, G. A., C. L. Kucera, and G. H. Wagner. 1987. Comparative analyses of carbon dynamics in native and cultivated ecosystems. *Ecology* **68:**2023–2031.

Colbourn, P., and R. J. Dowdell. 1984. Denitrification in field soils. *Plant Soil* **76:**213–226.

Davidson, E. A., and W. T. Swank. 1987. Factors limiting denitrification in soils from mature and disturbed southeastern hardwood forests. *Forest Sci.* **33:**135–144.

Firestone, M. K. 1982. Biological denitrification, p. 289–326. *In* F. J. Stevenson (ed.), *Nitrogen in Agricultural Soils*. American Society of Agronomy, Madison, Wis.

Focht, D. D., and W. Verstraete. 1977. Biochemical ecology of nitrification and denitrification, p. 135–214. *In* M. Alexander (ed.), *Advances in Microbial Ecology*. Plenum Press, New York.

Folonoruso, O. A., and D. E. Rolston. 1984. Spatial variability of field-measured denitrification gas fluxes and soil properties. *Soil Sci. Soc. Am. J.* **49:**1087–1093.

Goodroad, L. L., and D. R. Keeney. 1984. Nitrous oxide production in aerobic soil under varying pH, temperature and water content. *Soil Biol. Biochem.* **16:**39–43.

Groffman, P. M., and J. M. Tiedje. 1989a. Denitrification in north temperate forest soils: relationships between denitrification and environmental factors at the landscape scale. *Soil Biol. Biochem.* **21:**621–626.

Groffman, P. M., and J. M. Tiedje. 1989b. Denitrification in north temperate forest soils: spatial and temporal patterns at the landscape and seasonal scales. *Soil Biol. Biochem.* **21:**613–620.

Groffman, P. M., J. M. Tiedje, D. L. Mokma, and S. Simkins. Submitted for publication.

Groffman, P. M., J. M. Tiedje, and C. W. Rice. Manuscript in preparation.

Groffman, P. M., J. M. Tiedje, G. P. Robertson, and S. Christensen. 1988. Denitrification at different temporal and geographical scales: proximal and distal controls, p. 174–192. *In* J. R. Wilson (ed.), *Advances in Nitrogen Cycling in Agricultural Ecosystems.* CAB International, Wallingford, U.K.

Hall, F. G., D. E. Strebel, and P. J. Sellers. 1988. Linking knowledge among spatial and temporal scales: vegetation, atmosphere, climate and remote sensing. *Landscape Ecol.* **2:**3–22.

Hole, F. D. 1978. An approach to landscape analysis with an emphasis on soils. *Geoderma* **21:**1–23.

Jeffers, J. N. R. 1988. Statistical and mathematical approaches to issues of scales in ecology, p. 47–56. *In* T. Rosswall, R. G. Woodmansee, and P. G. Risser (ed.), *Scales and Global Change: Spatial and Temporal Variability in Biospheric and Geospheric Processes.* John Wiley & Sons, Inc., New York.

Jenkinson, D. S., and D. S. Powlson. 1976. The effects of biocidal treatments on metabolism in soil. V. A method for measuring soil biomass. *Soil Biol. Biochem.* **8:**209–213.

Jenny, H. 1980. *The Soil Resource: Origin and Behavior.* Springer-Verlag, New York.

Keller, M., W. A. Kaplan, and S. C. Wofsy. 1986. Emissions of N_2O, CH_4 and CO_2 from tropical forest soils. *J. Geophys. Res.* **91:**11791–11802.

Knapp, A. K., and T. R. Seastedt. 1986. Detritus accumulation limits productivity of tallgrass prairie. *BioScience* **36:**662–668.

Lowrance, R., and P. M. Groffman. 1987. Impacts of low and high input agriculture on landscape structure and function. *Am. J. Altern. Agric.* **2:**175–183.

Matson, P. A., and R. C. Hariss. 1988. Prospects for aircraft-based gas exchange measurements in ecosystem studies. *Ecology* **69:**1318–1325.

Matson, P. A., and P. M. Vitousek. 1987. Cross-system comparisons of soil nitrogen transformations and nitrous oxide flux in tropical forest ecosystems. *Global Biogeochem. Cycles* **1:**163–170.

Matson, P. A., and P. M. Vitousek. 1990a. Ecosystem approach to a global nitrous oxide budget. *BioScience* **40:**677–672.

Matson, P. A., and P. M. Vitousek. 1990b. Remote sensing and trace gas fluxes, p. 157–167. *In* R. J. Hobbs and H. A. Mooney (ed.), *Remote Sensing of Biosphere Functioning.* Springer-Verlag, New York.

Matson, P. A., P. M. Vitousek, G. P. Livingston, and N. Swanberg. Sources of variation in nitrous oxide flux from Amazonian forests: effects of soil fertility and disturbance. *J. Geophys. Res.*, in press.

Matson, P. A., P. M. Vitousek, and D. S. Schimel. 1989. Regional extrapolation of trace gas flux based on soils and ecosystems, p. 97–108. *In* M. O. Andreae and D. S. Schimel (ed.), *Exchange of Trace Gases between Terrestrial Ecosystems and the Atmosphere.* Dahlem Konferenzen. John Wiley & Sons, Chichester, U.K.

McClaugherty, C. A., J. Pastor, J. D. Aber, and J. M. Melillo. 1985. Forest litter decomposition in relation to soil nitrogen dynamics and litter quality. *Ecology* **66:**266–275.

Mosier, A. R., W. J. Parton, and G. L. Hutchinson. 1983. Modelling nitrous oxide production from cropped and native soils, p. 229–242. *In* R. Hallberg (ed.), *Environmental Biogeochemistry.* Ecological Bulletins no. 35. Stockholm.

Myrold, D. D. 1988. Denitrification in ryegrass and winter wheat cropping systems of western Oregon. *Soil Sci. Soc. Am. J.* **52:**412–415.

Nieder, R., G. Schollmayer, and J. Richter. 1989. Denitrification in the rooting zone of cropped soils with regard to methodology and climate: a review. *Biol. Fertil. Soils* **8:**219–226.

O'Neill, R. V. 1988. Hierarchy theory and global change, p. 29–46. *In* T. Rosswall, R. G.

Woodmansee, and P. G. Risser (ed.), *Scales and Global Change: Spatial and Temporal Variability in Biospheric and Geospheric Processes*. John Wiley & Sons, Inc., New York.

O'Neill, R. V., D. L. DeAngelis, J. B. Waide, and T. F. H. Allen. 1986. *A Hierarchical Concept of the Ecosystem*. Princeton University Press, Princeton, N.J.

Parkin, T. B., J. L. Starr, and J. J. Meisinger. 1987. Influence of sample size on measurement of soil denitrification. *Soil Sci. Soc. Am. J.* **51:**1492–1501.

Parton, W. J., A. R. Mosier, and D. S. Schimel. 1988. Rates and pathways of nitrous oxide production in a shortgrass steppe. *Biogeochemistry* **6:**45–58.

Pastor, J., J. D. Aber, and C. A. McClaugherty. 1982. Geology, soils and vegetation of Blackhawk Island, Wisconsin. *Am. Midland Nat.* **108:**266–277.

Pastor, J., J. D. Aber, C. A. McClaugherty, and J. M. Melillo. 1984. Aboveground production and N and P cycling along a nitrogen mineralization gradient on Blackhawk Island, Wisconsin. *Ecology* **65:**256–268.

Pastor, J., and W. M. Post. 1986. Influence of climate, soil moisture, and succession on forest carbon and nitrogen cycles. *Biogeochemistry* **2:**3–27.

Poth, M., and D. D. Focht. 1985. ^{15}N kinetic analysis of N_2O production by *Nitrosomonas europaea*: an examination of nitrifier denitrification. *Appl. Environ. Microbiol.* **49:**1134–1141.

Reiners, W. A., L. Strong, P. A. Matson, I. Burke, and D. Ojima. Estimating biogeochemical fluxes across sagebrush-steppe landscapes with thematic mapper imagery. *Remote Sensing Environ.*, in press.

Ritchie, G. A. F., and D. J. D. Nicholas. 1972. Identification of the sources of nitrous oxide produced by oxidative and reductive processes in *Nitrosomonas europaea*. *Biochem. J.* **126:**1181–1191.

Robertson, G. P. 1986. Nitrogen: regional contributions to the global cycle. *Environment* **28:**16–21.

Robertson, G. P., and T. Rosswall. 1986. Nitrogen in West Africa: the regional cycle. *Ecol. Monogr.* **56:**43–72.

Robertson, G. P., and J. M. Tiedje. 1984. Denitrification and nitrous oxide production in old growth and successional Michigan forests. *Soil Sci. Soc. Am. J.* **48:**383–389.

Robertson, G. P., and J. M. Tiedje. 1988. Deforestation alters denitrification in a lowland tropical rain forest. *Nature* (London) **336:**756–759.

Running, S. W., and J. C. Coughlan. 1988. A general model of forest ecosystem processes for regional applications. I. Hydrological balance, canopy gas exchange and primary production processes. *Ecol. Modelling* **42:**125–154.

Sahrawat, K. L., and D. R. Keeney. 1986. Nitrous oxide emission from soils. *Adv. Soil Sci.* **4:**103–148.

Salati, E., R. S. Bradley, and R. L. Victoria. 1982. Regional gains and losses of nitrogen in the Amazon basin. *Plant Soil* **67:**367–376.

Schaetzl, R. J. 1986. Soilscape analysis of contrasting glacial terrains in Wisconsin. *Ann. Assoc. Am. Geographers* **76:**414–425.

Schimel, D. S., D. C. Coleman, and K. A. Horton. 1985a. Soil organic matter dynamics in paired rangeland and cropland toposequences in North Dakota. *Geoderma* **36:**201–214.

Schimel, D. S., S. Simkins, T. Rosswall, A. R. Mosier, and W. J. Parton. 1988. Scale and the measurement of nitrogen-gas fluxes from terrestrial ecosystems, p. 179–194. *In* T. Rosswall, R. G. Woodmansee, and P. G. Risser (ed.), *Scales and Global Change: Spatial and Temporal Variability in Biospheric and Geospheric Processes*. John Wiley & Sons, Inc., New York.

Schimel, D. S., M. A. Stillwell, and R. G. Woodmansee. 1985b. Biogeochemistry of C, N and P in a soil catena of the shortgrass steppe. *Ecology* **66:**276–282.

Schmidt, J., W. Seiler, and R. Conrad. 1988. Emission of nitrous oxide from temperate forest soils into the atmosphere. *J. Atmospher. Chem.* **6:**95–115.

Schmugge, T. 1983. Remote sensing of soil moisture with microwave radiometers. *Trans. Am. Soc. Agric. Eng.* **26:**748–753.

Sellers, P. J., F. G. Hall, G. Asrar, D. E. Strebel, and R. E. Murphy. 1988. The first ISLSCP field experiment (FIFE). *Bull. Am. Meteorol. Soc.* **69:**22–27.

Smith, M. S., and J. M. Tiedje. 1979. Phases of denitrification following oxygen depletion in soil. *Soil Biol. Biochem.* **11:**262–267.

Soil Conservation Service, USDA. 1975. Soil survey of Riley county and part of Geary county, Kansas. USDA Soil Conservation Service, Washington, D.C.

Swanson, F. J., T. K. Kratz, N. Caine, and R. G. Woodmansee. 1988. Landform effects on ecosystem patterns and process. *BioScience* **38:**92–98.

Tiedje, J. M. 1988. Ecology of denitrification and dissimilatory nitrate reduction to ammonium, p. 179–244. *In* A. J. B. Zehnder (ed.), *Biology of Anaerobic Microorganisms.* John Wiley & Sons, Inc., New York.

Tucker, C. J., and P. J. Sellers. 1986. Satellite remote sensing of primary productivity. *Internat. J. Remote Sensing* **7:**1395–1416.

Vitousek, P. M., and J. M. Melillo. 1979. Nitrate losses from disturbed forests: patterns and mechanisms. *Forest Sci.* **25:**605–619.

Wessman, C. A., J. D. Aber, D. L. Peterson, and J. M. Melillo. 1988. Remote sensing of canopy chemistry and nitrogen cycling in temperate forest ecosystems. *Nature* (London) **335:**154–156.

Woodmansee, R. G. 1988. Ecosystem processes and global change, p. 11–28. *In* T. Rosswall, R. G. Woodmansee, and P. G. Risser (ed.), *Scales and Global Change: Spatial and Temporal Variability in Biospheric and Geospheric Processes.* John Wiley & Sons, Inc., New York.

Zak, D. R., K. S. Pregitzer, and G. E. Host. 1986. Landscape variation in nitrogen mineralization and nitrification. *Can. J. Forest Res.* **16:**1258–1263.

Chapter 12

Fluxes of Nitrous Oxide and Nitric Oxide from Terrestrial Ecosystems

Eric A. Davidson

Atmospheric chemists might wonder why nitric oxide (NO) and nitrous oxide (N_2O) are addressed in the same chapter. Nitric oxide is very reactive in the troposphere, where it participates in numerous reactions affecting acid deposition and photochemical production of tropospheric ozone (Logan, 1983). In contrast, N_2O is stable in the troposphere and is radiatively important as a greenhouse gas (Ramanathan, 1988). In the stratosphere, N_2O participates in reactions leading to destruction of stratospheric ozone (Cicerone, 1987). Although their atmospheric consequences are very different, NO and N_2O should be studied together by biologists, because both are products of the same microorganisms. Figure 1 presents the major pathways of NO and N_2O production and consumption.

Chemoautotrophic nitrifying bacteria gain energy from the oxidation of NH_4^+ to NO_2^- and NO_3^-. A number of possible biochemical pathways exist for NO and N_2O production via nitrification (Firestone and Davidson, 1989). Denitrifying bacteria can both produce and consume NO and N_2O, as they use oxides of N for terminal electron acceptors in respiration when O_2 is limiting. A number of soil microorganisms besides nitrifying and denitrifying bacteria are capable of producing NO and N_2O (Robertson and Tiedje, 1987), but their significance in soils is unknown. Several abiological reactions involving NO and N_2O also occur in soil (Nelson, 1982). Self-decomposition of HNO_2 produces NO and NO_2, and reactions of soil organic matter with HNO_2 can produce NO and N_2O.

The objectives of this paper are (i) to relate observed patterns of fluxes to our understanding of cellular and ecosystem regulation of nitrification, denitrification, and abiological processes and (ii) to identify areas of research needed to improve global budgets and to assess the effects of present and future human activities on global budgets.

Eric A. Davidson • Woods Hole Research Center, P.O. Box 296, Woods Hole, Massachusetts 02543.

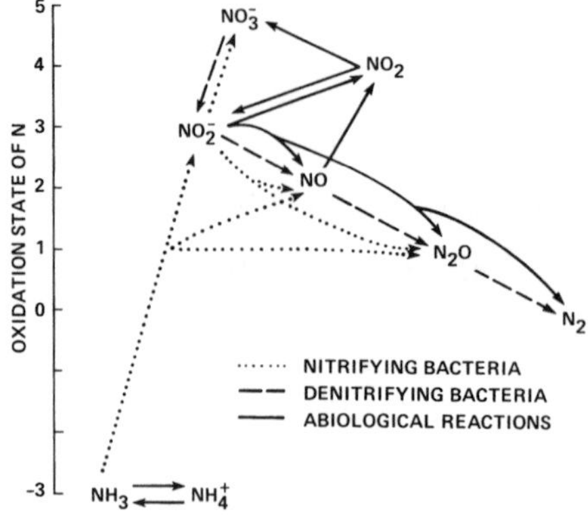

Figure 1. Biological and abiological processes of production and consumption of NO and N_2O.

GLOBAL BUDGETS

Nitrous Oxide

A budget published by McElroy and Wofsy (1986) and an update based on recent literature are given in Table 1. The original estimate by McElroy and Wofsy of 4 Tg of N_2O-N produced per year by combustion of fossil fuels is very close to the 3.5 Tg annual accumulation that has been observed in the atmosphere. Hence, it appeared that increased use of fossil fuels could account for increasing atmospheric N_2O concentrations. Hypothesized increases in biogenic soil emissions of N_2O due to increasing use of N fertilizers in modern agriculture (Council for Agricultural Science and Technology, 1976) or due to accelerating rates of deforestation in the tropics no longer appeared relevant. However, a recent discovery of an artifact in methods used to measure N_2O from industrial sources (Muzio et al., 1989) requires that the fossil fuel combustion estimate be lowered to a negligible source and has renewed interest in biogenic sources.

The budget of McElroy and Wofsy also indicated a very important biogenic source of N_2O from tropical terrestrial ecosystems. Based on only a few field studies, this preliminary estimate stimulated more research in the tropics. Although the data are still fewer than desirable, Matson and Vitousek (1990) have calculated new estimates from the tropics, using published flux data from over 30 sites. They stratified the land area of the humid tropical forest biome by taxonomic classification of soil orders. They then multiplied the area of each stratum by the arithmetic mean of flux estimates from studies conducted on soils of that taxonomic order. They have also observed higher N_2O fluxes from pastures as compared with adjacent intact forests and have calculated the increase in N_2O fluxes due to conversion of humid tropical forests to pastures. Unfortunately, nearly all the data for the tropics are from humid regions. Only one study can be used to represent fluxes from all seasonally dry tropical forests and pastures. The new estimates confirm the importance of tropical ecosystems, although the estimated total

Table 1. Global budget of nitrous oxide

Sinks and sources	N_2O budget (Tg of N_2O-N $year^{-1}$)	
	McElroy and Wofsy (1986)	Update
Sinks and accumulation		
Stratospheric photolysis and reactions	10.6	
Accumulation in the atmosphere	3.5	
Total	14 ± 3.5	
Sources		
Ocean	2	2
Combustion		
Coal and oil	4	< 0.1[a]
Biomass	0.7	0.1–0.3[b]
Fertilized agricultural land	0.8	0.5(0.2–2.1)[c]
Temperate grasslands	0.1	0.1
Boreal and temperate forests	0.1–0.5	0.3–1.5[d]
Tropical and subtropical forests		
Forests and woodlands	7.4	
Humid tropical forests		2.4[e]
Humid forest conversion to pasture		0.7[e]
Seasonal dry tropical forests		1.0[f]
Dry forest conversion		?
Total	15.3 ± 6.7	7.9 ± ?

[a] Calculated from effect of sampling artifact reported by Muzio et al. (1989).
[b] Crutzen and Andreae (1990).
[c] Eichner (1990).
[d] Higher estimates from Schmidt et al. (1988).
[e] Matson and Vitousek (1990).
[f] Vitousek et al. (1989).

contribution from the tropics is less than the original estimate by McElroy and Wofsy.

Only relatively minor changes are suggested for the other categories in the McElroy and Wofsy budget. A recent study of N_2O fluxes from European forests that have received significant inputs of N deposition via acid rain (Schmidt et al., 1988) indicates that the total flux from temperate forests may be somewhat larger than earlier estimates that were based on North American forests. The bottom line of the revised budget, however, shows that the global N_2O budget is unbalanced and that significant errors could exist in any of the current estimates.

Ronan et al. (1988) have suggested that contaminated aquifers may be a forgotten component of the global N_2O budget. Their measurements of N_2O dissolved in groundwater and aquifers of sites fertilized with sewage effluents, and similar measurements in agricultural drains (Dowdell et al., 1979) and in seeps from a clearcut forest (Bowden and Bormann, 1986), show that N_2O concentrations dissolved in the liquid phase are often significantly elevated. Concentrations of

Table 2. Global budget of nitric oxide (Logan, 1983)

Sinks and sources	NO budget (Tg of NO-N $year^{-1}$)
Sinks	
Precipitation	12–42
Dry deposition	12–22
Total	24–64
Sources	
Fossil fuel combustion	21 (14–28)
Biomass burning	12 (4–24)
Lightning	8 (2–20)
Microbial activity in soils	8 (4–16)[a]
Oxidation of ammonia	1–10[b]
Photolytic or biological processes in the ocean	<1
Input from the stratosphere	0.5
Total	25–99

[a] Note that this estimate is not in agreement with a new estimate for the soil emission NO source given in Table 3.
[b] Oxidation of ammonia may provide a sink for NO_x of similar magnitude.

N_2O in liquid phase have been 1 to 4 orders of magnitude higher than the N_2O concentration that would be expected if the soil water were in equilibrium with atmospheric N_2O concentration. To assess the global significance of these measurements, assumptions must be made regarding the representativeness of these estimates for defined categories of land areas, the areal extent of such categories, the rate at which groundwater and soil water enter surface streams, lakes, and seas, and the proportion of dissolved N_2O that escapes to the atmosphere. Davidson and Swank (1990) found that N_2O concentrations in soil water of a disturbed temperate hardwood forest were 30 times ambient, but when extrapolated globally based on a "worst-case" scenario of the above assumptions, the flux to the atmosphere was insignificant. Similar assessments are needed for agricultural systems and land irrigated with sewage effluent.

Nitric Oxide

The wide ranges of values given by Logan (1983) in her global NO budget (Table 2) reveal the uncertainty of NO budget estimates. The estimate for "microbial activity in soils" is based on global extrapolation of the median flux observed in one study of pastures in Australia. Penner et al. (1991) used this value for a global mean of soil emissions in a general circulation model (GCM), but weighted the soil source regionally so that NO fluxes were higher from tropical regions and lower from temperate regions. The model predicted global distribution of tropospheric NO that generally agreed well with observed NO concentrations. The

Table 3. Biogenic sources of nitric oxide by ecosystem type

Ecosystem type	Area[a] (10^{12} m^2)	Length of season (days)	Flux (ng of N cm^{-2} h^{-1})		No. of flux estimates	Global flux estimate (Tg of N $year^{-1}$)
			Mean	Range of fluxes		
Cultivated land						
Temperate	7	200	6.4	0.2–34	6	2.2
Tropical[b]	9	360	6.4		0	5.0
Temperate grassland	8	200	1.5	0.6–3.6	4	0.6
Temperate forest	9	200	0.2	0.1–0.4	3	0.1
Tropical rain forest	11	360	2.3	0.2–4.4	4	2.2
Tropical seasonally dry forest	5	360	0.8	0.1–57	1	0.4
Savanna	19	360	4.7	0.1–23	3	7.7
Chaparral and sclerophyll forest	4	360	5.8	3.6–7.6	1	2.0
Mangrove, swamp, marsh	2	360	< 0.1	0–0.04	1	0.0
Desert and semi-desert	30				0	
Peatland, mixed forest, tiaga	19				0	
Tundra, polar desert	8				0	
Total					23	20.2

[a] Area estimates are from Bolin et al. (1979).
[b] No estimates from tropical agricultural systems are available. The same daily flux rate observed in temperate agricultural systems is assumed.

purpose of this modeling exercise was to use NO data to test the validity of the GCM coupled with equations for photochemical reactions in the trophosphere. It would be instructive to biologists to take the reverse approach and to use GCMs to conduct a sensitivity analysis of variation in biogenic NO. We know that considerable variation in biogenic NO fluxes exists across ecosystem types, but we do not currently know how important this variation is for regional and global tropospheric chemistry.

Rather than list each of the numerous flux estimates published since Logan's 1983 budget, a global inventory of biogenic NO sources stratified by ecosystem type is offered in Table 3. The mean flux estimates for studies cited by Johansson (1989) and a few more that have since become available (see Tables 4 and 5 and Williams and Fehsenfeld, 1991) provide the data base. The arithmetic mean for each ecosystem type was calculated, and each mean was multipled by the respective areal extent of that ecosystem type.

Obviously, 23 flux estimates are too few to develop a reliable estimate of a global budget. About 44% of the earth's land is covered by ecosystem types for which NO flux data are not available. Nevertheless, compared with previous global estimates based on extrapolation of a single study, an estimate based on 23 studies should provide some improvement. Perhaps this inventory of biogenic NO sources based on incomplete data will help stimulate further research, as did McElroy and Wofsy's global budget of N_2O.

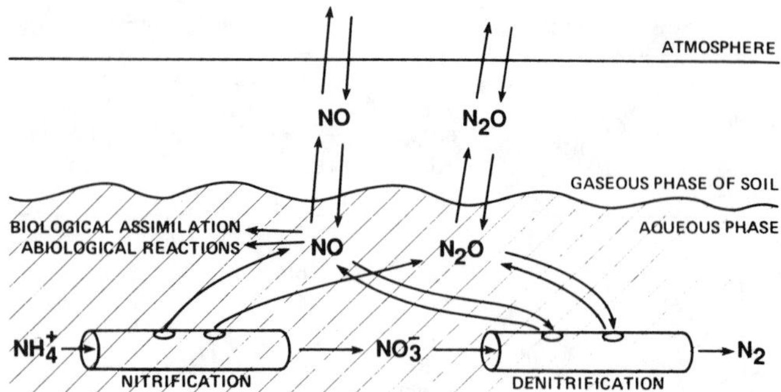

Figure 2. Three levels of regulation of NO and N_2O flux: (i) the rates of nitrification and denitrification (amount of N flowing through the pipes); (ii) the ratios of end products (the size of the holes in the pipes); and (iii) diffusion and consumption of N gases prior to escape from the soil to the atmosphere.

The global total for biogenic emissions of NO is over 2 times greater than Logan's estimate of 8 Tg of NO-N $year^{-1}$. However, this higher value should be regarded with caution. Most of the increase is due to high fluxes observed in a pair of studies in a Venezuelan savanna with a sandy loam soil (Johansson and Sanhueza, 1988; Johansson et al., 1988). Another Venezuelan savanna with a clay loam soil had very low NO emissions (Sanhueza et al., 1990). More data are needed from savannas to improve estimates of global emissions.

Relatively high fluxes have also been observed in chaparrals and seasonally dry tropical forests, which, like savannas, have prolonged dry seasons and distinct wet seasons. Seasonal variation in soil moisture and episodic wetting events may have important implications for N cycling processes in general and N trace gas production in particular. If dry areas that receive occasional precipitation are important NO sources, then the gap in data for deserts and semi-deserts should be filled.

REGULATION OF NO AND N_2O FLUXES

Firestone and Davidson (1989) have presented a "hole-in-the-pipe" model (Fig. 2) to show that three levels of regulation are relevant: (i) the factors affecting rates of nitrification and denitrification (the amount of N flowing through the pipe); (ii) the factors that affect relative proportions of end products produced (the size of the holes in the pipe); and (iii) the factors that affect gaseous diffusion through the soil and to the atmosphere.

Regulation of N Cycling Rates

An example of the first level of regulation is a correlation between net N mineralization and N_2O fluxes observed in several humid tropical forests grouped by soil order (Matson and Vitousek, 1987). At these sites, the net N mineralization assay (the net increase in soil inorganic N during a 10-day aerobic laboratory incubation) served as a good index of the rate of N cycling and soil fertility among humid tropical forest soils. Fluxes of N_2O increased along a soil fertility gradient of increasing net N mineralization. However, this relationship did not hold for nearby pastures, where N_2O fluxes were much higher than would be expected based on net mineralization assays (Matson et al., 1990). Hence, the net mineralization assay was valuable as long as apples were compared to apples (i.e., humid forests of varying fertility were compared to each other), but it lost its value when an orange was thrown in (i.e., a pasture). It is possible that the net mineralization assay does not provide the same information regarding N cycling rates within both forest and pasture soils. The net mineralization assay is an index of N cycling rates and is not a direct measure of actual gross rates of microbial mineralization and immobilization of N. More direct measures of gross rates of microbial transformations of soil N could help reveal differences in microbial processes among ecosystem types. The second and third levels of regulation, factors affecting the proportion of end products and diffusion of gases, might also account for differences in N_2O fluxes between intact forests and adjacent pastures.

A possible example of the first level of regulation (effect of N cycling rates) for NO flux reveals the importance of excess soil N relative to soil C. A correlation was observed between soil NO_3^- and NO fluxes across a range of forest and agricultural sites in the United States (Williams and Fehsenfeld, 1991). Presence of a large soil NO_3^- pool indicates that nitrification has occurred, but more importantly, it indicates that soil N is in excess relative to soil C. In agricultural soils, N is added as fertilizer, and soil C is depleted when tillage exposes soil organic matter and enhances rates of decomposition. Temperate forest soils, in contrast, generally are rich in organic C and are N limited. Excess NH_4^+ availability relative to C availability in agricultural soil results in "flow" of N through the nitrification and perhaps the denitrification "pipes," with concomitant production of NO. Soil NO_3^- was a good indicator of excess N relative to C in the soils of widely contrasting ecosytem types studied by Williams and Fehsenfeld (1991). However, differences in NO fluxes at sites within the same ecosystem type were poorly correlated with soil NO_3^-. Perhaps factors affecting end products (second or third levels of regulation) would help explain within-ecosystem variation.

Williams and Fehsenfeld (1991) also observed a significant diel effect: NO fluxes were positively correlated with soil temperature at each site. However, diel variation in NO fluxes within sites was less than 1 order of magnitude, whereas variation in mean NO fluxes across sites was about 3 orders of magnitude. While diel effects must be considered when making flux measurements, N cycling characteristics of ecosystems are probably of greatest significance for development of regional and global budgets.

Regulation of End Product Proportions

The second level of regulation occurs at the cellular level in the sense that inhibition of enzymes (e.g., nitrous oxide reductase) and relative availabilities of electron donors (usually organic C) and electron acceptors (O_2 and N oxides) affect the physiological functions of denitrifying bacteria (Firestone and Davidson, 1989). Similarly, acidity, pO_2, and [NH_4^+] at membrane surfaces affect activities of nitrifying bacteria. Differences in end product ratios could also be inherent properties of certain species of organisms. However, with the exception of identification of a few organisms that lack N_2O reductase, no species-specific characterization of end product ratios for nitrifying or denitrifying bacteria has been demonstrated.

Although much is understood about how environmental factors affect the physiology of N gas production at the cellular level (Firestone and Davidson, 1989), relating ecosystem properties to end product ratios remains challenging. For example, laboratory studies have shown that when availability of electron acceptors (NO_3^-) is high relative to the availability of electron donors (organic C) the primary end product of denitrification is N_2O; when the converse is true, N_2 is the dominant end product (Firestone et al., 1980). Although this relationship has been elegantly demonstrated in the laboratory, predicting availability of organic C and NO_3^- at the microsite scale of a denitrifying bacterium in heterogeneous soils in situ is complicated by numerous sources and sinks, spatial heterogeneity, and diffusional constraints. Similarly, increasing acidity inhibits N_2O reductase in the laboratory, but the importance of variation in soil acidity to N_2O:N_2 ratios from denitrification has not been demonstrated in the field.

Drying and wetting cycles probably cause important temporal changes in end products. Numerous studies have demonstrated increased fluxes of NO and N_2O following precipitation and irrigation events (Table 4; Anderson et al., 1988; Davidson and Swank, 1986; Hao et al., 1988; Johansson et al., 1988; Johansson and Sanhueza, 1988; Levine et al., 1988; Slemr and Seiler, 1984; Williams et al., 1987). Nitrate and NO_2^- can accumulate in dry soil due to low rates of nitrification occurring in the absence of plant and microbial sinks of inorganic N (Davidson et al., 1990). Both organic C and inorganic N become rapidly available when dry soil is wetted, due to turnover of microbial biomass that died from desiccation and starvation in dry soil (Bottner, 1985). Nitrifying and denitrifying bacteria can become active within minutes of wetting of very dry soil (Davidson et al., 1990; Rudaz et al., in press). Immediately following wetting, a pulse of microbial respiration reduces the partial pressure of O_2 within the soil. Soil moisture, porosity, drainage, and rate of respiration affect O_2 diffusion and consumption. Variation in degree of anaerobiosis within the soil permits both nitrification and denitrification to occur at different microsites. Denitrification dominates in very wet soils, whereas nitrification dominates at more moderate soil moisture (Linn and Doran, 1984). In the field, the dominant trace gas product of denitrification is probably N_2O, whereas NO fluxes often exceed N_2O fluxes from nitrification (Davidson et al., in press; Rudaz et al., in press; Tortoso and Hutchinson, 1990).

Consumption of N Gases

A third level of regulation shown in Fig. 2 illustrates the effects of gaseous diffusion on possible consumption of gaseous products before they escape the soil.

Table 4. Seasonal comparisons of NO and N_2O fluxes

Ecosystem type	Flux (ng of N cm^{-2} h^{-1})	
	Dry season	Wet season
Humid forest		
NO	4.4[a]	0.8[b]
N_2O	1.5–3.5[c]	1.5–3.5[c]
Humid pasture		
NO	ND[d]	ND
N_2O	3.0–4.0[c]	6.0–11.0[c]
Dry seasonal forest		
NO	0.3[e]	1.0[e]
	50 (wet-up)[e]	0.7 (wet-up)[e]
N_2O	0.2[e,f]	0.9[f]
	5–12 (wet-up)[e,f]	0.9 (wet-up)[f]
Savanna		
NO	2.9[g]	1.8 (dry soil)[h]
	30–60 (wet-up)[g]	54–90 (after rain)[h]
N_2O	0.4[i]	ND
	1.6 (wet-up)[i]	ND

[a] Kaplan et al. (1988).
[b] Bakwin et al. (1990).
[c] Matson et al. (1990).
[d] ND, no data.
[e] Davidson et al., in press; "wet-up" means an experimental wetting.
[f] Vitousek et al. (1989).
[g] Johansson et al. (1988).
[h] Johansson and Sanhueza (1988).
[i] Hao et al. (1988).

Biogenic NO and N_2O must move from aqueous solution to the gaseous phase of the soil and then must diffuse through gas-filled pores before being emitted to the atmosphere. The probability of consumption increases as impediments to diffusion increase. Because more possible consumptive fates exist for NO than for N_2O, this third level of regulation may be particularly important for NO. In a laboratory study, net NO release from a soil sample was affected by the thickness of the sample, while N_2O was unaffected (Remde et al., 1989). Apparently, NO consumption increased as the diffusional path length of the soil sample increased. Unlike N_2O, which contains a double bond between two N atoms, NO might be easily assimilated by plants and microorganisms as an N source. Furthermore, denitrifying bacteria tend to utilize more oxidized forms, NO_2^- and NO_3^-, before utilizing N_2O and, similarly, may utilize NO preferentially to N_2O.

A possible example of the third level of regulation (gaseous diffusion) is provided by a comparison of NO and N_2O fluxes from clayey and sandy soils of humid tropical forests of Brazil and savannas of Venezuela (Table 5). Fluxes of NO exceed N_2O fluxes in the sandy soils, but the opposite is true for the clayey soils. Because gaseous diffusion is less restricted in sandy soil than in the clayey soil, NO consumption within sandy soil may not be important. In contrast, consumption of

Table 5. N-gas fluxes and soil texture

Ecosystem type	Flux (ng of N cm^{-2} h^{-1}) in soil texture:	
	Clayey	Sandy
Humid tropical forests		
Nitric oxide	0.8[a]	2.8[a]
Nitrous oxide	2.0[b]	0.3[b]
Savannas		
Nitric oxide	0.1[c]	2.9[d]
Nitrous oxide	0.5[c]	0.4[e]

[a] Bakwin et al. (1990).
[b] Matson et al. (1990).
[c] Sanhueza et al. (1990).
[d] Johansson et al. (1988).
[e] Hao et al. (1988).

NO along the tortuous diffusional pathway of a clayey soil could be very significant. Hence, the gross production rates of NO and N_2O could be similar in the clayey soil, but greater NO consumption could account for lower fluxes of NO from the soil surface. In short, we cannot always distinguish between the effects of differences in rates of gross production and gross consumption.

Soil moisture can also impede gaseous diffusion and thus affect NO consumption. Fluxes of N_2O did not vary seasonally in a humid tropical forest, but NO fluxes were higher in the dry season than in the wet season (Table 4). Gross production rates of NO could be higher in the dry season, or restricted diffusion and higher consumption rates could account for lower fluxes in the wet season.

Gross NO production rates were 1 to 2 orders of magnitude greater under anaerobic conditions than under aerobic conditions in a laboratory study (Remde et al., 1989). However, net release of NO was strongly dependent on the flow rate of gas purging the flask containing the soil, indicating that diffusion and consumption of NO were significant factors affecting net release. Sterilization eliminated the ability of the soil to consume NO, indicating that the consumptive process was biological. Hence, while NO production via denitrification may occur, net NO fluxes may be relatively low from wet soils with high rates of denitrification. The larger fluxes of NO observed from sandy soils and during the dry season (Tables 4 and 5) probably result from modest rates of gross production under conditions that permit NO escape.

IDENTIFICATION OF SOURCES

Higher field fluxes of N_2O when fertilizer is added as NO_3^- rather than NH_4^+ have been given as evidence for denitrification as the major source of N_2O (Bakwin et al., 1990; Keller et al., 1988; Livingston et al., 1988). These results prove that a potential for increased denitrification exists, but they do not prove that N_2O from unfertilized plots is necessarily from denitrification.

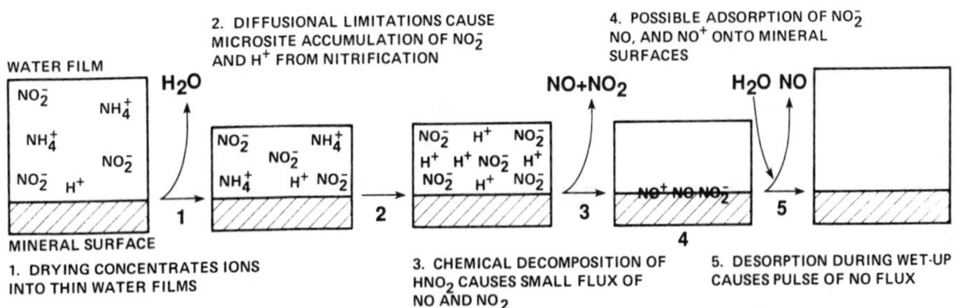

Figure 3. Conceptualization of hypothesized interactions of nitrification and abiological reactions that produce NO in dry and wetted soil.

In laboratory studies, modest production of N_2O via nitrification has been demonstrated by observing partial inhibition of N_2O production in moderately moist soils in the presence of acetylene (C_2H_2), a potent inhibitor of nitrification (Davidson et al., 1986; Rudaz et al., in press). However, when soil moisture was above field capacity (i.e., soil macropores were water filled), N_2O production rates were 1 to 2 orders of magnitude higher and were not inhibited by C_2H_2, indicating that denitrification was the dominant source. Although flux rates may be highest via denitrification, well-drained soils probably produce N_2O at high rates only during brief periods following episodic wetting events. More modest production rates via nitrification between wetting events could still be important relative to total annual emissions.

Fluxes of NO have increased following addition of NO_3^- (Johansson et al., 1988) and NH_4^+ and urea (Slemr and Seiler, 1984). Production of NO in dry soils strongly suggests that a nondenitrifying source is important at least some of the time.

Occurrence of NO and N_2O production via nitrification might result from a combination of NH_4^+ oxidation to NO_2^- by nitrifying bacteria and subsequent abiological reactions of HNO_2 with organic matter or HNO_2 with itself. These abiological reactions are concentration dependent, and thus are affected by soil [NO_2^-], soil pH, and the amount of soil organic matter (Blackmer and Cerrato, 1986). It is interesting to note that soil NO_2^- was remarkably high (2 to 4 μg of NO_2^--N g^{-1} dry soil) at the savanna sites of Venezuela where very high NO fluxes have also been observed (Johansson and Sanhueza, 1988).

A conceptual model is proposed in Fig. 3 whereby drying and wetting events could affect biological NO production and flux. Soil drying causes concentration of ions, including H^+ and NO_2^-, in thin water films. Nitrification occurs in dry soil, albeit at low rates, and both H^+ and NO_2^- are produced by NH_4^+-oxidizing bacteria. Although soil NO_2^- pools are usually very low, microsite accumulations could be important when diffusion of ions is limited. Abiological production of NO may occur where NO_2^- and H^+ are concentrated in thin water films near sites of NH_4^+ oxidation. If NO_2^- and H^+ are adsorbed to clay surfaces, then NO production might occur when they are desorbed following wetting. Adsorption

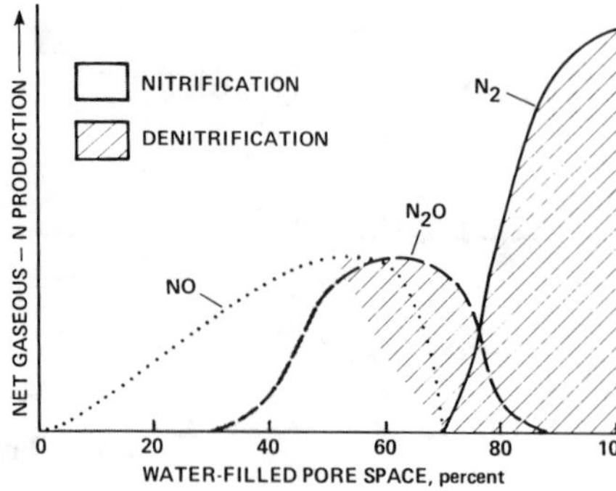

Figure 4. Model of the relationship between WFPS (water-filled pore space) of soil and relative fluxes of N gases.

and desorption of NO and NO^+ are also possible, but very little is known about the potential for NO adsorption onto soil surfaces. The model presented in Fig. 3 is highly speculative and is offered in hope that it will stimulate much-needed research on these processes.

A useful expression of soil moisture for predicting fluxes of gases is water-filled pore space (WFPS). Data from a variety of sites show that denitrification becomes increasingly important as WFPS increases above 60% (Davidson et al., 1986; Linn and Doran, 1984; Parton et al., 1988; Rolston et al., 1984). When WFPS exceeds 80%, N_2O consumption occurs and N_2 becomes the major end product. Optimal moisture contents for nitrification range between about 30% and 70% WFPS (Davidson et al., 1986; Linn and Doran, 1984; Parton et al., 1988), with N_2O production occurring towards the wetter end of that range. Similar data for NO are lacking. However, significant NO fluxes have been measured in dry soil and increase upon wetting (Table 4), but soil moisture above field capacity has reduced NO fluxes (Anderson and Levine, 1987). A synthesis of these data that is consistent with observed flux measurements and with our understanding of cellular regulation of nitrification and denitrification end products is offered in Fig. 4. Production of NO via nitrification or a combination of nitrification and abiological reactions dominates in dry soils. At intermediate moisture contents, both NO and N_2O are produced by both nitrification and denitrification, and much of the gaseous production results in net flux from the soil. Under moderately wet conditions, more of the NO produced is consumed before escaping the soil and N_2O either gradually diffuses from the soil or remains in aqueous phase until gaseous diffusion occurs as the soil dries. Under very wet conditions only N_2 is produced. The relative flux rates and the moisture contents at which inflection points of these curves occur are very speculative. Further development of this model may help predict seasonal variation of NO and N_2O fluxes within sites.

EFFECTS OF HUMAN DISTURBANCE OF ECOSYSTEMS

Table 1 includes an estimate by Matson and Vitousek (1990) of enhanced N_2O fluxes due to conversion of humid tropical forest to pastures. Their estimate is equal

Table 6. Effects of burning on inorganic-N and N-gas fluxes

Site	Inorganic N (μg of N g^{-1} soil)		N gas flux (ng of N cm^{-2} h^{-1})	
	NH_4^+	NO_3^-	NO	N_2O
Venezuelan savanna[a]				
Preburn	33	0.7	3–15	ND[b]
Postburn	38	1.9	30–100	ND
Californian chaparral[c]				
Preburn, dry	13–16	1–6	9.7	ND
Preburn, watered	ND	ND	21.4	ND
Postburn, dry	22–72	2–5	13.3	<5
Postburn, watered	ND	ND	60.7	53.9

[a] Johansson et al. (1988).
[b] ND, no data.
[c] Anderson et al. (1988).

to about 20% of the observed annual increase in atmospheric N_2O. Interestingly, they did not find similar increases in N_2O fluxes from forest clearings. Apparently, destruction of the forest and increased solar radiation impinging on the soil are not the important factors, but rather some (as yet unidentified) characteristic of a tropical pasture ecosystem appears critical. Furthermore, fluxes from intact forests did not exhibit seasonal variation, whereas fluxes from the pastures were highest during the wet season and significantly lower during the dry season (Luizao et al., 1989). Preliminary data from forests and pastures in Costa Rica fail to confirm that conversion of humid tropical forest to pasture increases N_2O fluxes, but rather indicate that N_2O fluxes can be lower in some pastures (Keller and Reiners, 1990). Too little is known about the N cycling processes in these ecosystems to speculate about mechanistic explanations for these observations. More data are also needed from seasonally dry tropical forests and savannas, where human habitation has had the greatest impact on landscapes.

Biomass burning is recognized as a source of NO and N_2O (Tables 1 and 2). In addition to emissions from the fire itself, however, enhanced fluxes of NO and N_2O from the soil after the burn are probably important sources that are usually omitted from estimates of burning emissions. Burning has been shown to increase inorganic N pools and NO and N_2O flux rates (Table 6). The combination of burning followed by wetting produced the greatest observed N_2O flux. Fire temporarily reduces the plant sinks for inorganic N, and it also reduces the microbial sink for N by removing litter with high C:N ratios. This lack of sinks of inorganic N results in favorable conditions for both nitrification and denitrification. Increased burning to clear tropical forests, where the potential for high NO and N_2O fluxes has already been shown in studies of unburned areas, could result in an important contribution to regional and global N trace gas budgets.

Changing patterns of agricultural practices in the tropics are also worthy of note. In the dry seasonal forest area of western Mexico, maize is grown for only 1 or 2 years following forest clearing, and then the site is used for pasture. Although fluxes of NO and N_2O were three times higher in maize fields than in intact forests, most of the pastures did not exhibit elevated fluxes relative to the forest (Davidson et al., in press; Vitousek et al., 1989). Hence, the long-term effect of deforestation in regions of seasonally dry tropical forests is unclear. In the humid tropics of Central America, denitrification rates were lower in the aggrading forests of abandoned agricultural fields than in either cleared fields or intact primary forest (Robertson and Tiedje, 1988). If aggrading forests in abandoned agricultural fields become an important land use classification, then emissions of N trace gases from the region could be lower as a result of cutting primary forest. Hence, the period of agricultural use, the type of agriculture practiced, and the ultimate fate of the land must be considered to assess the long-term impact of shifting agriculture on N-gas fluxes.

RESEARCH NEEDS

Obviously, more flux data are needed to improve regional and global budgets. The tropics seem to be the major biogenic source of N_2O, although uncertainty remains in all estimates. The potential for contaminated aquifers to contribute to the global atmospheric N_2O budget needs assessment by hydrologic models and global information on aquifer quality. High NO flux rates from savannas and other ecosystems with a prolonged dry season need confirmation.

Flux measurements are time consuming and expensive, especially in remote areas. While more flux data are needed, other approaches should also be pursued. Better mechanistic models will aid efforts to extrapolate from existing data sets to areas not sampled and to areas that are undergoing human disturbance. Progress has been made by identifying and focusing upon different levels of regulation of each mechanistic process, but applying these conceptual models to variation in the field remains challenging. Although N cycling rates are clearly important, measures or indices of N cycling that are applicable across broad ranges of ecosystem types have not been developed. Relating ecosystem and soil properties to our understanding of cellular control of end product ratios of nitrification and denitrification will be especially difficult. The processes of gaseous diffusion in soil, abiological reactions, sorption phenomena of NO_2^-, NO, and NO^+, and diffusional constraints of microbial processes in dry soil deserve more attention.

Whether better estimates are derived from more field flux measurements or from predictive mechanistic models, application of these estimates to global scales will require better land use data. Better data on rates of forest cutting and fates of cleared land could be obtained with existing remote sensing technology. Future developments in remote sensing might also provide information about soil moisture, soil type, and vegetation characteristics that characterize N cycling rates. Ideally, mechanistic models of N gas production would be driven by remotely sensed data.

Finally, microbial production of radiatively important trace gases provides an excellent example of the need to bridge scales of investigation and disciplines of study. This paper has attempted to relate microbial processes that occur on the scale of a microbe with a cross-sectional area of 10^{-12} m^2 to generalizations about vegetative biomes that span over 10^{12} m^2. This range in scale of investigation roughly corresponds to a range in scientific disciplines, from soil microbiologist, to ecosystems ecologist, to global modeler. Although many references have been cited, data sets from interdisciplinary studies that are designed to address extrapolation across scales are few. Hence, the generalizations offered in the present paper are tenuous at best. Identifying these weaknesses, however, should help investigators design research that fosters integration across scales and disciplines.

Acknowledgments. I thank Mary Firestone, Pamela Matson, and Peter Vitousek for many hours of discussion that contributed to the ideas presented here. I also thank the National Research Council Associateship Program for support at the NASA Ames Research Center while preparing this paper.

REFERENCES

Anderson, I. C., and J. S. Levine. 1987. Simultaneous field measurements of biogenic emissions of nitric oxide and nitrous oxide. *J. Geophys. Res.* **92:**965–976.

Anderson, I. C., J. S. Levine, M. A. Poth, and P. J. Riggan. 1988. Enhanced biogenic emissions of nitric oxide and nitrous oxide following surface biomass burning. *J. Geophys. Res.* **93:**3893–3898.

Bakwin, P. S., S. C. Wofsy, S. Fan, M. Keller, S. Trumbore, and J. M. daCosta. 1990. Emission of nitric oxide (NO) from tropical forest soils and exchange of NO between the forest canopy and atmospheric boundary layers. *J. Geophys. Res.* **95:**16755–16764.

Blackmer, A. M., and M. E. Cerrato. 1986. Soil properties affecting formation of nitric oxide by chemical reactions of nitrite. *Soil Sci. Soc. Am. J.* **50:**1215–1218.

Bolin, B., E. T. Degens, P. Duvigneaud, and S. Kempe. 1979. The global biogeochemical carbon cycle, p. 1–53. *In* B. Bolin, E. T. Degens, S. Kempe, and P. Ketner (ed.), *The Global Carbon Cycle.* John Wiley & Sons, Inc., New York.

Bottner, P. 1985. Response of microbial biomass to alternate moist and dry conditions in a soil incubated with ^{14}C- and ^{15}N-labelled plant material. *Soil Biol. Biochem.* **17:**329–337.

Bowden, W. B., and F. H. Bormann. 1986. Transport and loss of nitrous oxide in soil water after forest clear-cutting. *Science* **233:**867–869.

Cicerone, R. J. 1987. Changes in stratospheric ozone. *Science* **237:**35–42.

Council for Agricultural Science and Technology. 1976. Effect of increased nitrogen fixation on stratospheric ozone. Report no. 53. Iowa State University, Ames.

Crutzen, P. J., and M. O. Andreae. 1990. Biomass burning in the tropics: impact on atmospheric chemistry and biogeochemical cycles. *Science* **250:**1669–1678.

Davidson, E. A., J. M. Stark, and M. K. Firestone. 1990. Microbial production and consumption of nitrate in an annual grassland. *Ecology* **71:**1968–1975.

Davidson, E. A., and W. T. Swank. 1986. Environmental parameters regulating gaseous-N losses from two forested ecosystems via nitrification and denitrification. *Appl. Environ. Microbiol.* **52:**1287–1292.

Davidson, E. A., and W. T. Swank. 1990. Nitrous oxide dissolved in soil solution: an insignificant pathway of nitrogen loss from a southeastern hardwood forest. *Water Resour. Res.* **26:**1687–1690.

Davidson, E. A., W. T. Swank, and T. O. Perry. 1986. Distinguishing between nitrification and denitrification as sources of N_2O from soil. *Appl. Environ. Microbiol.* **52:**1280–1286.

Davidson, E. A., P. M. Vitousek, P. A. Matson, R. Riley, G. Garcia-Mendez, and J. M. Maass. Soil emissions of nitric oxide in a seasonally dry tropical forest of Mexico. *J. Geophys. Res.*, in press.

Dowdell, R. J., J. R. Burford, and R. Crees. 1979. Losses of nitrous oxide dissolved in drainage water from agricultural land. *Nature* (London) **278:**342–343.

Eichner, M. J. 1990. Nitrous oxide emissions from fertilized soils: summary of available data. *J. Environ. Qual.* **19:**272–280.

Firestone, M. K., and E. A. Davidson. 1989. Microbiological basis of NO and N_2O production and consumption in soil, p. 7–21. *In* M. O. Andreae and D. S. Schimel (ed.), *Exchange of Trace Gases between Terrestrial Ecosystems and the Atmosphere*. John Wiley & Sons, Inc., New York.

Firestone, M. K., R. B. Firestone, and J. M. Tiedje. 1980. Nitrous oxide from soil denitrification: factors controlling its biological production. *Science* **208:**749–751.

Hao, W. M., D. Scharffe, P. J. Crutzen, and E. Sanhueza. 1988. Production of N_2O, CH_4, and CO_2 from soils in the tropical savanna during the dry season. *J. Atmos. Chem.* **7:**92–105.

Johansson, C. 1989. Fluxes of NO_x above soil and vegetation, p. 229–246. *In* M. O. Andreae and D. S. Schimel (ed.), *Exchange of Trace Gases between Terrestrial Ecosystems and the Atmosphere*. John Wiley & Sons, Inc., New York.

Johansson, C., H. Rodhe, and E. Sanhueza. 1988. Emission of NO in a tropical savanna and a cloud forest during the dry season. *J. Geophys. Res.* **93:**7180–7192.

Johansson, C., and E. Sanhueza. 1988. Emission of NO from savanna soils during rainy season. *J. Geophys. Res.* **93:**14193–14198.

Kaplan, W. A., S. C. Wofsy, M. Keller, and J. M. DaCosta. 1988. Emission of NO and deposition of O_3 in a tropical forest system. *J. Geophys. Res.* **93:**1389–1395.

Keller, M., W. A. Kaplan, S. C. Wofsy, and J. M. DaCosta. 1988. Emissions of N_2O from tropical forest soils: response to fertilization with NH_4^+, NO_3^-, and PO_4^{3-}. *J. Geophys. Res.* **93:**1600–1604.

Keller, M., and W. A. Reiners. 1990. Emissions of nitrous oxide and nitric oxide from wet tropical forest soils. *EOS* **71:**1259.

Levine, J. S., W. R. Cofer III, and D. I. Sebacher. 1988. The effects of fire on biogenic soil emissions of nitric oxide and nitrous oxide. *Global Biogeochem. Cycles* **2:**445–449.

Linn, D. M., and J. W. Doran. 1984. Effect of water-filled pore space on carbon dioxide and nitrous oxide production in tilled and nontilled soils. *Soil Sci. Soc. Am. J.* **48:**1267–1272.

Livingston, G., P. M. Vitousek, and P. A. Matson. 1988. Nitrous oxide fluxes and nitrogen transformations across a landscape gradient in Amazonia. *J. Geophys. Res.* **93:**1593–1599.

Logan, J. 1983. Nitrogen oxides in the troposphere: global and regional budgets. *J. Geophys. Res.* **88:**10785–10807.

Luizao, F., P. Matson, G. Livingston, R. Luizao, and P. Vitousek. 1989. Nitrous oxide flux following tropical land clearing. *Global Biogeochem. Cycles* **3:**281–285.

Matson, P. A., and P. M. Vitousek. 1987. Cross-system comparisons of soil nitrogen transformations and nitrous oxide flux in tropical forest ecosystems. *Global Biogeochem. Cycles* **1:**163–170.

Matson, P. A., and P. M. Vitousek. 1990. Ecosystem approach to a global nitrous oxide budget. *Bioscience* **40:**667–672.

Matson, P. A., P. M. Vitousek, G. P. Livingston, and N. A. Swanberg. 1990. Sources of variation in nitrous oxide flux from amazonian ecosystems. *J. Geophys. Res.* **95:**16789–16798.

McElroy, M. B., and S. C. Wofsy. 1986. Tropical forests: interactions with the atmosphere, p. 33–60. *In* G. T. Prance (ed.), *Tropical Rain Forests and the World Atmosphere*. Westview, Boulder, Colo.

Muzio, L. J., M. E. Teague, J. C. Kramlich, J. A. Cole, J. M. McCarthy, and R. K. Lyon. 1989. Errors in grab sample measurements of N_2O from combustion sources. *J. Air Pollut. Control Assoc.* **39:**287–293.

Nelson, D. W. 1982. Gaseous losses of nitrogen other than through denitrification, p. 327–364. *In* F. J. Stevenson (ed.), *Nitrogen in Agricultural Soils*. American Society of Agronomy, Madison, Wis.

Parton, W. J., A. R. Mosier, and D. S. Schimel. 1988. Rates and pathways of nitrous oxide production in a shortgrass steppe. *Biogeochemistry* **6:**45–58.

Penner, J. E., C. S. Atherton, J. Dignon, S. J. Ghan, J. J. Walton, and S. Hameed. 1991. Tropospheric nitrogen: a three-dimensional study of sources, distributions, and deposition. *J. Geophys. Res.* **96:**959–991.

Ramanathan, V. 1988. The radiative and climatic consequences of the changing atmospheric composition of trace gases, p. 159–186. *In* F. S. Rowland and I. S. A. Isaksen (ed.), *The Changing Atmosphere*. John Wiley & Sons, Inc., New York.

Remde, A., F. Slemr, and R. Conrad. 1989. Microbial production and uptake of nitric oxide in soil. *FEMS Microb. Ecol.* **62:**221–230.

Robertson, G. P., and J. M. Tiedje. 1987. Nitrous oxide sources in aerobic soils: nitrification, denitrification and other biological processes. *Soil Biol. Biochem.* **19:**187–193.

Robertson, G. P., and J. M. Tiedje. 1988. Deforestation alters denitrification in a lowland tropical rain forest. *Nature* (London) **336:**756–759.

Rolston, D. E., P. S. C. Rao, J. M. Davidson, and R. E. Jessup. 1984. Simulation of denitrification losses of nitrate fertilizer applied to uncropped, cropped, and manure-amended field plots. *Soil Sci.* **137:**270–279.

Ronan, D., M. Magaritz, and E. Almon. 1988. Contaminated aquifers are a forgotten component of the global N_2O budget. *Nature* (London) **335:**57–59.

Rudaz, A., E. A. Davidson, and M. K. Firestone. Production of nitrous oxide immediately after wetting dry soil. *FEMS Microb. Ecol.*, in press.

Sanhueza, E., W. M. Hao, D. Scharffe, L. Donoso, and P. J. Crutzen. 1990. N_2O and NO emissions from soils of the northern part of the Guayana Shield, Venezuela. *J. Geophys. Res.* **95:**22481–22488.

Schmidt, J., W. Seiler, and R. Conrad. 1988. Emission of nitrous oxide from temperate forest soils into the atmosphere. *J. Atmos. Chem.* **6:**95–115.

Slemr, R., and W. Seiler. 1984. Field measurements of NO and NO_2 emissions from fertilized and unfertilized soils. *J. Atmos. Chem.* **2:**1–24.

Tortoso, A. C., and G. L. Hutchinson. 1990. Contributions of autotrophic and heterotrophic nitrifiers to soil NO and N_2O emissions. *Appl. Environ. Microbiol.* **56:**1799–1805.

Vitousek, P. M., P. A. Matson, C. Volkmann, J. M. Maass, and G. Garcia. 1989. Nitrous oxide flux from dry tropical forests. *Global Biogeochem. Cycles* **3:**375–382.

Williams, E. J., and F. C. Fehsenfeld. 1991. Measurement of soil nitrogen oxide emissions at three North American ecosystems. *J. Geophys. Res.* **96:**1033–1042.

Williams, E. J., D. D. Parrish, and F. C. Fehsenfeld. 1987. Determination of nitrogen oxide emissions from soils: results from a grassland site in Colorado, United States. *J. Geophys. Res.* **92:**2173–2179.

Chapter 13

Cycling of NO_x in Tropical Forest Canopies

Daniel J. Jacob and Peter S. Bakwin

Nitrogen oxides (NO_x = NO + NO_2) play a central role in the chemistry of the troposphere. They regulate the photochemical production of ozone and the abundance of the hydroxyl radical (OH), which is the main oxidant for a number of trace gases including methane. Sources of NO_x to the troposphere include fossil fuel combustion, biomass burning, biogenic emissions from soils, and lightning (Logan, 1983; Table 1). Atmospheric oxidation of NO_x takes place on a time scale on the order of 1 day and produces nitric acid which is removed by deposition; this deposition is a major source of nitrogen and acidity to terrestrial ecosystems.

Simulation of NO_x is currently a top priority in the development of global models for atmospheric chemistry (Brost et al., 1988; Levy and Moxim, 1989). However, as shown in Table 1, the size of the biogenic source is a serious uncertainty. Biogenic emissions could dominate the budget of NO_x over the continental tropics, where soil NO_x fluxes are high and anthropogenic emissions are relatively low (Johansson et al., 1988; Johansson and Sanhueza, 1988; Jacob and Wofsy, 1988, 1990). Proper accounting of this source requires consideration not only of the magnitude of soil NO_x emissions (Davidson, this volume), but also of the fraction of those emissions that is lost by deposition to vegetation during transport from the soil to canopy top. The latter effect has been so far neglected in the construction of atmospheric source inventories, but we will argue here that it can reduce considerably the export of NO_x from forested canopies (by about a factor of 4). As a result, current global estimates of the biogenic source of NO_x to the atmosphere may be seriously exaggerated.

BIOSPHERE-ATMOSPHERE EXCHANGE OF NO_x

The biosphere is both a source and a sink for atmospheric NO_x. Microbes in soils emit NO_x (mainly as NO), while vegetation scavenges NO_x (mainly as NO_2). The mechanisms for NO_x deposition to vegetation have been reviewed recently by Johansson (1989). The principal route appears to be uptake of NO_2 by the plant stomata, followed by reduction to nitrite at the mesophyll and assimilation via

Daniel J. Jacob and Peter S. Bakwin • Department of Earth and Planetary Sciences and Division of Applied Sciences, Harvard University, 29 Oxford Street, Cambridge, Massachusetts 02138.

Table 1. Global source inventory for NO_x in the troposphere[a]

Source	NO_x (Tg of N year^{-1})
Fossil fuel combustion	21 (14–28)
Biomass burning	12 (4–24)
Soil emissions	8 (4–16)
Lightning	8 (2–20)
Atmospheric oxidation of NH_3	1–10
Oceans	<1
Stratospheric input	<1
Total	25–99

[a] Data from Logan (1983). Numbers are best estimates and possible ranges.

nitrite reductase (Rogers et al., 1979). Uptake of NO is negligibly slow compared to uptake of NO_2. Johansson (1989) stressed the need to consider the balance between soil emissions of NO on the one hand and NO_2 deposition to vegetation and to the ground on the other hand in assessing the flux of NO_x between the biosphere and the atmosphere. Indeed, measurements in polluted regions have documented net downward fluxes of NO_x to vegetation, due to the uptake of NO_x advected from anthropogenic sources upwind (Delany and Davies, 1983; Delany et al., 1986).

We wish to examine how the biosphere-atmosphere exchange of NO_x over tropical forests should be treated in regional and global models for atmospheric chemistry. These models generally use the top of the canopy as the lower boundary. The biogenic source of NO_x in the model is defined by the ventilation flux, F_v (molecules per square centimeter per second), of biogenic NO_x at canopy top, viz.:

$$F_v = \alpha E_{NO} \quad (1)$$

where E_{NO} is the soil emission flux of NO (molecules per square centimeter per second) and α is an "export efficiency" ($0 \leq \alpha \leq 1$) representing the fraction of NO_x emitted by soil that is ventilated to the atmosphere above the canopy. Estimates for α will be presented below. The ventilation flux, F_v, is balanced by a downward flux, F_d (molecules per square centimeter per second), of NO_x from the atmosphere to the canopy, which is usually expressed in terms of a deposition velocity, V_d (centimeters per second):

$$F_d = V_d(NO_x) \quad (2)$$

where (NO_x) is the concentration of NO_x (molecules per cubic centimeter) immediately above the canopy. Estimates for V_d will also be presented below. The net upward flux of NO_x at canopy top (i.e., at the lower boundary of the model) is the difference $F_v - F_d$.

Rapid chemical cycling between NO and NO_2 takes place inside a forest canopy (Fig. 1; Table 2). NO is oxidized to NO_2 by O_3 subsiding from aloft and by

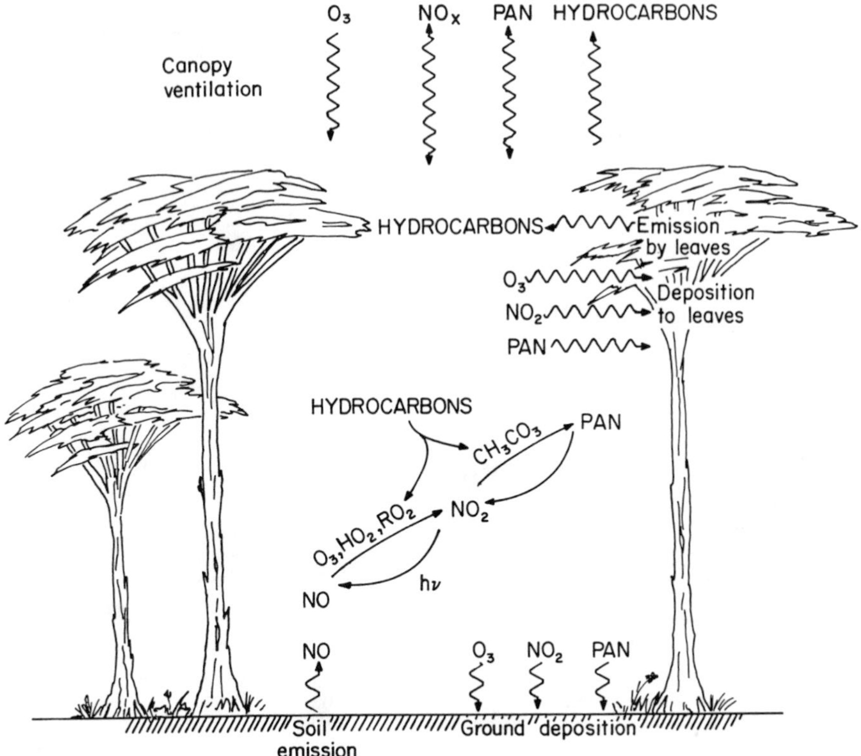

Figure 1. Chemical cycling of NO_x in a forest canopy.

peroxy radicals produced within the canopy from decomposition of biogenic hydrocarbons (e.g., isoprene and terpenes). The oxidation of NO takes place on a time scale of minutes, less than typical time scales for canopy ventilation (Trumbore et al., 1990). In the daytime NO_2 is photolyzed back to NO, and a chemical equilibrium between NO and NO_2 is established where the NO_2/NO molar ratio is of order unity. At night there is no mechanism for reaction of NO_2 back to NO, so that NO may be quantitatively converted to NO_2.

The NO_x budget inside the canopy is complicated by conversion of NO_2 to $CH_3C(O)OONO_2$ (peroxyacetylnitrate, abbreviated as PAN). This conversion is driven by CH_3CO_3 (peroxyacetyl) radicals generated locally from photochemical oxidation of biogenic hydrocarbons (Jacob and Wofsy, 1990). The lifetime of PAN against thermal decomposition back to NO_2 is relatively short at the temperatures found in tropical forest canopies (e.g., 35 min at 300 K), so that a chemical equilibrium may be established between NO_2 and PAN. Depending on the efficiency of PAN deposition to vegetation and to the ground, the NO_2/PAN equilibrium could affect the export of biogenic NO_x out of the canopy.

Table 2. Important reactions cycling NO_x inside a forest canopy

Reaction	Rate constant (cm^3 molecule^{-1} s^{-1} or s^{-1})
$NO + O_3 \longrightarrow NO_2$ or O_2	$2.2 \times 10^{-12} e^{-1,430/T}$ [a]
$NO_2 + h\nu \xrightarrow{O_2} NO + O_3$	5.6×10^{-3} [b]
$O_3 + h\nu \xrightarrow{H_2O} 2\ OH + O_2$	4.3×10^{-6} [b,c]
Hydrocarbons $+ OH \xrightarrow{O_2} RO_2$	$2.5 \times 10^{-11} e^{409/T}$ [d]
$NO + RO_2 \longrightarrow NO_2 + RO$	$4.2 \times 10^{-12} e^{180/T}$
$RO + O_2 \longrightarrow R'CHO, R'O_2, HO_2$	Fast[e]
$NO + HO_2 \longrightarrow NO_2 + OH$	$3.7 \times 10^{-12} e^{240/T}$
$CH_3COCHO + h\nu \xrightarrow{2O_2} CH_3CO_3 + HO_2 + CO_2$	8.4×10^{-4} [b,f]
$NO_2 + CH_3CO_3 \longrightarrow CH_3C(O)OONO_2$	4.7×10^{-12}
$CH_3C(O)OONO_2 \longrightarrow NO_2 + CH_3CO_3$	$1.95 \times 10^{16} e^{-13,543/T}$

[a] T is temperature in units of K.
[b] Photolysis rate constant at noon at canopy top, computed at the equator with a Rayleigh scattering code assuming 30% opaque cloud cover overhead, an O_3 column of 7.1×10^{18} molecules cm^{-2}, and a canopy albedo of 0.1. The photolysis rate constants inside the canopy are reduced because of extinction of light by leaves; correction factors are given in the text.
[c] For $P_{H_2O} = 24.7$ mb.
[d] Oxidation of hydrocarbons by OH produces organic peroxy radicals (RO_2). The rate constant is strongly dependent on hydrocarbon type; the value shown here is for oxidation of isoprene.
[e] Depending on the nature of the RO radical, products from this reaction may include HO_2, organic peroxy radicals, aldehydes, ketones, and dicarbonyls.
[f] Model calculations by Jacob and Wofsy (1990) indicate that this reaction (where methylglyoxal is produced from oxidation of isoprene) accounts for $\approx 90\%$ of CH_3CO_3 production in the Amazon forest canopy.

The data base of NO_x flux measurements is limited, due in part to the difficulty of measuring NO_x concentrations accurately (Johansson, 1989). In particular there are at this time no reliable data for NO_x fluxes over forest canopies. Ideally, a field experiment designed to study the biosphere-atmosphere exchange of NO_x over a forest should include the following measurements: (i) soil emission fluxes of NO, (ii) turbulent fluxes of NO_x at canopy top, (iii) concentrations of trace gases inside and just above the canopy (NO, NO_2, O_3, PAN, and isoprene are key species), (iv) leaf and ground resistances to deposition, and (v) rates of vertical mass exchange within the canopy and at canopy top. The data set collected during the NASA/ABLE-2B expedition to the Amazon forest in the wet season of 1987 (Harriss et al., 1990) comes closest at this time to meeting the above requirements. As part of this expedition, extensive data were collected from a 40-m-high tower erected in a terra firme forest 20 km northeast of Manaus, Brazil. Trace gas concentrations at high altitudes were measured by aircraft. We will see below that these data afford important constraints on the export of biogenic NO_x out of the Amazon forest canopy.

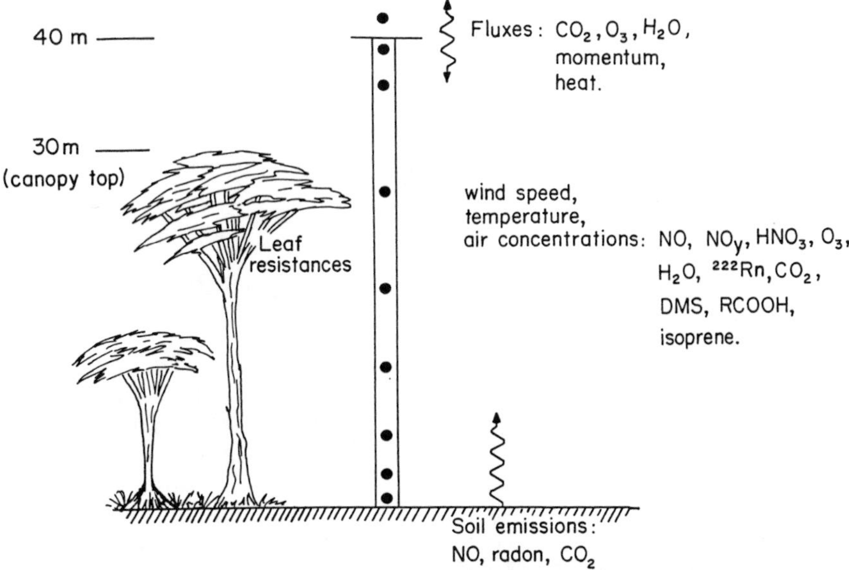

Figure 2. Ensemble of measurements made at the Ducke Forest Reserve, 20 km northeast of Manaus, Brazil, during the ABLE-2B expedition in April–May 1987. Fluxes and air concentrations were measured from a 40-m-high tower erected through the forest canopy. Fluxes were measured at the top of the tower by the eddy correlation method. Air concentrations were measured at several altitudes, indicated by the solid circles. Soil emissions were measured using static (closed) chambers. Leaf resistances were measured by Roberts et al. (1990) during earlier expeditions at the site. Further details on investigators and methods are given by Harriss et al. (1990).

OBSERVATIONAL CONSTRAINTS ON THE NO_x BUDGET IN THE AMAZON FOREST CANOPY

Overview of the ABLE-2B Data

Figure 2 gives an overview of the measurements available from the ABLE-2B tower site. The forest canopy was 30 m high, with emergents up to 35 m and a leaf area index of 7 (the leaf area index is the total leaf area per unit area of air column, counting only one side of leaf; it is a dimensionless measure of vegetation loading). Soil fluxes of NO from the predominant clay soils averaged 8.9×10^9 molecules cm^{-2} s^{-1} with no significant diurnal variation (Bakwin et al., 1990b). Concentrations of NO inside the canopy decreased sharply with altitude (Fig. 3), reflecting the emission from soil and the rapid oxidation to NO_2 (Bakwin et al., 1990a). Maximum concentrations of NO were observed at night when canopy ventilation was restricted and O_3 concentrations were near zero. Aircraft measurements at 100 to 300 m altitude indicated NO concentrations of 12 $\pm$ 7 parts per trillion by volume (ppt), much lower than inside the canopy.

Concentrations of NO_y, representing the sum of all reactive nitrogen oxides

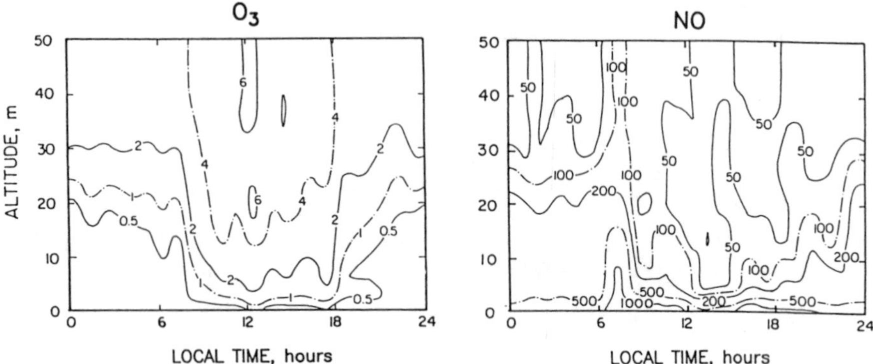

Figure 3. Average concentrations of O_3 and NO measured at the ABLE-2B tower site as a function of time of day. Units are parts per billion by volume (ppb) for O_3 and parts per trillion by volume (ppt) for NO.

(including NO_x, HNO_3, PAN, and other organic nitrates), were measured at 19 and 39 m altitude by Bakwin et al. (1990a) (Fig. 4). Values at 19 m averaged 400 ppt with little diurnal variation; values at 39 m were slightly higher, implying a net downward flux of NO_y to the canopy. The principal components of NO_y were not clearly identified. Concentrations of HNO_3 were only 20 to 50 ppt (Talbot et al., 1990), and concentrations of PAN measured from aircraft at 100 to 300 m altitude were less than 20 ppt (Singh et al., 1990). Concentrations of NO_2 were not measured. Model calculations (Jacob and Wofsy, 1990) suggest that NO_2 could have accounted for most of the NO_y in the canopy at night, but not in the daytime.

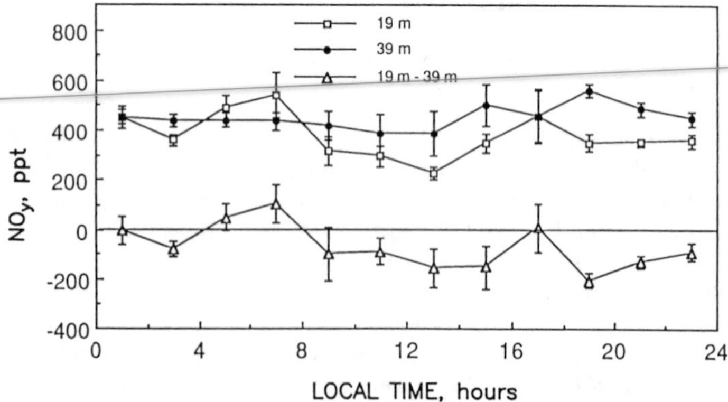

Figure 4. Average concentrations of NO_y measured at the ABLE-2B tower site as a function of time of day at 19 and 39 m altitude. Figure adapted from Bakwin et al. (1990a).

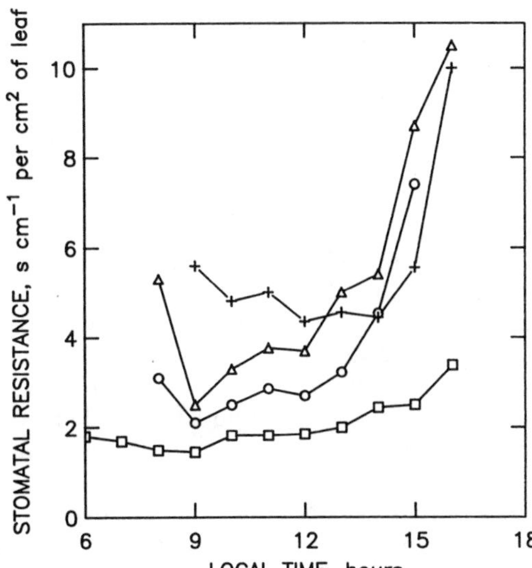

Figure 5. Average stomatal resistances measured at the ABLE-2B tower site as a function of time of day for four species present at different altitudes. Symbols: □, *Piptadenia suaveolens* at 33 m; ○, *Naucleopsis glabra* at 17 m; △, *Gustavia angusta* at 8 m; +, *Scheelea* sp. at 0 to 3 m. Figure adapted from Roberts et al. (1990).

Organic nitrates subsiding from aloft were probably the major contributors to daytime NO_y.

Concentrations of O_3 within the canopy (Fig. 3) were very low compared to values commonly observed in surface air (Logan, 1985), reflecting the rapid deposition of O_3 to the forest vegetation. Deposition velocities of O_3 measured at 40 m altitude averaged 1.8 cm s^{-1} in the daytime (Fan et al., 1990), a factor of 2 higher than typical values observed over mid-latitudes forests (Lenschow et al., 1982). At night the deposition velocities for O_3 were lower but still substantial (averaging 0.3 cm s^{-1}).

Deposition of O_3, NO_2, and other gases to vegetation is thought to take place primarily at the stomata of leaves (Hicks et al., 1987). Stomatal and boundary (leaf-atmosphere) resistances to water vapor transfer were measured at the tower site by Roberts et al. (1990) during expeditions in 1983 to 1985 which preceded ABLE-2B. The boundary resistances were always small compared to the stomatal resistances. The stomatal resistances were themselves quite low (Fig. 5), consistent with observations in other tropical forests (J. Roberts, personal communication, 1989), and presumably reflecting the high insolation and the availability of water. Lowest values for the stomatal resistances were observed for trees in the upper canopy; these increased gradually from mid-morning to mid-afternoon, possibly as a result of water stress.

The low stomatal resistances of the Amazon forest vegetation, together with the high leaf area index, indicate the potential for rapid uptake of trace gases. One can derive the stomatal resistances to uptake of a particular gas by using the measured values for water vapor and scaling by the ratio of molecular diffusivities.

Fan et al. (1990) explained their observed daytime deposition velocities for O_3 during ABLE-2B by using a standard leaf resistance model (Hicks et al., 1987) constrained with the stomatal resistance data of Roberts et al. (1990). The deposition of O_3 observed at night (when the stomata were closed) suggests that significant O_3 uptake took place also at the outer (cuticular) surfaces of leaves (Fan et al., 1990).

A key variable regulating the export of biogenic NO_x out of the canopy is the canopy ventilation rate. Much of the canopy ventilation at the ABLE-2B site occurred by episodic large-scale downdrafts flushing the entire canopy (Fitzjarrald and Moore, 1990; Fitzjarrald et al., 1990). This ventilation mechanism is typical of forest environments in general (Gao et al., 1989) and cannot be properly described with standard mixing-length models for atmospheric turbulence. Vertical mass exchange rates during ABLE-2B can nevertheless be constrained quite well by using the observed vertical distributions and fluxes of three chemical tracers: radon-222, CO_2, and O_3 (Trumbore et al., 1990; Fan et al., 1990). Radon-222 is a particularly useful tracer as it is released at a relatively constant and uniform rate by the soil, and it is removed from the atmosphere solely by radioactive decay (half-life, 3.8 days). Trumbore et al. (1990) used the radon-222 and CO_2 data from ABLE-2B to derive air residence times (τ) within the 0–40-m column of 5.5 h at night and ≤ 1 h in the daytime. Sharp vertical gradients of radon-222 concentration were observed in the lowest 2 m above ground, but above that altitude the concentrations were relatively uniform.

A Box Model for the Amazon Forest Canopy

The ABLE-2B data place some important constraints on the export of biogenic NO_x from the forest to the atmosphere. These constraints can be expressed in a simple way with a steady-state box model for the 0–40-m air column, including no explicit chemistry. In the next section we will present a more detailed process-based model which includes temporal and vertical resolution, as well as a full description of photochemistry; we will see that results from the box model capture to a good approximation the main features of the complicated model. The upper boundary of the box model is chosen at 40 m because the canopy ventilation times (τ) given by Trumbore et al. (1990) are defined with respect to that altitude. We define "biogenic NO_x" as the NO_x supplied to the 0–40-m column by soil emission, in contrast to "atmospheric NO_x" supplied to the column from aloft.

The ventilation flux (F_v) of biogenic NO_x through the top of the 0–40-m air column is given by:

$$F_v = \frac{\Delta Z}{\tau}(NO_x) \tag{3}$$

where (NO_x) is the concentration in the column and $\Delta Z = 40$ m. Replacing in equation 1, we obtain an expression for the export efficiency α of biogenic NO_x out of the canopy:

$$\alpha = \frac{(NO_x)}{E_{NO}} \frac{\Delta Z}{\tau} \tag{4}$$

where E_{NO} is the soil emission flux. Although (NO_x) was not measured in ABLE-2B, an upper limit is imposed by the NO_y concentrations shown in Fig. 4: $(NO_y) \approx 400$ ppt implies that biogenic $(NO_x) < 400$ ppt. At night when $\tau \approx 5.5$ h, this upper limit on biogenic (NO_x) places a severe constraint on α. Inserting the average observed value $E_{NO} = 8.9 \times 10^9$ molecules cm^{-2} s^{-1} into equation 4, and converting (NO_x) to units of molecules per cubic centimeter (1 ppt $= 2.43 \times 10^7$ molecules cm^{-3}), we find $\alpha < 0.22$. In the daytime, by contrast, the NO_y data afford no constraint on α; replacing $\tau = 1$ h into equation 4 yields the useless result $\alpha < 1.2$. Most of the NO_y in daytime was contributed by non-NO_x species, as opposed to nighttime (see Overview of the ABLE-2B Data, above). The upper limit, $(NO_x) < 400$ ppt, is evidently not sufficiently stringent to constrain α in the daytime.

An estimate for α in the daytime can, however, be derived from the deposition flux of biogenic NO_x to vegetation. Let R (seconds per centimeter) be the total leaf resistance to NO_2 deposition per square centimeter of air column, and let (NO_2) be the concentration of biogenic NO_2 in the 0–40-m column. The deposition flux of biogenic NO_x is given by $(NO_2)/R$ and represents the balance between the soil emission flux, E_{NO}, and the ventilation flux, F_v:

$$\frac{(NO_2)}{R} = E_{NO} - F_v \tag{5}$$

By combining equations 3 through 5 we obtain:

$$\alpha = \frac{1}{1 + \dfrac{\tau\,(NO_2)}{R \Delta Z (NO_x)}} \tag{6}$$

The ratio $(NO_2)/(NO_x)$ in daytime is of order 0.7 (see below). An upper limit for R can be estimated by assuming that deposition of NO_2 is restricted to the leaf stomata. We then decompose R as the sum of a boundary resistance (R_b), a stomatal resistance (R_s), and a mesophyllic resistance (R_m) placed in series (Hicks et al., 1987):

$$R = \frac{R_b + \beta R_s + R_m}{L} \tag{7}$$

Here $L = 7$ is the leaf area index, $\beta \approx 1.6$ is the ratio of the molecular diffusivities of water vapor and NO_2, and R_b, R_s, and R_m are in units of seconds per centimeter per square centimeter of leaf. Assuming $R_m = 0$ (Wesely, 1989), mean values for

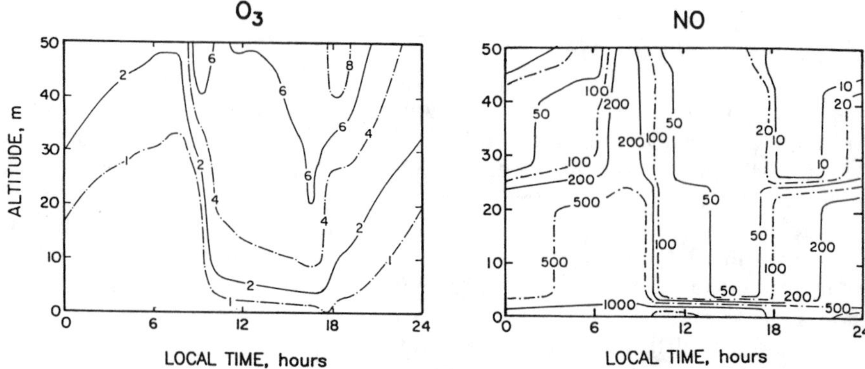

Figure 6. Concentrations of O_3 (ppb) and NO (ppt) simulated by the model of Jacob and Wofsy (1990), as a function of altitude and time of day. Results are in good agreement with observations (compare with Fig. 3). Figure adapted from Jacob and Wofsy (1990).

R_b and R_s of 0.5 and 3 s cm^{-1} per cm^2 of leaf, respectively (Roberts et al., 1990), and $\tau = 1$ h, we obtain $\alpha = 0.56$. Averaging the daytime and nighttime results yields a 24-h mean value of $\alpha \leq 0.39$. It appears therefore that only a small fraction of NO emitted by soil is exported to the atmosphere above the canopy.

ONE-DIMENSIONAL MODEL OF NO_x CYCLING IN THE AMAZON FOREST CANOPY

The box model presented above suffers from obvious flaws, in particular the assumption of uniform NO_x concentrations and the neglect of deposition to the ground and to the cuticles of leaves. An improved estimate of the NO_x budget in the Amazon forest canopy can be obtained by using the one-dimensional process-based model of Jacob and Wofsy (1990). This model was developed to simulate the atmospheric chemistry observed during the ABLE-2B expedition; it has proven successful at reproducing various observations from tower and aircraft, in particular the NO and O_3 concentrations inside the canopy (Fig. 6). The full model extends from the ground to 2,000 m altitude and includes (i) vertical transport rates constrained by observations for radon-222, O_3, and CO_2; (ii) detailed photochemistry describing the oxidation of biogenic hydrocarbons and the cycle of reactive nitrogen oxides; and (iii) biosphere-atmosphere exchange regulated by local vegetation density, leaf resistances, insolation, and temperature.

We consider here a truncated model domain, extending from 0 to 40 m altitude only, with fixed concentrations at 40 m as upper boundary conditions. The domain is subdivided into four grid cells (Table 3); cells 1 to 3 are in the canopy (0 to 30 m), and cell 4 is above the canopy (30 to 40 m). Each cell is assumed to be individually well mixed and exchanges air with adjacent cells by turbulent diffusion. The

Table 3. Structure of the Jacob and Wofsy (1990) model for the Amazon forest canopy[a]

Grid cell no.[b]	Atmospheric column[b] (m)	Leaf area index[c]	Exchange velocity at grid cell top[d] ($cm\ s^{-1}$)	
			0900–1600	1600–0900
1	0–2	1	0.13	0.4
2	2–20	2	2	0.5
3	20–30	4	15	1
4	30–40	0	2	0.2

[a] The model is one-dimensional in the vertical and simulates concentrations and fluxes of chemically reactive trace gases in the air column between 0 and 40 m altitude. The top of the forest canopy is at 30 m altitude.
[b] The atmosphere extending from 0 m (ground level) to 40 m altitude is subdivided into four grid cells (1 through 4), which are assumed to be individually well mixed and to exchange mass with adjacent grid cells.
[c] The leaf area index in each grid cell measures the local density of vegetation (square centimeters of leaf area per square centimeter of air column, counting only one side of the leaf). The sum of leaf area indices for all grid cells is 7 and represents the total leaf area index of the canopy.
[d] The exchange velocities define the rates of turbulent mass transfer between grid cells, as given by equation 8 in the text. We distinguish between a daytime regime (0900 to 1600 local time) when turbulence is vigorous and a nighttime regime (1600 to 0900 local time) when vertical motions are suppressed.

upward vertical flux, $F_{i,j}$ (molecules per square centimeter per second), of species i through the top of grid cell j is given by:

$$F_{i,j} = V_j \left[\frac{n_{i,j}}{N_j} - \frac{n_{i,j+1}}{N_{j+1}} \right] \frac{N_j + N_{j+1}}{2} \qquad (8)$$

where V_j (centimeters per second) is an exchange velocity, $n_{i,j}$ (molecules per cubic centimeter) is the concentration of species i in cell j, and N_j (molecules per cubic centimeter) is the air density in cell j. The exchange velocities measure the intensity of turbulence and are specified to fit the observed concentrations and vertical fluxes of radon-222, O_3, and CO_2 (Table 3). The residence time in the 0–40-m model air column for an inert tracer emitted at the ground is 0.6 h at noon and 5.0 h at midnight, consistent with the radon-222 data of Trumbore et al. (1990).

Upper boundary concentrations for the model at 40 m are specified from ABLE-2B observations. Very low concentrations of NO_x and PAN are imposed at 40 m in order to suppress the downward flux of atmospheric NO_x and thus isolate F_v from F_d. The simulations are iterated to a diurnal steady state, i.e., to a solution where concentrations show no net change over a 24-h cycle. The time scale to reach this steady state is less than 1 day, and hence we believe it to be a good approximation of atmospheric conditions.

The total leaf area index of the canopy ($L = 7$) is apportioned among the three canopy grid cells (0 to 30 m) as shown in Table 3. Half of the total leaf area is in the upper canopy. Extinction of light by vegetation is computed by treating the canopy as a grey absorber and assuming a uniform angular distribution of leaves. The canopy optical depth at any altitude is then equal to half the leaf area index of the vegetation overhead (Verstraete, 1987). The resulting insolation near the ground at noon (θ = 13°C) is only 2.8% of the value at canopy top; the optical depth of the vegetation has thus a major effect on photochemical rates (in particular NO_2 photolysis).

Natural hydrocarbons (mainly isoprene and acetaldehyde) are emitted by vegetation in each grid cell at a rate dependent on vegetation density, temperature, and insolation, as described by Jacob and Wofsy (1990). The hydrocarbon emission fluxes peak in the upper canopy at noon; 24-h average emission fluxes at canopy top are 9.1×10^{10} molecules cm^{-2} s^{-1} for isoprene and 4.0×10^{9} molecules cm^{-2} s^{-1} for acetaldehyde. Deposition of O_3, NO_2, and other reactive species is simulated in each grid cell using equation 7, with additional terms to describe deposition to the ground and to the leaf cuticles. For O_3 and NO_2 we assume R_m = 0, cuticular resistances of 10 s cm^{-1} per cm^2 of leaf (Fan et al., 1990), and resistances to deposition at the ground of 2 s cm^{-1} in the lowest grid cell (Johansson et al., 1988; Wesely, 1989). Values of R_b and R_s are taken from Roberts et al. (1990) and vary with altitude and time of day (see Fig. 5).

Figure 7 shows the main pathways for NO_x cycling within the 0–40-m air column at noon and at midnight. The NO_x budget reflects largely a balance between soil emissions of NO, deposition of NO_2, and ventilation. Conversion of NO_2 to PAN turns out to be a minor process; the concentrations of PAN inside the canopy at noon are only $\approx$ 10% those of NO_2, and the upward flux of PAN at 40 m is only 4% that of NO_x. Production of PAN is inhibited by the optical thickness of the canopy, which suppresses the photochemical decomposition of biogenic hydrocarbons and hence the source of CH_3CO_3 radicals.

The NO_x concentrations in the canopy are higher at night than in the daytime, partly because of the higher resistance to NO_2 deposition and partly because of the lower canopy ventilation rate. At night NO is oxidized to NO_2 by O_3 and there is no conversion of NO_2 back to NO, so that NO_x reaching the top of the canopy is $\approx$ 90% NO_2. In the daytime, oxidation of NO is facilitated by the higher concentrations of O_3 (Fig. 6) and by the presence of peroxy radicals produced from the decomposition of isoprene (which account for about half of total NO oxidation). Photolysis of NO_2 maintains a daytime NO_2/NO_x ratio in the range 0.5 to 0.9 inside the canopy.

Inspection of Fig. 7 indicates that the uptake of NO_2 by vegetation is roughly evenly distributed with altitude and that the ground accounts for 30% of total deposition. The export efficiency (α) of NO_x out of the canopy is 0.39 at noon and 0.17 at midnight; the 24-h average value is 0.25. The lower value of α at night follows mainly from the restricted ventilation, which is only partially compensated by the higher resistance to deposition. The values of α obtained here are somewhat lower than in the box model described above, principally due to the inclusion of ground deposition.

Deposition to the Amazon forest of atmospheric NO_x supplied from aloft (equation 2) was investigated in a separate simulation in which we assumed zero

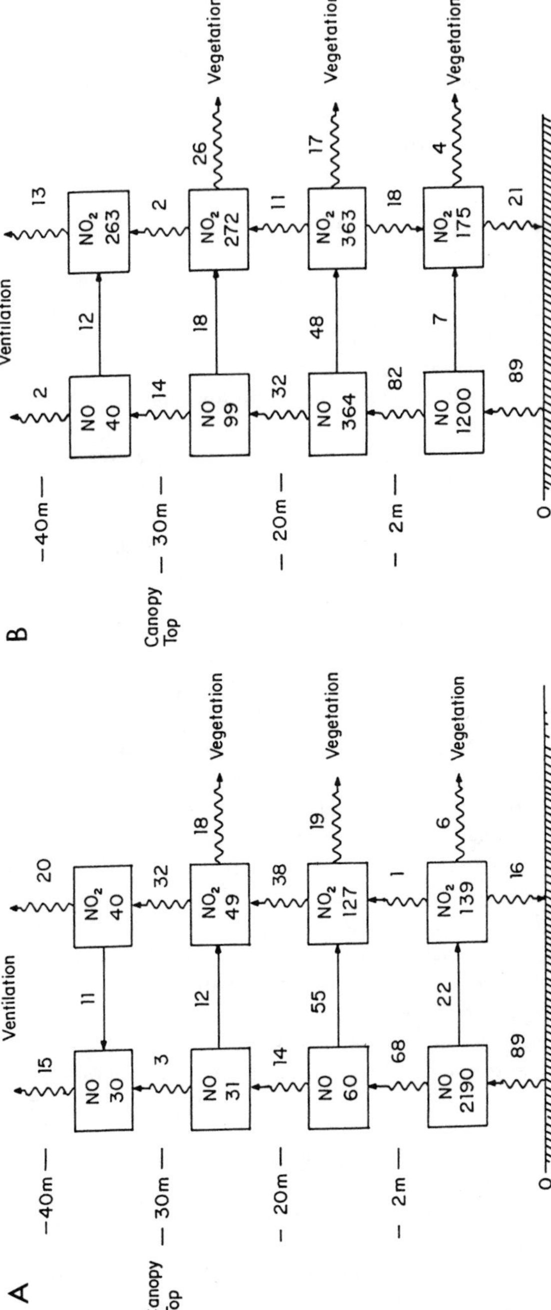

Figure 7. Cycling of biogenic NO_x in the Amazon forest canopy, as simulated by the model at noon (A) and at midnight (B). Concentrations in each grid cell (boxes) are in units of ppt. Fluxes and transformation rates per unit area of air column (arrows) are in units of 10^8 molecules cm^{-2} s^{-1}. Conversion factor: 10^8 molecules cm^{-2} s^{-1} $\approx$ 150 ppt m h^{-1}.

soil emissions of NO and a fixed NO_x concentration at 40 m of 26 ppt (taken from Jacob and Wofsy [1990]). We found deposition velocities (V_d) for NO_x at 40 m of 0.78 cm s^{-1} at noon and 0.13 cm s^{-1} at midnight; these values were insensitive to the speciation of NO_x assumed at 40 m because of the rapid chemical cycling between NO and NO_2. Over 70% of total deposition of atmospheric NO_x took place in the upper canopy (20 to 30 m).

EXTENSION TO OTHER TROPICAL FOREST ENVIRONMENTS

The fraction α of biogenic NO_x emitted by soil that is ventilated to the atmosphere above the forest (export efficiency) depends on a number of environmental variables including (i) the magnitude of soil emission, (ii) the O_3 concentration above the canopy, (iii) the leaf resistances to NO_2 and O_3 deposition, (iv) the canopy ventilation rate, and (v) the leaf area index. Figure 8 shows results from sensitivity simulations in which values for each of these variables were individually modified from the values used in the standard simulation described in the previous section. The range of perturbations was chosen to span the conditions likely to be encountered in tropical forests. The perturbations to R and τ were applied by multiplying all leaf and ground resistances, or all exchange velocities, by a given factor relative to the values in the standard simulation.

An increase in the O_3 concentration above the canopy increases the NO_2/NO_x ratio and hence the scavenging of NO_x by vegetation. At O_3 concentrations of >30 parts per billion by volume (ppb), the NO_2/NO_x ratio approaches unity through most of the canopy and a lower limit for α is reached which is dependent solely on the relative rates of NO_2 deposition and ventilation. An increase in NO emission, by contrast, decreases the NO_2/NO_x ratio and hence increases α. High soil fluxes of NO could result theoretically in complete titration of O_3 inside the canopy and consequently in very high values of α. However, it seems unlikely that titration of O_3 could ever occur since high soil fluxes of NO would foster photochemical production of O_3 in the atmosphere above the forest, thus increasing the supply of O_3 to the canopy from aloft. Such a situation was indeed encountered during the dry season ABLE-2A expedition, which operated from the same site as ABLE-2B (Harriss et al., 1988). Soil emission fluxes of NO in the dry season averaged 5.2×10^{10} molecules cm^{-2} s^{-1}, six times higher than in the wet season, but O_3 concentrations averaged $\approx$20 ppb at 40 m and 10 to 20 ppb inside the canopy (Kaplan et al., 1988). Model calculations for the dry season (Jacob and Wofsy, 1988) indicate that the concentrations of biogenic NO_x above the canopy were sufficiently high to promote photochemical production of O_3, explaining in part the relatively high O_3 levels (NO_x from biomass burning was an additional explanation). From the average dry season values for E_{NO} and O_3 concentrations we derive a 24-h mean α of 0.22, slightly lower than in the wet season (0.25).

Changes in the resistance to deposition, the canopy ventilation rate, or the leaf area index have remarkably little effect on α because of negative feedbacks. For example, an increase in the resistance to deposition hinders the uptake of NO_2 but also of O_3; it facilitates penetration of O_3 inside the canopy and hence increases the

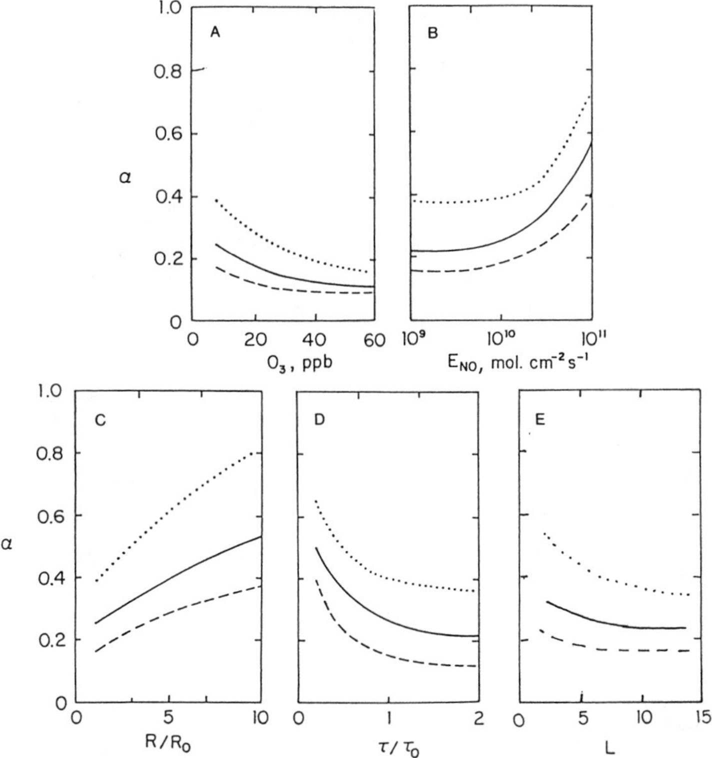

Figure 8. Export efficiency α of biogenic NO_x out of the forest canopy as a function of environmental variables. The export efficiency represents the fraction of NO_x emitted by soil that is exported to the atmosphere above the canopy. The environmental variables include (A) O_3 concentration at 40 m altitude, (B) soil emission flux of NO (E_{NO}), (C) total resistance to deposition (R), (D) residence time of air inside the 0–40-m column (τ), and (E) canopy leaf area index (L). The perturbations to R and τ are applied by multiplying the individual resistances and exchange velocities used in the standard simulation (corresponding to standard values R_0 and τ_0) by a given factor. The figure shows 24-h average values for α (solid line) and values at noon and midnight (dashed and dotted lines, respectively).

NO_2/NO_x ratio. Similar arguments can be made to explain the weak dependences of α on canopy ventilation rate and on leaf area index.

CONCLUSIONS

Results from a process-based model for the Amazon forest atmosphere indicate that only 25% of biogenic NO_x emitted by soil is exported to the atmosphere above the canopy (35% in the daytime, 17% at night). The balance is deposited to vegetation during transport from the ground to canopy top. The deposited NO_x may be assimilated into the plant material, thus recycling nitrogen within the ecosystem. The role of the canopy filter in limiting the export of biogenic NO_x to the atmosphere appears to be important over a broad range of conditions encountered in tropical forest environments. Neglect of this effect in atmospheric chemistry models may lead to serious overpredictions of NO_x concentrations above tropical continents.

Extension of our results to non-forest canopies is difficult because canopy ventilation rates and leaf surface resistances may be vastly different. High soil emissions of NO have been observed from wet savannas (Johansson and Sanhueza, 1988) and from fertilized cropland (Williams et al., 1988), but the ventilation of such open canopies could be much faster than for tropical forests, and the seasonality of leaf activity would need to be considered. Open chamber flux measurements by Slemr and Seiler (1984) indicate lower NO emissions from grassland than from adjacent bare soil, suggesting uptake of NO_2 by grass. However, Johansson and Granat (1984) observed the same effect when using a static chamber where O_3 (and hence NO_2) should be depleted due to excess NO. More research is needed to explain these observations.

Acknowledgments. This work was funded by the National Science Foundation (NSF-ATM88-58974 and NSF-ATM-89-21119), the Host Foundation, and the Packard Foundation.

REFERENCES

Bakwin, P. S., S. C. Wofsy, and S.-M. Fan. 1990a. Measurements of reactive nitrogen oxides (NO_y) within and above a tropical forest canopy in the wet season. *J. Geophys. Res.* **95:**16765–16772.

Bakwin, P. S., S. C. Wofsy, S.-M. Fan, M. Keller, S. Trumbore, and J. M. da Costa. 1990b. Emission of nitric oxide (NO) from tropical forest soils and exchange of NO between the forest canopy and atmospheric boundary layers. *J. Geophys. Res.* **95:**16755–16764.

Brost, R. A., R. B. Chatfield, J. P. Greenberg, P. L. Haagenson, B. G. Heikes, S. Madronich, B. A. Ridley, and P. R. Zimmerman. 1988. Three-dimensional modeling of transport of chemical species from continents to the Atlantic Ocean. *Tellus* **40:**358–379.

Delany, A. C., and T. D. Davies. 1983. Dry deposition of NO_x to grass in rural East Anglia. *Atmos. Environ.* **17:**1391–1394.

Delany, A. C., D. L. Fitzjarrald, D. H. Lenschow, R. Pearson, Jr., G. J. Wendel, and B. Woodruff. 1986. Direct measurements of nitrogen oxides and ozone fluxes over grasslands. *J. Atmos. Chem.* **4:**429–444.

Fan, S.-M., S. C. Wofsy, P. S. Bakwin, D. J. Jacob, and D. R. Fitzjarrald. 1990. Atmosphere-biosphere exchange of CO_2 and O_3 in the central Amazon forest. *J. Geophys. Res.* **95:**16851–16864.

Fitzjarrald, D. R., and K. E. Moore. 1990. Mechanisms of nocturnal exchange between the rain forest and the atmosphere. *J. Geophys. Res.* **95:**16839–16850.

Fitzjarrald, D. R., K. E. Moore, O. M. R. Cabral, J. Scolar, A. O. Manzi, and L. D. de Abreu Sa. 1990. Daytime turbulent exchange between the Amazon forest and the atmosphere. *J. Geophys. Res.* **95:**16825–16838.

Gao, W., R. H. Shaw, and K. T. Paw U. 1989. Observation of organized structure in turbulent flow within and above a forest canopy. *Boundary-Layer Meteorol.* **47:**349–377.

Harriss, R. C., et al. 1988. The Amazon Boundary Layer Experiment (ABLE 2A): dry season 1985. *J. Geophys. Res.* **93:**1351–1360.

Harriss, R. C., et al. 1990. The Amazon Boundary Layer Experiment: wet season 1987. *J. Geophys. Res.* **95:**16721–16736.

Hicks, B. B., D. D. Baldocchi, T. P. Meyers, D. R. Matt, and R. P. Hosker. 1987. A multiple preliminary resistance routine for deriving dry deposition velocities from measured quantities. *Water Air Soil Pollut.* **36:**311–330.

Jacob, D. J., and S. C. Wofsy. 1988. Photochemistry of biogenic emissions over the Amazon forest. *J. Geophys. Res.* **93:**1477–1486.

Jacob, D. J., and S. C. Wofsy. 1990. Budgets of reactive nitrogen, hydrocarbons, and ozone over the Amazon forest during the wet season. *J. Geophys. Res.* **95:**16737–16754.

Johansson, C. 1989. Fluxes of NO_x above soil and vegetation, p. 229–246. *In* M. O. Andreae and D. S. Schimel (ed.), *Exchange of Trace Gases between Terrestrial Ecosystems and the Atmosphere.* John Wiley & Sons, Inc., New York.

Johansson, C., and L. Granat. 1984. Emission of nitric oxide from arable land. *Tellus* **36:**25–37.

Johansson, C., H. Rodhe, and E. Sanhueza. 1988. Emission of NO in a tropical savanna and a cloud forest during the dry season. *J. Geophys. Res.* **93:**7180–7192.

Johansson, C., and E. Sanhueza. 1988. Emission of NO from savanna soils during rainy season. *J. Geophys. Res.* **93:**14193–14198.

Kaplan, W. A., S. C. Wofsy, M. Keller, and J. M. Da Costa. 1988. Emission of NO and deposition of O_3 in a tropical forest system. *J. Geophys. Res.* **93:**1389–1395.

Lenschow, D. H., R. Pearson, Jr., and B. B. Stankov. 1982. Measurements of ozone vertical flux to ocean and forest. *J. Geophys. Res.* **87:**8833–8837.

Levy, H., II, and W. J. Moxim. 1989. Simulated global distribution and deposition of reactive nitrogen emitted by fossil fuel combustion. *Tellus* **41:**256–271.

Logan, J. A. 1983. Nitrogen oxides in the troposphere: global and regional budgets. *J. Geophys. Res.* **88:**10785–10807.

Logan, J. A. 1985. Tropospheric ozone: seasonal behavior, trends, and anthropogenic influence. *J. Geophys. Res.* **90:**10463–10482.

Roberts, J., O. M. R. Cabral, and L. F. De Aguiar. 1990. Stomatal and boundary-layer conductances in an Amazonian Terra Firme rain forest. *J. Appl. Ecol.* **27:**336–353.

Rogers, H. H., J. C. Campbell, and R. J. Volk. 1979. Nitrogen-15 dioxide uptake and incorporation by *Phaseolus vulgaris* (L.). *Science* **206:**333–335.

Singh, H. B., D. Herlth, D. O'Hara, L. Salas, A. L. Torres, G. L. Gregory, G. W. Sachse, and J. F. Kasting. 1990. Atmospheric peroxyacetyl nitrate measurements over the Brazilian Amazon Basin during the wet season: relationships with nitrogen oxides and ozone. *J. Geophys. Res.* **95:**16945–16954.

Slemr, F., and W. Seiler. 1984. Field measurements of NO and NO_2 emissions from fertilized and unfertilized soils. *J. Atmos. Chem.* **2:**1–24.

Talbot, R. W., R. C. Harriss, M. O. Andreae, H. Berresheim, P. Artaxo, M. Garstang, R. C. Harriss, K. M. Beecher, and S.-M. Li. 1990. Aerosol chemistry during the wet season in central Amazonia: the influence of long-range transport. *J. Geophys. Res.* **95:**16955–16970.

Trumbore, S. E., M. Keller, S. C. Wofsy, and J. M. da Costa. 1990. Measurements of soil and canopy exchange rates in the Amazon rain forest using ^{222}Rn. *J. Geophys. Res.* **95:**16865–16874.

Verstraete, M. M. 1987. Radiation transfer in plant canopies: transmission of direct solar radiation and the role of leaf orientation. *J. Geophys. Res.* **92:**10985–10995.

Wesely, M. L. 1989. Parameterization of surface resistance to gaseous dry deposition in regional-scale numerical models. *Atmos. Environ.* **23:**1293–1304.

Williams, E. J., D. T. Parrish, M. P. Buhr, and F. C. Fehsenfeld. 1988. Measurements of soil NO_x emissions in central Pennsylvania. *J. Geophys. Res.* **93:**9539–9546.

Chapter 14

Aspects of the Marine Nitrogen Cycle with Relevance to the Dynamics of Nitrous and Nitric Oxide

Douglas G. Capone

The nitrogen cycle of the oceans has been a subject of scientific interest for almost a century (Johnstone, 1908). Studies of most aspects of the marine nitrogen (N) cycle accelerated during the 1960s when N was generally accepted as the nutrient most often limiting marine phytoplankton production (Ryther and Dunstan, 1971; Thomas, 1966). Recurring nuisance algal blooms in inshore and coastal waters further intensified efforts to identify the role of nitrogen in such phenomena (Ryther and Dunstan, 1971). The 1960s and 1970s also brought initial concerns about the effects of certain atmospheric trace gases, including nitrous oxide (N_2O), on stratospheric ozone (Crutzen, 1981). Substantial effort has been directed towards characterizing the oceanic concentrations and distributions of N_2O in an effort to establish whether the oceans are a net sink or source to the atmosphere for this trace gas (Hahn, 1981).

The detection of increasing N_2O in the atmosphere (Khalil and Rasmussen, 1983; Weiss, 1981) and the recognition of its high capacity for absorbing solar radiation (Dickenson and Cicerone, 1986) have stimulated renewed interest in N_2O fluxes from major ecosystems. Unfortunately, our present knowledge of oceanic N_2O cycling is sparse, and that of nitric oxide (NO) is virtually nonexistent. However, a large body of information is available on "mainstream" aspects of the marine N cycle. An understanding of the major pathways of N in the sea provides information directly relevant to understanding the fluxes of biologically active trace gases such as N_2O and NO (Capone, in press).

GENERAL ASPECTS OF THE MARINE NITROGEN CYCLE

The distribution of major organic and inorganic N species and the predominant biological transformation of N have been relatively well characterized over a broad range of oceanic environments. Comprehensive reviews of various aspects can be

Douglas G. Capone • Center for Environmental and Estuarine Studies, Chesapeake Biological Laboratory, University of Maryland, Solomons, Maryland 20688-0038.

found in the volumes edited by Carpenter and Capone (1983) and Blackburn and Sorensen (1988).

Marine Environments

The oceans can be categorized, with respect to N, by distance from land and water column depth. With respect to distance offshore, three distinct zones are apparent: the open ocean, coastal waters, and nearshore/inshore areas. Distinct depth zones include the upper euphotic layer, subeuphotic/aphotic waters, and the bottom waters in contact with the sediments.

Nearshore waters are generally shallow and well mixed down to the bottom sediments. Biogeochemical processes in the water column and underlying sediments are often coupled (Nixon, 1981). Furthermore, nearshore waters can be heavily influenced by adjacent land masses and are commonly enriched in nutrients from input of land runoff. As one moves offshore and the depth of the water column increases, there can develop two major layers in the water column, separated by a thermal gradient referred to as the thermocline. The direct influence of the terrestrial environment and the coupling of benthic and pelagic processes decrease seaward.

Nitrogen Inventories and Distributions

Inventories of the major forms of N in each of the major zones of the sea and in the ocean overall are given in Table 1. Dinitrogen gas (N_2) is by far the most abundant form of N, but is nutritionally available only to certain prokaryotic, N_2-fixing (diazotrophic) organisms. Nitrate represents the next largest pool of N. N_2O occurs in relatively small amounts. No attempt was made to estimate NO inventories because of insufficient information.

In general, pools of N (Table 1) and biological transformations (Table 2) in the open ocean quantitatively dominate the oceanic N cycle. However, because of intensive activities in coastal and nearshore areas, particular processes in these zones can also have global relevance (see below).

Nitrogen Transformations

There is intense interest in and research concerning the N cycle in marine environments because N is often the primary factor limiting biological productivity in many marine systems (Howarth, 1988). Most of the known biological transformations of N occur in oceanic environments; many of these can have direct bearing on the pools of N_2O and NO (Fig. 1). The most quantitatively important component of the marine N cycle, on a local or global basis, is the flux between inorganic (NH_4^+ and NO_3^-) and organic (living biomass and detritus) forms of N. A stoichiometric amount of N is required by marine autotrophs for primary productivity to proceed. Over relatively short time and space scales, there is probably a balance between autotrophic N uptake and the subsequent release of organically

Table 1. Inventory for several oceanic nitrogen pools by volumetric extrapolation[a]

Environment	Area (10^6 km^2)	Vol (10^{18} liters)	Inventory (Tg)				
			N_2	N_2O	NO_3^-	NH_4^+	Organic N
Open ocean							
Euphotic zone[b]		64	716,800	17.9	1,859	1,926	4,838
Subeuphotic zone		1,376	22,153,600	809	674,240	4,648	57,792
Sediments[c]		0.016		0.045	2.24	46.5	46.5
Total	320	1,440	22,870,400	827	676,101	6,621	62,677
Coastal							
Euphotic zone[d]		1.8	22,680	1	37.8	30.2	151
Subeuphotic zone		5.4	68,040	3.02	1,134	907	454
Sediments[c]		0.0018		0.01		1.26	1.26
Total	36	7	90,720	4	1,172	938	606
Nearshore/estuary							
Water column[e]		0.125	1,575	0.07	175	350	35
Sediments[c]		0.00025		0.004	0	0.7	0.7
Total	5	0.125	1,575	0.074	175	351	36
Total ocean							
This compilation			22,962,695	831	677,448	7,910	63,318
Soderlund and Svensson (1976)			22,000,000	200	570,000	7,000	530,000
Delwiche (1981)			20,000,000		100,000		45,000

[a] Average concentrations for each oceanic zone were taken from Sharp (1983); areas and volumes of oceanic basins were those of Kossina (1921) as given in Sverdrup et al. (1942).
[b] Assumed a mixed layer depth of 200 m.
[c] Inventories estimated for top 10 cm of sediment and assuming a sediment porosity of 50%.
[d] Assumed a euphotic zone depth of 50 m.
[e] Assumed a euphotic zone depth of 25 m.

Table 2. Oceanic nitrogen fluxes

Process	Tg of N $year^{-1}$	Reference
Inputs		
Runoff	13–24	Soderlund and Svensson, 1976
Atmospheric deposition	29–82	Soderlund and Svensson, 1976
N_2 fixation		
Water column	10	Capone and Carpenter, 1982
Benthic, shelf	20	Capone and Carpenter, 1982
Total	72–136	
New production[a]		
Open ocean (6%)	296	
Ocean upwell (18%)	19	
Coast upwell (40%)	9	
Shelf (30%)	303	
Nearshore (46%)	65	
Total	692	
Outputs		
Sedimentation	20–38	Soderlund and Svensson, 1976; McElroy, 1983
Denitrification		
Water column, Atlantic and Pacific	50–60	Hattori, 1983
Arabian Sea	25	Codispoti, 1989
Benthic, deep sea	4–7	Hattori, 1983
Benthic, shelf	50–75	Christensen et al., 1987
Benthic, other	6–12	Seitzinger, 1988
Other	10	Liu, 1979
Total	165–227	

[a] Based on productivity estimates derived by DeVooys (1979) and the percentage of new production from different zones from Eppley and Peterson (1979) (in parentheses). See Capone (in press) for details.

bound N during heterotrophic catabolism. With regard to N assimilation, there is some evidence that N_2O may arise directly from assimilatory NO_3^- reduction (e.g., Weathers, 1984).

Biological N_2 fixation, i.e., the ability of some bacteria to reduce N_2 to NH_4^+, occurs in open ocean waters and shallow nearshore areas (Capone and Carpenter, 1982). Interestingly, N_2O is among the substrates reduced by the nitrogenase enzyme system (Rivera-Ortiz and Burris, 1975), although the importance of this in the overall N cycle is not known.

Nitrification, the two-step oxidation of NH_4^+ through NO_2^- to NO_3^- by certain bacteria, occurs in oxygenated oceanic waters and sediments (Kaplan, 1983). Nitrification may yield N_2O as a by-product or as an alternate product of the first step (Poth and Focht, 1985; Ritchie and Nicholas, 1972). Denitrification, the reduction of oxidized inorganic forms of N to gaseous end products during

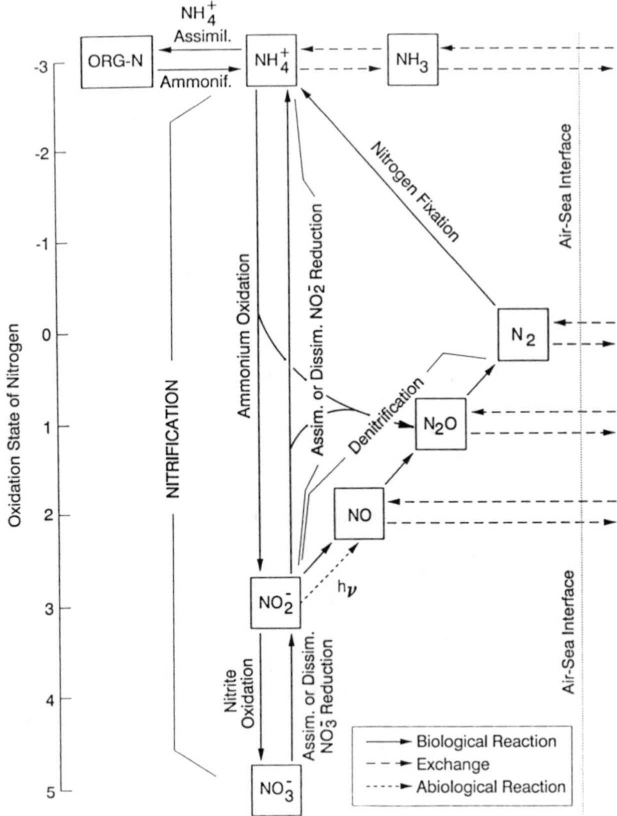

Figure 1. Biological nitrogen transformations indicating the oxidation state of nitrogen and gaseous species of nitrogen which may exchange across the air/sea interface.

anaerobic respiration by certain bacteria, is also an important process in the sea (Hattori, 1983). NO and N_2O are sequentially produced in denitrification (Fig. 1): NO is probably an enzyme-bound intermediate, whereas N_2O may be evolved or further reduced to N_2 (Payne, 1981). A very close coupling appears to exist between nitrification and denitrification in many environments, including nearshore, coastal, and shelf sediments (Codispoti and Christensen, 1985; Jenkins and Kemp, 1984; Horrigan and Capone, 1985). Indeed, denitrification may be limited in such systems by the rate of NO_3^- production through nitrification. In anaerobic environments, dissimilatory NO_3^- reduction to NH_4^+ may compete with denitrification for available NO_3^- (Hattori, 1983; Koike and Sorensen, 1988).

New versus Recycled Production

Conceptually, there are two kinds of photoautotrophic or "primary" production in the sea (Dugdale and Goering, 1967; Eppley and Peterson, 1979). "New

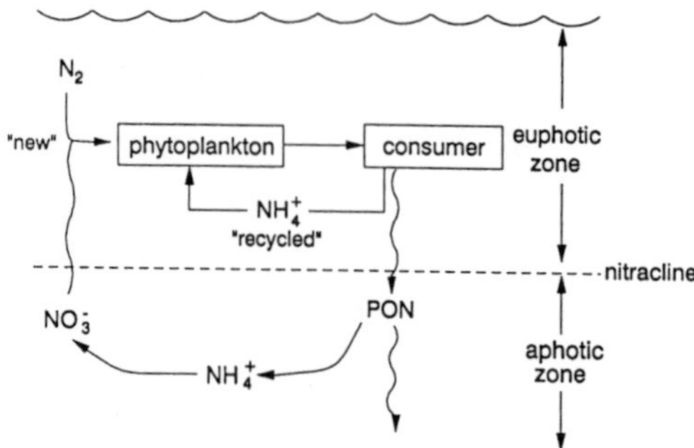

Figure 2. Conceptual diagram illustrating the sources of new and recycled nitrogen and the balance between sedimentation and subeuphotic zone ammonification, nitrification, and nitrate infiltration into the euphotic zone. PON, particulate organic nitrogen.

production'' is dependent upon N imported into the system (e.g., NO_3^- flux across the thermocline, N_2 fixation) (Dugdale and Goering, 1967; Eppley and Peterson, 1979) (Fig. 2). The amount of ''new N'' coming into a system sets a limit on the amount of net primary production that can be exported from the system (e.g., through sedimentation out of the euphotic zone) if a dynamic steady state exists. ''Recycled'' production is the production resulting from rapidly recycled forms of N, such as NH_4^+, within the system. The concept of new and recycled production has been used to set independent constraints on the amount of excess atmospheric CO_2 that may be biologically sequestered through autotrophic uptake and subsequent sedimentation. The new production concept also bears directly on N_2O dynamics in the sea (see below; also Capone, in press).

Organic material sedimented out of the euphotic zone (new production or ''J-flux'') may be recycled in the lower water column or deposited in the sediments. There is a strong negative relationship between water column depth and sediment deposition rate (Suess, 1980): in deeper water, less material makes it to the bottom. Low concentrations of NH_4^+ and the relatively high concentrations of NO_3^- in the subeuphotic zone indicate that in this region reduced N released through regeneration is relatively rapidly oxidized by nitrification.

Assuming a steady state for NO_3^- pools in supra- and subthermocline waters, nitrification at depth should therefore be roughly equal to NO_3^- flux across the thermocline and, in turn, the assimilation of NO_3^- in the euphotic zone. That is, the amount of nitrification integrated over depth should approximately equal the amount of new production.

Open Ocean

In the vast expanses of the open ocean, a thin surface layer (typically 100 to 200 m in depth), called the mixed layer and containing the euphotic zone, is separated from a much deeper (average, 4,000 m) aphotic zone by a permanent thermocline. Deep ocean sediments represent a third distinct zone of the open ocean.

The euphotic zone of much of the open ocean is often termed ''oligotrophic'' and is characterized by exceedingly low quantities of inorganic N species (e.g., NO_3^-, NH_4^+) (Fig. 3). The euphotic zone is a site of active photoautotrophy and

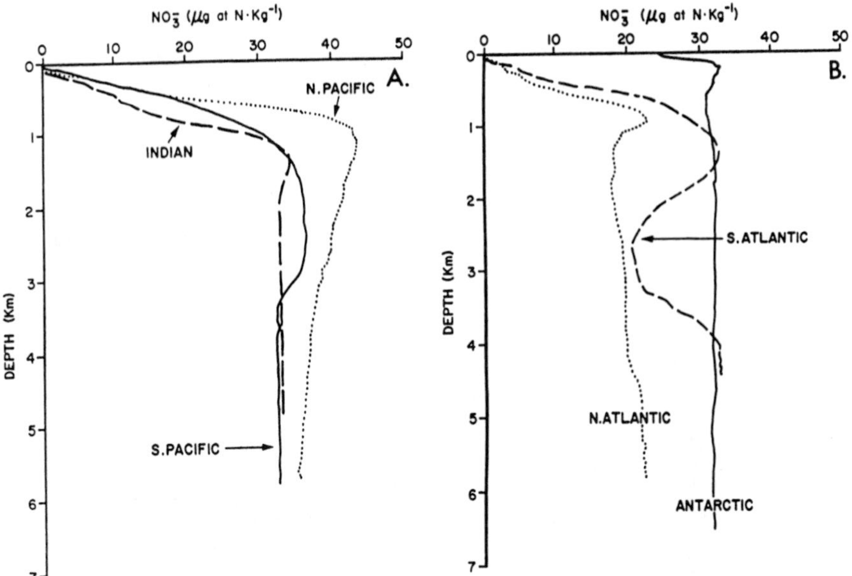

Figure 3. Typical profiles of nitrate in different ocean basins, taken from GEOSECS data (from Sharp, 1983).

associated N assimilation by planktonic microalgae (phytoplankton). About 5,000 Tg of N is assimilated each year by open ocean primary producers (Fig. 4). In contrast to coastal and nearshore waters, areal rates of primary productivity are quite low in the open ocean and largely dependent upon internal recycling of inorganic nitrogen within the euphotic zone (see below).

A source of "new" N in tropical and subtropical euphotic zone waters is N_2 fixation by planktonic cyanobacteria (Capone and Carpenter, 1982). While some open ocean cyanobacteria appear to satisfy their nutritional requirement for N through N_2 fixation (Carpenter, 1983), the total amount of N provided by this pathway is very small relative to overall N assimilation by all open ocean phytoplankton (Fig. 4, Table 2; Capone and Carpenter, 1982). Furthermore, N_2 fixation is rarely (if ever) observed in more nutrient-rich coastal and nearshore waters. The relatively small amount of N_2 fixation occurring in an "N-limited" ecosystem has been noted (Capone and Carpenter, 1982), and hypotheses relating to secondary limitation of N_2 fixation by trace metals (Capone and Carpenter, 1982; Howarth and Cole, 1985) or other factors (Paerl et al., 1987) have been put forth. There is also current concern that the extent of N_2 fixation in the open ocean may have been previously underestimated, either through methodological artifact or by simply having overlooked other N_2-fixing organisms in the open ocean (Legendre and Gosselin, 1989).

To meet the demand for recycled N, most of the organic production of phytoplankton in the euphotic zone must be consumed within that zone ("recy-

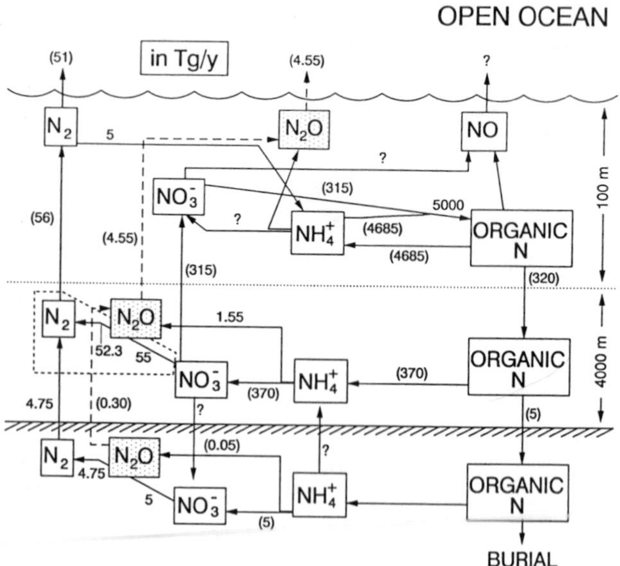

Figure 4. Estimates for annual transfers between major nitrogen pools in the open ocean. All values are in teragrams ($= 10^{12}$ g) per year. Values in parentheses refer to inferred or estimated rates of transfer as described by Capone (in press), in contrast to values taken from the literature.

cled'' production; estimated to be about 90% of total production) (Fig. 4). The remainder, which is equal to the amount exported by sedimentation through the thermocline (the ''new'' production), is predominantly consumed by heterotrophic organisms in the subeuphotic zone before reaching the deep sediments. As mentioned above, NO_3^- levels sharply increase below the thermocline (Fig. 3), a result of the heterotrophic degradation and release of organically bound N (ammonification) and its subsequent oxidation (nitrification).

A common feature of the open ocean is an NO_2^- maximum near the base of the euphotic zone. Otherwise, NO_2^- is generally undetectable throughout the water column; the absence of detectable NO_2^- is probably a result of the closely coupled activities of the two groups of nitrifiers, the NH_4^+ oxidizers and the NO_2^- oxidizers (Kaplan, 1983). Both groups are inhibited in the light and appear to be only minimally active through most of the euphotic zone (Olson, 1981; Ward et al., 1989). However, NH_4^+ oxidizers are somewhat less inhibited by light than NO_2^- oxidizers. The primary NO_2^- maximum may occur because NH_4^+ oxidation commences at a shallower depth (and higher light level) than does NO_2^- oxidation, resulting in a zone where NO_2^- production can outpace its consumption (Olson, 1981).

Higher concentrations of NO_3^- are found at depth in the Pacific Ocean, compared to the Atlantic (Fig. 3), a result of the large-scale circulation of deep waters (Broecker and Peng, 1982). Deep ocean water is formed by extreme cooling at the surface in the North Atlantic and around Antarctica; after sinking, it travels through the Indian Ocean into the Pacific. It surfaces again, "upwelling" in areas of the Eastern Pacific (Broecker and Peng, 1982). The longer a parcel of water is at depth, the greater its encounter with sinking, decaying particles, and thus the higher its content of regenerated inorganic constituents (i.e., NO_3^-, PO_4^{3-}, and HCO_3^-) and concurrent depletion of O_2 through microbial respiration. In fact, one may predict with considerable accuracy the amount of inorganic species in a parcel of deep water from the amount of oxygen consumption in that water relative to the amount of oxygen present when the water parcel first chilled and sank (Redfield et al., 1963). The oldest deep water may be almost completely depleted of the O_2 it initially took to depth, and large volumes of the North Pacific are known for their hypoxic condition. Where O_2 concentrations in the water column are strongly depleted, there are often significant NO_3^- "anomalies" in that NO_3^- is deficient relative to that predicted by water age and O_2 depletion. This is indicative of active denitrification in these zones. Indeed, in deep waters of the eastern tropical north Pacific, denitrification is of global importance, amounting to about 50 to 60 Tg of N year^{-1} (Fig. 4, Table 2; Hattori, 1983). Codispoti (1989) has recently concluded that the Arabian Sea is another quantitatively important site of denitrification and has estimated annual denitrification there to amount to 25 Tg.

Pools of dissolved organic N (DON), until recently thought to be relatively uniform through the upper and lower water column, are considerably higher in the euphotic zone and decrease rapidly with depth, only becoming uniform below the euphotic zone (Suzuki and Sugimura, 1985). The fraction removed, probably by the action of heterotrophic bacterioplankton, represents labile forms of N (e.g., amino acids) derived from upper water column production. The values given in Table 1 include both dissolved and particulate forms of organic N, in which DON usually predominates by a factor of 10 (Sharp, 1983). However, these values are taken from the older literature on DON and will likely be revised considerably upwards based on these new observations. An emerging view of DON is that it is a very dynamic pool in the upper water column of the oceans (Ward et al., 1989; Suzuki and Sugimura, 1985). Furthermore, export of DON from the euphotic zone may occur and would necessitate increasing our estimate of new production (Toggweiler, 1989).

Because of technological and resource constraints, we are only just beginning to acquire generalized knowledge about the pools of N and the intensity of N transformations in deep ocean sediments. Because of their depth and the paucity of organic material reaching these sediments from surface waters, as well as the vast volume of water per unit area of deep sediment, it is unlikely that deep ocean sediments have a major short-term impact on the oceanic N cycle. Inferences of biological N transformations have been made based on analysis of N_2 and NO_3^- profiles (Bender et al., 1977; Wilson, 1978). To a limited extent, incubation chambers have been deployed at depth to determine exchange across the sediment/water interface (e.g., Christensen and Rowe, 1984). Retrieved cores have also been

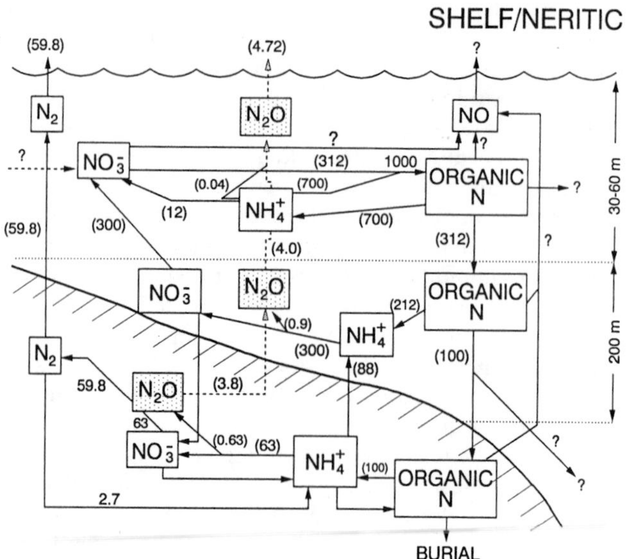

Figure 5. Estimates for annual transfers between major nitrogen pools in coastal waters. Same as Fig. 4.

used to assess directly rates of some nitrogen transformations (e.g., Hartwig and Stanley, 1978; Sorensen and Wilson, 1984). However, until the areal extent of the recently detected biological communities of the deep-sea hydrothermal vents is fully evaluated, we should reserve any final conclusions concerning the role of deep sediments in oceanic nutrient cycles.

Coastal Waters

The continental shelves represent an area only about 10% of that covered by the open ocean (Sverdrup et al., 1942). Like the open ocean, coastal waters are a three-layered system, consisting of a euphotic water column, an aphotic water column, and underlying sediments (Fig. 5). However, the depth to the sediments is considerably less (<200 m) than in the open ocean; hence, more material produced in the euphotic zone may reach the sediments. Waters over the continental shelves, termed coastal or neritic, usually have greater concentrations of inorganic N than the open ocean. Furthermore, while the euphotic zone is substantially shallower (typically 30 to 60 m), algal densities and areal productivities are higher than in the open ocean.

Even though coastal zones (shelf and coastal upwelling) represent a small fraction of the total surface area of the sea, productivity here accounts for uptake of about 1,000 Tg of N year^{-1} (Fig. 5). Compared to offshore waters, algal production in the coastal euphotic zone is less dependent on rapid recycling within the euphotic zone for its N requirement, but uses N outwelled from adjacent nearshore waters, regenerated and released from underlying sediments or upwelled from deep offshore waters. Hence, similar absolute amounts of new production occur in coastal and offshore waters (Table 2; Berger et al., 1989).

Higher organic input to sediments leads to higher nitrogen pools and metabolic activities in shelf sediments. Nutrients regenerated during heterotrophic metabolism in shelf sediments can be released from the sediments and may be mixed up into productive waters over short (seasonal) time scales. On the basis of investigations and analysis by Christensen et al. (1987), denitrification in shelf sediments is also a quantitatively important process with respect to the oceanic N cycle, amounting to about 50 Tg of N year^{-1} (Table 2, Fig. 5).

Nearshore, Estuaries

Nearshore and estuarine waters represent only a small fraction of the total ocean. However, such areas are important because of human activities. Compared to coastal or open ocean waters, the highest concentrations of inorganic N have been reported for nearshore waters (Sharp, 1983). Areal productivity by phytoplankton populations in such areas is among the highest noted anywhere (De-Vooys, 1979). Furthermore, the shallowness of nearshore areas results in essentially a two-layered system (well-mixed water column and sediments), and light penetration to the sediments may at times support very large populations of benthic macrophytes (macroalgae and sea grasses) that contribute to high areal rates of primary productivity. Much of the productivity of nearshore/inshore areas depends on de novo inputs of N (e.g., riverine, anthropogenic), rather than on N recycled in the water column (Table 2, Fig. 6).

Sediments of nearshore areas or estuaries, while very heterogeneous, may be organically enriched, be metabolically active, and have large N effluxes (Nixon, 1981). Anaerobic conditions often found in such sediments make them important sites of denitrification and, in vegetated areas, N_2 fixation (Table 2, Fig. 6).

DISTRIBUTIONS AND DYNAMICS OF NO

Relatively little is known about nitric oxide in marine ecosystems. Sorensen (1978) reported concentrations of up to 200 μM NO in nearshore sediments. At times, maximum concentrations of NO were coincident with zones of maximum denitrification, while in other cases the NO concentration maximum occurred below the peak in denitrification.

Zafiriou and McFarland (1981) reported concentrations of NO in surface waters that appeared to be a product of photolysis of NO_2^-. The partial pressure of NO (pNO) during light periods could be up to 10^3 times greater than the atmospheric pNO, suggesting that the sea may be a net source of NO. The half-life of NO pools was extremely short, on the order of seconds.

More recently, Ward and Zafiriou (1988) determined NO and nitrification profiles in eastern North Pacific waters. NO concentrations of up to 60 pM were

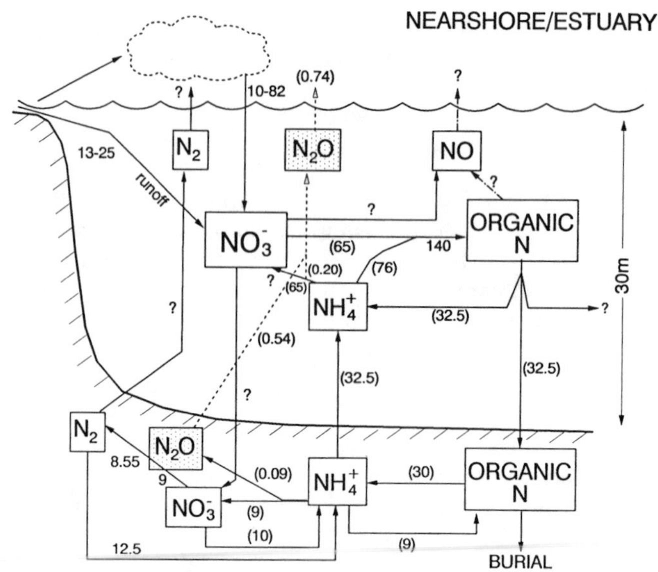

Figure 6. Estimates for annual transfers between major nitrogen pools in nearshore and estuarine areas. Same as Fig. 4.

noted, with maxima for NO concentration and production occurring at the bottom of the euphotic zone and coincident with zones of active nitrification (Fig. 7). At times, rates of NO production could account for up to 13% of the nitrification rate.

At present, it is impossible to make any generalized statements concerning the significance of oceanic NO cycling. This awaits future research that will more broadly and quantitatively define the relationships between light, nitrite levels, and

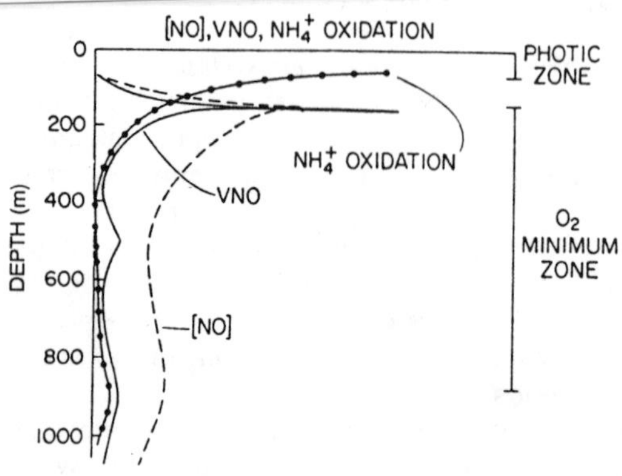

Figure 7. Conceptual diagram for vertical distribution of NO, NO production, and ammonium oxidation (from Ward and Zafiriou, 1988). VNO, NO production rate.

NO production in the upper water column and nitrification and NO production throughout the water column.

N_2O DISTRIBUTIONS AND DYNAMICS

Relative to NO, we have a better understanding of the general distributions of N_2O in the oceans. Much of our current knowledge comes from intensive studies of the oceanic concentrations and distributions of N_2O in the 1960s and 1970s (Hahn, 1981; Scranton, 1983).

N_2O is a trace constituent of seawater. It is often found at levels in excess of atmospheric equilibrium: a slight supersaturation of N_2O is present in most oceanic waters and concentration often increases with depth, being maximal at mid-depths (Cohen and Gordon, 1979; Hahn, 1981; Fig. 8A). Recently, Law and Owens (1990) found very high levels (up to 246%) of surface supersaturation over extensive areas of the Arabian Sea. River and estuarine waters, particularly those with substantial anthropogenic inputs of inorganic nitrogen, may have very high concentrations of N_2O during periods of low flow (Deck, 1980; McElroy et al., 1978). Nearshore sediments can contain very high levels of N_2O, as much as 50 times air-saturated concentrations (Jensen et al., 1984; Sorensen, 1978). These observations are generally interpreted as indicating that there is a net efflux of N_2O from the oceans.

While some abiological reactions yield N_2O, a biological source in the oceans is usually assumed. However, debate continues over which biological processes are quantitatively most important in the production or consumption of N_2O. Two decades ago, when research on environmental N_2O fluxes was first seriously initiated, N_2O was well established as both an intermediate and an end product of denitrification (Payne, 1981). Hence, denitrification was generally considered the primary source of N_2O (Hahn, 1981). However, biological nitrification has also long been known as a source of N_2O, either as a result of the breakdown of an unstable enzyme-bound intermediate during NH_4^+ oxidation (Ritchie and Nicholas, 1972) or through a reductive detoxification pathway during periods of low O_2 and nitrite accumulation (Poth and Focht, 1985; Ritchie and Nicholas, 1972).

Evidence for which biological processes might account for most N_2O in the sea was first gleaned from N_2O distributions. N_2O vertical profiles in the open ocean tend to show one to two peaks, increasing with depth in both the Atlantic and Pacific oceans (Fig. 8A). Nitrate distributions are positively correlated with the N_2O distribution, and there is generally a negative correlation between O_2 level and N_2O (Fig. 8B) that does not extend to very low O_2 levels. Denitrification was thought not to commence until O_2 reached very low levels, typically less than 0.2 mg of O_2 liter^{-1} (6.3 μmol liter^{-1}) (Hattori, 1983). In fact, highly anoxic waters appear to be net sinks for N_2O (Cohen, 1978; Codispoti and Christensen, 1985). Given the laboratory studies with nonmarine ammonium oxidizers which found increasing yield of N_2O with decreasing O_2 concentration (e.g., Ritchie and Nicholas, 1972), the negative relationship between O_2 and N_2O and the positive relationships between NO_3^- and N_2O suggested nitrification, rather than denitrification, as the primary source of N_2O in the oceans.

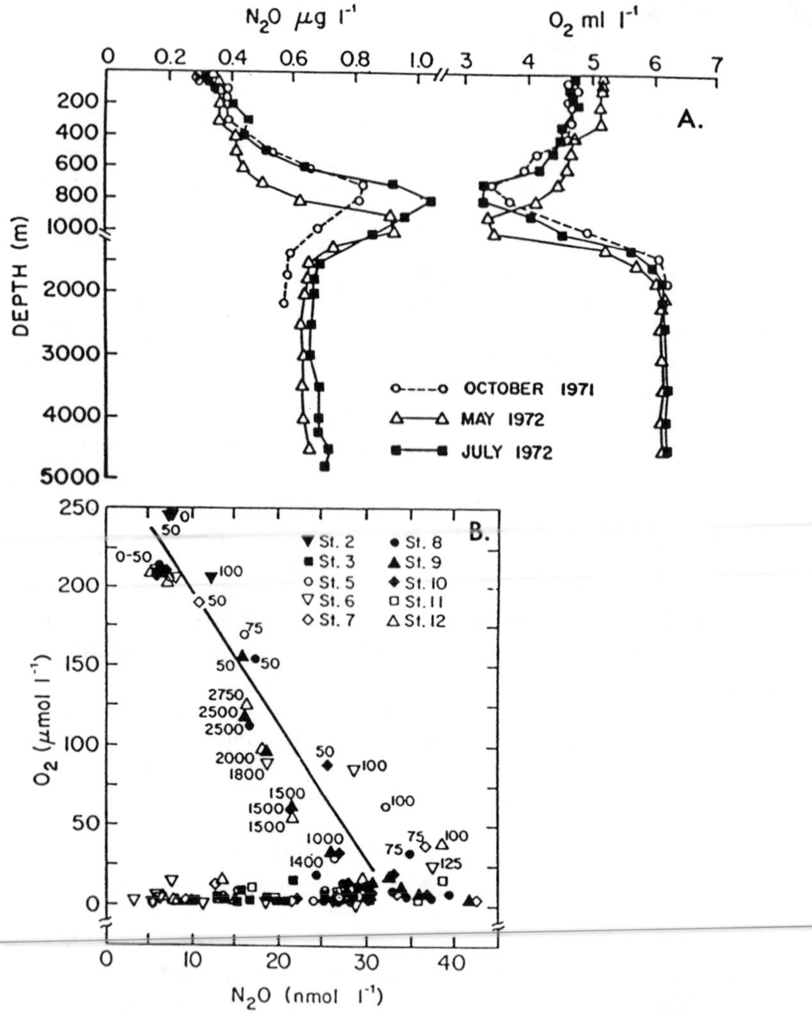

Figure 8. (A) Typical vertical distributions of N_2O (from Yoshinari, 1976); (B) relationship between N_2O and O_2 in the eastern tropical North Pacific (from Cohen and Gordon, 1979).

The results of laboratory culture studies of a marine nitrifier (Goreau et al., 1980) corroborated the field observations and strengthened the general consensus in the 1970s that nitrification rather than denitrification was probably the main biological source of N_2O, while denitrification was possibly a net sink (Elkins et al., 1978; Hahn, 1981). However, over the past 5 years evidence has accumulated in several systems to suggest that denitrification is indeed an important source of N_2O

Table 3. Biologically constrained estimate of annual N_2O production[a]

System	Process	Nitrogen flux (Tg of N year^{-1})	N_2O yield (%)	N_2O produced (Tg of N year^{-1})
Open ocean	Nitrification (new production)[b]	315	0.30	0.95
	Denitrification	85	5.00	4.25
	Nitrification (denitrification)[c]	60	1.00	0.60
Coastal/upwelling	Nitrification (new production)	312	0.30	0.94
	Denitrification	63	5.00	3.15
	Nitrification (denitrification)	63	1.00	0.63
Nearshore	Nitrification (new production)	65	0.30	0.20
	Denitrification	9	5.00	0.45
	Nitrification (denitrification)	9	1.00	0.09
Total				11.25[d]

[a] Adapted from Capone (in press).
[b] Yield of N_2O from nitrification, assuming a quantity of annual nitrification equivalent to new production in the water column.
[c] Yield of N_2O from nitrification, assuming a quantity of nitrification equivalent to denitrification.
[d] Compare with geochemically modeled estimates of: Cohen and Gordon (1979), 10 Tg; Singh et al. (1979), 12.7 to 19.2 Tg; Hahn (1981), 30 to 45 Tg; Elkins et al. (1978), 5 to 7.1 Tg; and Butler et al. (1989), 1.6 Tg of N year^{-1}.

under certain circumstances (Jorgensen et al., 1984; Yoshida et al., 1989). Further research is required to substantiate the generality of either of these conclusions in various environments. The factors regulating the amount of N_2O produced by either pathway need to be better defined.

N_2O Fluxes from the Sea

Estimates of N_2O flux (ocean to atmosphere) have most often been derived using a geochemical approach which considers the positive gradient of N_2O with depth into the sea, the diffusion rate of the gas, and the boundary conditions at the air/sea interface (Hahn, 1981). Such estimates range from 0.9 to 23 μmol m^{-2} day^{-1} (Hahn, 1981). Several investigators have directly measured the flux of N_2O from shallow marine environments (Jensen et al., 1984; Nishio et al., 1983; Seitzinger et al., 1984; Seitzinger, 1987). In general, positive fluxes of about 0.5 to 84 μmol m^{-2} day^{-1} have been reported for these environments. Flux estimates for open ocean regions extrapolated to ocean-wide values range from 1.6 Tg of N year^{-1} during the 1987 El Nino/Southern Ocean Oscillation event (Butler et al., 1989) to a high value of 45 Tg of N year^{-1} (Table 3).

I have recently taken another approach to estimating N_2O fluxes from the ocean (Capone, in press). Because we now have a reasonable understanding of many of the important processes in the marine nitrogen cycle that are of direct or indirect relevance to N_2O fluxes, one may estimate N_2O production from these processes. Nitrogen uptake, N_2 fixation, and denitrification have all been well characterized in a number of marine systems, and their ocean-wide importance has been estimated (Table 2). Extensive surveys of marine primary production have also been conducted, and the distributions of production have been mapped and integrated for the world's oceans (Berger et al., 1989; DeVooys, 1979).

For marine nitrification, which may contribute directly to N_2O fluxes, there is relatively little direct information. However, available quantitative estimates for processes coupled to nitrification allow us to approximate the extent of nitrification if we assume a dynamic steady state. Nitrate flux (by diffusion or upwelling) from deeper water across the thermocline into the euphotic zone is probably the primary source of "new" nitrogen to the surface layers of the ocean (Ward et al., 1989). At steady state, the ultimate source of this NO_3^- is sedimenting organic material from which reduced nitrogen has been recycled and oxidized via nitrification (Fig. 2). Thus, using the information available on oceanic primary production and current estimates of the portion of production dependent on new or regenerated forms of nitrogen, one may calculate the annual demand for new nitrogen in various oceanic environments (Table 2). This may be extended to an estimate of the magnitude of nitrification. If new production is substantially underestimated due to production and export of DON (Toggweiler, 1989), we would also need to revise our estimate of nitrification upwards.

Similarly, denitrification, itself a potential source of N_2O, is ultimately dependent upon nitrification for its substrate (Fig. 1). As mentioned, evidence from a variety of environments suggests that these processes are intimately coupled at times. Therefore, with the possible exception of nearshore regions where large exogenous inputs of NO_3^- may be present, oceanic nitrification must also accommodate denitrification demands in addition to the demands of new production. Quantitative estimates of denitrification can be used to place a second constraint on nitrification rates.

Deriving an estimate of oceanic nitrification (Table 3), one may go on to place boundaries on N_2O fluxes in the sea. To use effectively the available and inferred information on oceanic new production, nitrification, and denitrification in estimating N_2O fluxes, we still require information on the net amount of N_2O evolved during nitrification and denitrification. Unfortunately, the close coupling of the two processes in many systems makes it difficult to separate the contribution of each process to N_2O production.

Several field studies of shallow sediment systems have empirically assessed N_2O fluxes relative to N_2 fluxes or denitrification rates. Ratios of N_2O to N_2 in dissolved gas fluxes were 0.13% in an unpolluted estuary in Florida (Seitzinger, 1987) and 0.07 to 5.6% for a eutrophic estuary in Rhode Island (Seitzinger et al., 1984). For shallow sediments in a Danish estuary, N_2O flux accounted for 0.7 to 13% of the concurrently measured denitrifying activity (Jensen et al., 1984). A 0.35% yield of N_2O relative to N_2 emission was found in sediments of a Japanese estuary (Nishio et al., 1983). The N_2O fluxes from each of these studies were probably a result of both nitrification and denitrification.

Studies of pure cultures of marine NH_4^+ oxidizers have found that the N_2O yield from nitrification increased at lower partial pressures of O_2 (Goreau et al.,

1980). Under fully aerobic conditions (20.3 kPa), N_2O yields were about 0.25% of NH_4^+ oxidation but increased to 10% at a pO_2 of 0.5 kPa. Similarly, nitrification in sediment suspensions yielded little N_2O at 0.5 kPa of O_2, but shunted up to 25% of NH_4^+ oxidation to N_2O at 0.1 kPa of O_2 (Jorgensen et al., 1984). This latter study found that over a broad range of pO_2, denitrification was generally the most important source of N_2O, with 5 to 10% of denitrification yielding N_2O at low pO_2 (<0.1 to 0.2 kPa), increasing to 50% N_2O yield at >0.5 kPa. Thus, except over a very narrow range between 0.1 to 0.2 kPa of O_2, where nitrification could account for up to 60% of the total N_2O flux, denitrification was the main source of N_2O. Thus, these two studies indicate that for both nitrification and denitrification, the partial pressure of O_2 is a critical determinant of the extent of N_2O evolution.

Table 3, presenting data from Capone (in press), provides the summed estimates for N_2O production. Nitrification in support of oceanic new production includes estimates for both the open ocean and regions of oceanic upwelling. Similarly, the estimate for coastal nitrification balancing new production in that zone sums coastal upwelling and nonupwelling areas. I chose an N_2O yield of 0.3% for the oceanic nitrification rate, assuming that it occurs under relatively aerobic conditions. In contrast, nitrification linked to denitrification likely occurs in low-O_2 waters or sediments; hence, I have assumed an N_2O yield of 1%. For denitrification, an N_2O yield of 5% was assumed. I thus derived an estimate of 11.25 Tg of N year^{-1}. This is within the range of several of the geochemically based estimates previously given (Table 3).

The rationale and details of this calculation are more fully given elsewhere (Capone, in press). One area clearly deserving more research attention and which would add credibility to such calculations is the analysis of factors affecting N_2O yields from nitrification and denitrification.

CONCLUSIONS

Nitrogenous inputs to the sea, and oceanic productivity, have varied considerably and naturally over the relatively recent geological past (Berger et al., 1989; McElroy et al., 1978). The observed short-term changes in global geochemical cycles and their perceived interaction in climate change are prompting the current cross-disciplinary effort to resolve and define the extent and importance of such interactions. Microbiological studies of the ecology and physiology of nitrification and denitrification in marine ecosystems will be crucial in any attempt to understand, and potentially manage, global N_2O cycling.

Acknowledgments. I thank John Rogers and William Steen for the invitation to the workshop which provided the impetus for this effort. I thank Kelly Cunningham, Linda Duguay, George McManus, and Ken Tenore for commenting on the manuscript and Fran Younger for her help with drafting. Support was provided by National Science Foundation grants OCE 84-17595, 85-15886, and 90-12199.

REFERENCES

Bender, M., K. A. Ganning, P. M. Froelich, G. R. Heath, and V. Maynard. 1977. Interstitial nitrate profiles and oxidation of sedimentary organic matter in the eastern equatorial Atlantic. *Science* **198:**605–609.

Berger, W. H., V. S. Smetacek, and G. Wefer. 1989. Ocean productivity and paleoproductivity—an overview, p. 1–34. *In* W. H. Berger, V. S. Smetacek, and G. Wefer (ed.), *Productivity in the Ocean: Present and Past*. John Wiley & Sons, Inc., New York.

Blackburn, T. H., and J. Sorensen (ed.). 1988. *Nitrogen Cycling in Coastal Marine Environments*. John Wiley & Sons, Inc., New York.

Broecker, W., and T.-H. Peng. 1982. *Tracers in the Sea*. Eldigio Press, New York.

Butler, J., J. W. Elkins, T. M. Thompson, and K. B. Egan. 1989. Tropospheric and dissolved N_2O of the West Pacific and East Indian Oceans during the El Nino Southern Oscillation Event of 1987. *J. Geophys. Res.* **94:**14865–14877.

Capone, D. G. A biologically constrained estimate of oceanic N_2O flux. *In* D. P. Adams, S. P. Seitzinger, and P. M. Crill (ed.), *Cycling of Reduced Gases in the Hydrosphere*. E. Schweizerbart'sche Verlagsbuchhandlungen, Stuttgart, in press.

Capone, D. G., and E. J. Carpenter. 1982. Nitrogen fixation in the marine environment. *Science* **217:**1140–1142.

Carpenter, E. J. 1983. Physiology and ecology of marine *Oscillatoria* (*Trichodesmium*). *Mar. Biol. Lett.* **4:**69–85.

Carpenter, E. J., and D. G. Capone (ed.). 1983. *Nitrogen in the Marine Environment*. Academic Press, Inc., New York.

Christensen, J. P., J. W. Murray, A. H. Devol, and L. A. Codispoti. 1987. Denitrification in continental shelf sediments has a major impact on the oceanic nitrogen budget. *Global Biogeochem. Cycles* **1:**97–116.

Christensen, J. P., and G. Rowe. 1984. Nitrification and oxygen consumption in northwest Atlantic deep-sea sediments. *J. Mar. Res.* **42:**1099–1116.

Codispoti, L. 1989. Phosphorus versus nitrogen limitation of new and export production, p. 377–394. *In* W. H. Berger, V. S. Smetacek, and G. Wefer (ed.), *Productivity in the Ocean: Present and Past*. John Wiley & Sons, Inc., New York.

Codispoti, L. A., and J. P. Christensen. 1985. Nitrification, denitrification and nitrous oxide cycling in the eastern tropical South Pacific Ocean. *Mar. Chem.* **16:**277–300.

Cohen, Y. 1978. Consumption of dissolved nitrous oxide in an anoxic basin, Saanich Inlet, British Columbia. *Nature* (London) **272:**235–237.

Cohen, Y., and L. Gordon. 1978. Nitrous oxide in the oxygen minimum of the eastern tropical North Pacific: evidence for its consumption during denitrification and possible mechanisms for its production. *Deep-Sea Res.* **25:**509–524.

Cohen, Y., and L. Gordon. 1979. Nitrous oxide production in the ocean. *J. Geophys. Res.* **84:**347–353.

Crutzen, P. 1981. Atmospheric chemical processes of the oxides of nitrogen, including nitrous oxide, p. 17–44. *In* C. C. Delwiche (ed.), *Denitrification, Nitrification, and Atmospheric Nitrous Oxide*. John Wiley & Sons, Inc., New York.

Deck, B. 1980. Nutrient element distribution in the Hudson estuary. Ph.D. thesis. Columbia University, New York.

Delwiche, C. C. 1981. The nitrogen cycle and nitrous oxide, p. 1–15. *In* C. C. Delwiche (ed.), *Denitrification, Nitrification, and Atmospheric Nitrous Oxide*. John Wiley & Sons, Inc., New York.

DeVooys, C. G. N. 1979. Primary productivity in aquatic environments, p. 259–292. *In* B. Bolin (ed.), *The Global Carbon Cycle*. John Wiley & Sons, Inc., New York.

Dickenson, R. E., and R. J. Cicerone. 1986. Future global warming from atmospheric trace gases. *Nature* (London) **319:**109–115.

Dugdale, R. C., and J. J. Goering. 1967. Uptake of new and regenerated forms of nitrogen in primary productivity. *Limnol. Oceanogr.* **12:**196–206.

Elkins, J. W., S. C. Wofsy, M. B. McElroy, C. E. Kolb, and W. Kaplan. 1978. Aquatic sources and sinks for nitrous oxide. *Nature* (London) **275:**602–606.

Eppley, R. W., and B. J. Peterson. 1979. Particulate organic matter flux and planktonic new production in the deep ocean. *Nature* (London) **282:**677–680.

Goreau, T. J., W. A. Kaplan, J. C. Wofsy, M. B. McElroy, F. W. Valois, and S. W. Watson.

1980. Production of NO_2 and N_2O by nitrifying bacteria at reduced concentrations of oxygen. *Appl. Environ. Microbiol.* **40:**526–532.

Hahn, J. 1981. Nitrous oxide in the oceans, p. 191–241. *In* C. C. Delwiche (ed.), *Denitrification, Nitrification, and Atmospheric Nitrous Oxide.* John Wiley & Sons, Inc., New York.

Hartwig, E. O., and S. O. Stanley. 1978. Nitrogen fixation in Atlantic deep-sea and coastal sediments. *Deep-Sea Res.* **25:**411–417.

Hattori, A. 1983. Denitrification and dissimilatory nitrate reduction, p. 191–232. *In* E. J. Carpenter and D. G. Capone (ed.), *Nitrogen in the Marine Environment.* Academic Press, Inc., New York.

Horrigan, S. G., and D. G. Capone. 1985. Rates of nitrification and nitrate reduction in nearshore marine sediments under varying environmental conditions. *Mar. Chem.* **16:**317–327.

Howarth, R., and J. Cole. 1985. Molybdenum availability, nitrogen availability and plankton growth in natural waters. *Science* **229:**653–655.

Howarth, R. W. 1988. Nutrient limitation of net primary production in marine ecosystems. *Annu. Rev. Ecol.* **19:**89–110.

Jenkins, M. C., and W. M. Kemp. 1984. The coupling of nitrification and denitrification in two estuarine sediments. *Limnol. Oceanogr.* **29:**609–619.

Jensen, H. B., K. S. Jorgensen, and J. Sorensen. 1984. Diurnal variation of nitrogen cycling in coastal, marine sediments. II. Nitrous oxide emission. *Mar. Biol.* **83:**177–183.

Johnstone, J. 1908. *Conditions of Life in the Sea.* Cambridge University Press, London. (Reprinted by Arno Press, New York, 1977.)

Jorgensen, K., H. B. Jensen, and J. Sorensen. 1984. Nitrous oxide production from nitrification and denitrification in various sediments at low oxygen concentrations. *Can. J. Microbiol.* **30:**1073–1078.

Kaplan, W. A. 1983. Nitrification, p. 139–190. *In* E. J. Carpenter and D. G. Capone (ed.), *Nitrogen in the Marine Environment.* Academic Press, Inc., New York.

Khalil, M. A. K., and R. A. Rasmussen. 1983. Increase and seasonal cycles of nitrous oxide in the earth's atmosphere. *Tellus* **35B:**161–169.

Koike, I., and J. Sorensen. 1988. Nitrate reduction and denitrification in marine sediments, p. 251–273. *In* T. H. Blackburn and J. Sorensen (ed.), *Nitrogen Cycling in Coastal Marine Environments.* John Wiley & Sons, Inc., New York.

Law, C. S., and N. J. P. Owens. 1990. Significant flux of atmospheric nitrous oxide from the northwest Indian Ocean. *Nature* (London) **346:**826–828.

Legendre, L., and M. Gosselin. 1989. New production and export of organic matter to the deep ocean: consequences of some recent discoveries. *Limnol. Oceanogr.* **34:**1374–1380.

Liu, K. K. 1979. Geochemistry of inorganic nitrogen compounds in two marine environments: the Santa Barbara Basin and the ocean off Peru. Ph.D. thesis. University of California, Los Angeles.

McElroy, M., J. W. Elkins, S. C. Wofsy, C. E. Kolb, A. P. Duran, and W. A. Kaplan. 1978. Production and release of N_2O from the Potomac Estuary. *Limnol. Oceanogr.* **23:**1168–1182.

McElroy, M. B. 1983. Marine biological controls on atmospheric CO_2 climate. *Nature* (London) **302:**328–329.

Nishio, T., I. Koike, and A. Hattori. 1983. Estimates of denitrification and nitrification in coastal and estuarine sediments. *Appl. Environ. Microbiol.* **45:**444–450.

Nixon, S. 1981. Remineralization and nutrient cycling in coastal marine ecosystems, p. 111–138. *In* B. J. Nielson and L. E. Cronin (ed.), *Estuaries and Nutrients.* Humana Press, Clifton, N.J.

Olson, R. J. 1981. Differential photoinhibition of marine nitrifying bacteria: a possible mechanism for the formation of the primary nitrite maximum. *J. Mar. Res.* **39:**227–238.

Paerl, H. W., K. M. Crocker, and L. E. Prufert. 1987. Limitation of N_2 fixation in coastal marine waters: relative importance of molybdenum, iron, phosphorus and organic matter availability. *Limnol. Oceanogr.* **32:**525–536.

Payne, W. J. 1981. The status of nitric oxide and nitrous oxide as intermediates in denitrification, p. 85–104. *In* C. C. Delwiche (ed.), *Denitrification, Nitrification, and Atmospheric Nitrous Oxide*. John Wiley & Sons, Inc., New York.

Poth, M., and D. D. Focht. 1985. ^{15}N kinetic analysis of N_2O production by *Nitrosomonas europaea*: an examination of nitrifier denitrification. *Appl. Environ. Microbiol.* **49:**1134–1141.

Redfield, A., B. H. Ketchum, and F. A. Richards. 1963. The influence of organisms on the composition of sea water, p. 26–77. *In* M. N. Hill (ed.), *The Sea*, vol. 2. Academic Press, Inc., New York.

Ritchie, G. A. F., and D. J. D. Nicholas. 1972. Identification of the sources of nitrous oxide produced by oxidative and reductive processes in *Nitrosomonas europaea*. *Biochem. J.* **126:**1181–1191.

Rivera-Ortiz, J. M., and R. H. Burris. 1975. Interactions among substrates and inhibitors of nitrogenase. *J. Bacteriol.* **123:**537–545.

Ryther, J., and W. M. Dunstan. 1971. Nitrogen, phosphorus and eutrophication in the marine environment. *Science* **171:**1008–1013.

Scranton, M. 1983. Gaseous nitrogen compounds in the marine environment, p. 37–64. *In* E. J. Carpenter and D. G. Capone (ed.), *Nitrogen in the Marine Environment*. Academic Press, Inc., New York.

Seitzinger, S., S. W. Nixon, and M. E. Q. Pilson. 1984. Denitrification and nitrous oxide production in coastal marine ecosystem. *Limnol. Oceanogr.* **29:**73–83.

Seitzinger, S. P. 1987. Nitrogen biogeochemistry in an unpolluted estuary: the importance of benthic denitrification. *Mar. Ecol. Prog. Ser.* **37:**65–73.

Seitzinger, S. P. 1988. Denitrification in freshwater and coastal marine ecosystems: ecological and geochemical significance. *Limnol. Oceanogr.* **33:**702–724.

Sharp, J. H. 1983. The distributions of inorganic nitrogen and dissolved and particulate organic nitrogen in the sea, p. 1–35. *In* E. J. Carpenter and D. G. Capone (ed.), *Nitrogen in the Marine Environment*. Academic Press, Inc., New York.

Singh, H. B., L. J. Salas, and H. Shigeishi. 1979. The distribution of nitrous oxide (N_2O) in the global atmosphere and the Pacific Ocean. *Tellus* **31:**313–320.

Soderlund, R., and B. H. Svensson. 1976. The global nitrogen cycle, p. 23–73. *In* B. Svensson and R. Soderlund (ed.), *Nitrogen, Phosphorus and Sulphur—Global Cycles*. SCOPE Report no. 7. Ecological Bulletin no. 21. NFR, Stockholm.

Sorensen, J. 1978. Occurrence of nitric and nitrous oxide in a coastal marine sediment. *Appl. Environ. Microbiol.* **36:**809–813.

Sorensen, J., and T. R. S. Wilson. 1984. A headspace technique for oxygen measurements in deep-sea sediment cores. *Limnol. Oceanogr.* **29:**650–652.

Suess, E. 1980. Particulate organic carbon flux in the ocean—surface productivity and oxygen utilization. *Nature* (London) **288:**260–262.

Suzuki, Y., and Y. Sugimura. 1985. A catalytic oxidation method for the determination of total nitrogen dissolved in sea-water. *Mar. Chem.* **16:**83–97.

Sverdrup, H. U., M. W. Johnson, and R. H. Fleming. 1942. *The Oceans: Their Physics, Chemistry, and General Biology*. Prentice-Hall, Englewood Cliffs, N.J.

Thomas, W. H. 1966. Surface nitrogenous nutrients and phytoplankton in the northeastern tropical Pacific Ocean. *Limnol. Oceanogr.* **15:**393–400.

Toggweiler, J. R. 1989. Is the downward dissolved organic matter (DOM) flux important in carbon transport?, p. 65–83. *In* W. H. Berger, V. S. Smetacek, and G. Wefer (ed.), *Productivity in the Ocean: Present and Past*. John Wiley & Sons, Inc., New York.

Ward, B. B., K. A. Kilpatrick, E. H. Renger, and R. W. Eppley. 1989. Biological nitrogen cycling in the nitracline. *Limnol. Oceanogr.* **34:**493–513.

Ward, B. B., and O. C. Zafiriou. 1988. Nitrification and nitric oxide in the oxygen minimum of the eastern tropical North Pacific. *Deep-Sea Res.* **35:**1127–1142.

Weathers, P. J. 1984. N_2O evolution by green algae. *Appl. Environ. Microbiol.* **48:**1251–1253.

Weiss, R. F. 1981. The temporal and spatial distribution of tropospheric nitrous oxide. *J. Geophys. Res.* **86:**7185–7195.

Wilson, T. R. S. 1978. Evidence for denitrification in aerobic pelagic sediments. *Nature* (London) **274:**354–356.

Yoshida, N., H. Morimoto, M. Hirano, I. Koike, S. Matsuo, E. Wada, T. Saino, and A. Hattori. 1989. Nitrification rates and ^{15}N abundances of N_2O and NO_3^- in the Western Pacific Ocean. *Nature* (London) **342:**895–897.

Yoshinari, T. 1976. Nitrous oxide in the sea. *Mar. Chem.* **4:**189–202.

Zafiriou, O. C., and M. McFarland. 1981. Nitric oxide from nitrite photolysis in the Central Equatorial Pacific. *J. Geophys. Res.* **86:**3173–3182.

Chapter 15

Formation of Halogenated Gases by Natural Sources

R. Wever

There is a growing interest in the release of halogenated gases in the atmosphere by natural sources, in particular since it was realized that these chemicals may affect the earth's climate and/or the chemical composition of the atmosphere (Barrie et al., 1988). For a study concerning the potential climatic effect of these trace gases, the reader is referred to Ramanathan et al. (1985), and a review concerning the sources and sinks of halogens in the atmosphere has been published by Cicerone (1981). In this paper the various natural sources of volatile halogenated gases will be discussed. As far as is known, the most important halogenated gases in terms of production on a global scale are methyl iodide, methyl bromide, and methyl chloride. Other volatile brominated compounds are also formed, but in substantially lower amounts. For these halohydrocarbons, however, more detailed information is available regarding their biosynthetic pathways, the enzymes involved, and their producers. As a consequence, the source of brominated compounds will be discussed in (more) depth.

This review is restricted to volatile halogenated compounds; the more complex natural halohydrocarbons will not be discussed here. For this, the interested reader is referred to several reviews (Faulkner, 1984; Fenical, 1979, 1982; Neidleman and Geigert, 1986).

METHYL IODIDE

As the sea seems to be the main source of methyl iodide, it may be a key compound in the natural cycle of iodine between the seas and the land (Lovelock, 1975; Whitehead, 1984). The first estimates of the annual production rate by Lovelock et al. (1973) yielded a value of 40 Tg per year. Based on a larger atmospheric residence time, a minimum release of CH_3I of 1 to 2 Tg per year was calculated by Zafiriou (1974). This value is the same as the global flux of 1.3 Tg per year reported by Rasmussen et al. (1982); however, Singh et al. (1983) estimated

R. Wever • E. C. Slater Institute for Biochemical Research and Biotechnological Center, University of Amsterdam, Plantage Muidergracht 12, 1018 TV Amsterdam, The Netherlands.

values of 0.3 to 0.5 Tg per year, in line with the estimate by Liss and Slater (1974). Table 1 summarizes the sources and amounts of this and other methyl halides.

The actual mechanism and the source of methyl iodide formation are poorly understood. There are, however, clear indications that macroalgae and plankton may contribute. Lovelock (1975) demonstrated early that marine macroalgae are a potential source of methyl iodide since seawater near kelp beds (*Laminaria digitata*) contained 10^3-fold more methyl iodide than found in open oceans. Regions with high algal biomass (coasts of Iceland and South Africa) appear to be mainly responsible for methyl iodide production (Rasmussen et al., 1982), although these waters represent only about 10% of the global oceanic area.

Depth profiles measured by Singh et al. (1983) show that oceanic methyl halides are most abundant in the top mixed layer of the ocean. This may be related to the observations of Korzh (1984). Using a laboratory model, he was able to show that organically bound iodine present in seawater may play an essential role in iodine enrichment of aerosols. Elementary iodine combines with some organic compounds present in seawater, and organo-iodine compounds concentrate at the water/air interface. It may be that not only algae are responsible for producing CH_3I: Singh et al. (1983) reported high methyl iodide levels in the eastern Pacific Ocean when the sampling vessel was surrounded by jellyfish.

The biosynthetic pathway by which methyl iodide is formed is still elusive. Three possibilities can be considered. It has been proposed (White, 1982a) that a reaction may occur between dimethyl sulfonium compounds present in algae and iodide according to:

$$(CH_3)_2S^+ - R + I^- \rightarrow CH_3I + CH_3SR$$

Most macroalgae (green, brown, and red) and marine phytoplankton contain dimethyl sulfonium compounds (White, 1982a). Further, algae are known to concentrate halides like I^-.

Methyl iodide may also be formed by iodination of organic molecules in algae or organic matter in seawater. The iodinated compounds then may decompose to methyl iodide. This requires, however, that algae contain iodoperoxidases which catalyze the peroxidation of iodide to iodine. Indeed, most brown and red algae do contain iodoperoxidases (De Boer et al., 1986a; Kylin, 1929, 1930; Vilter et al., 1983). An alternative possibility is that iodide is methylated via a methyltransferase (Wuosmaa and Hager, 1990).

METHYL BROMIDE

Data on the presence and formation of methyl bromide are scarce (Singh et al., 1983; Cicerone et al., 1988). Measurements in and over the eastern Pacific by Singh et al. (1983) show a dominant oceanic source of methyl halides. For methyl bromide, the global oceanic source was estimated to be 0.3 Tg per year. Interestingly, a common source for both methyl chloride and methyl bromide seems to exist (Singh et al., 1983), although it is as yet unidentified. Table 1 shows that the anthropogenic contribution is much smaller than the natural source.

Table 1. Global inputs of methyl halides and bromoform to the atmosphere

Compound	Source	Amt[a] (g/year)	References
CH_3I	Natural (oceanic)	0.3×10^{12}–2.0×10^{12}	Liss and Slater, 1974; Rasmussen et al., 1982; Singh et al., 1983; Zafiriou, 1974
CH_3Br	Natural (oceanic)	0.3×10^{12}	Singh et al., 1983
CH_3Br	Anthropogenic (fumigant)	0.05×10^{12}–0.08×10^{12}	Penkett et al., 1985; Singh et al., 1983
CH_3Cl	Natural (oceanic)	2.5×10^{12}–5×10^{12}	Rasmussen et al., 1980; Singh et al., 1979; Singh et al., 1983
CH_3Cl	Anthropogenic	0.03×10^{12}–0.3×10^{12}	Cicerone, 1981; Harper, 1985; Lovelock, 1975
$CHBr_3$	Natural (oceanic)	10^{12}	Penkett et al., 1985
$CHBr_3$	Natural (marine macroalgae)	About 10^{10}	Gschwend et al., 1985
$CHBr_3$	Anthropogenic (chlorination of sea/fresh water)	6×10^{8}–6×10^{9}	Gschwend et al., 1985

[a] The minimum and maximum values reported by the various authors are given.

METHYL CHLORIDE

The current environmental estimates of CH_3Cl input into the biosphere are about 5 Tg per year (Cicerone, 1981; Lovelock, 1975; Rasmussen et al., 1980; Singh et al., 1979) (Table 1). Unlike the other methyl halides, methyl chloride seems to have both an oceanic and a terrestrial source. Several fungal species, like mushrooms and common wood-rotting fungi, produce CH_3Cl as a secondary metabolite (Harper, 1985, 1986; Turner et al., 1975). The wood-decaying fungus *Phellinus pomaceus* also produces significant amounts of methyl iodide and methyl bromide when chloride is replaced by iodide or bromide, respectively (Harper, 1985, 1986). The biosynthesis pathway by which these gases are produced is not well understood. Data from White (1982b) suggest methylation of chloride. Indeed, in line with this a novel pathway for the biosynthesis has been described (Wuosmaa and Hager, 1990). A methyltransferase was isolated from a plant that catalyzes the methylation of chloride in the presence of *S*-adenosylmethionine. This discovery makes it less likely that lignin peroxidase is involved in the biosynthesis. Most wood-decaying fungi produce lignin peroxidase, which, in the presence of bromide and H_2O_2, is able to brominate a variety of aromatic substrates (Renganathan et al., 1987). However, chloride is not oxidized by lignin peroxidase.

With regard to the oceanic source of methyl chloride, it is still not known whether CH_3Cl is formed directly by biosynthesis or is the product of a nucleophilic reaction of Cl^- with methyl iodide (Zafiriou, 1975). However, the data of Singh et al. (1983), which show the coexistence of high concentrations of methyl iodide with relatively low concentrations of methyl chloride and vice versa, do not support the hypothesis that methyl chloride is produced from methyl iodide by chloride ion reactions. Rather, there is evidence for a common source for methyl chloride and methyl bromide. The recent work by Wuosmaa and Hager (1990) also suggests strongly that CH_3Cl is directly formed by a biosynthetic pathway. These authors showed that 24 of 44 randomly collected marine algae produced methyl chloride. It is interesting to note that a higher atmospheric level of CH_3Cl has been detected over equatorial oceanic areas (Rasmussen et al., 1980). The amount of CH_3Cl produced industrially which reaches the atmosphere is about 10% of the biogenic source (Table 1).

The reason why methyl halides are produced in nature is unclear. However, these compounds have antiparasitic and toxic properties and are bacterial mutagens (Singh et al., 1982). The possibility exists that these chemicals play a role in defense systems of organisms. It is also obvious that very little is known about the origin of the methyl halides, their biosynthetic pathways, and the organisms involved. Considering the amounts and the impact of these compounds on atmospheric processes such as the reaction with ozone in the troposphere (Barrie et al., 1988) and stratosphere (Cicerone, 1981), there is a clear need to understand these sources better.

VOLATILE ORGANOHALIDES

According to Penkett et al. (1985), the global yearly flux of bromoform produced by oceans is in the same order of magnitude as that of the methyl halides

(Table 1). In comparison, the quantities of volatile halogenated compounds released by marine brown, red, and green seaweeds are small. For the marine brown algae *Ascophyllum nodosum*, *Fucus vesiculosis*, *Fucus sargassum*, and *Laminiaria laminaria* (Class et al., 1986; Gschwend et al., 1985), the main components are dibromomethane (CH_2Br_2), bromoform ($CHBr_3$), bromodichloromethane ($CHBrCl_2$), and dibromochloromethane ($CHBr_2Cl_2$). A number of minor components have also been identified such as iodomethane (CH_3I), diiodomethane (CH_2I_2), and iodoethane (C_2H_5I). For the red seaweed *Gigertina stellata*, release of similar compounds has been observed (Gschwend et al., 1985). The essential oil of the edible red alga *Asparagopsis taxiformis* is composed of mainly bromine and iodine-containing haloforms of which 80% consist of bromoform and 5% are dibromoiodomethane (Burreson et al., 1976). The green seaweeds *Ulva lacta* and *Enteromorpha linza* also produce volatile halogenated organic compounds. From the rate of release of the halometabolites by these lower plants (1 to 10 μg of bromine per g of seaweed per day) and a global biomass of 10^{13} g (Waaland, 1981), a global input in the biosphere of 0.01 Tg per year is estimated. This is of the same order of magnitude as the reported anthropogenic input of organobromides (Gschwend et al., 1985). Thus these volatile halogenated compounds also contribute to the halohydrocarbon burden in the atmosphere.

Recently, a group of vanadium-containing bromoperoxidases has been discovered in the brown seaweeds *Ascophyllum nodosum*, *Laminaria saccharina*, *Alaria esculenta*, *Fucus distichus*, and *Chorda filum* (De Boer et al., 1986a; De Boer et al., 1986b; De Boer and Wever, 1987; Wever et al., 1988) and in the red seaweeds *Corallina pilulifera* (Krenn et al., 1989a) and *Ceramium rubrum* (Krenn et al., 1987). These enzymes produce free hypobromous acid (HOBr and Br_2) in the presence of H_2O_2 and Br^- (De Boer and Wever, 1988). The bromoperoxidases are probably involved in the biosynthesis of the volatile halogenated metabolites by bromination of nucleophilic acceptors. It is very likely that these acceptors are carboxymethyl ketones, since halogenated ketones have been detected in seaweeds and, further, it was shown that bromination of ketones by a bromoperoxidase led to a variety of halogenated ketones which decayed via the classical haloform reaction to form the volatile brominated methanes (Burreson et al., 1976; Moore, 1977; Theiler et al., 1978). Why these halometabolites are formed by the algae is not clear. However, these compounds have biocidal properties and their formation may simply be part of a defense system of the algae.

In the brown seaweed *Ascophyllum nodosum* at least two different vanadium bromoperoxidases are found. One is located inside the thallus, particularly around the conceptacles, and the presence of this enzyme shows seasonal variation (Vilter et al., 1983; Wever, 1988): maximal activity is found in winter and spring, when fruiting bodies are present, and a rapid decrease in activity is observed in summer. Another bromoperoxidase is located at the thallus surface and can be considered to be extracellular (Krenn et al., 1989b). This location is in line with the observations by Gschwend et al. (1985), who concluded that the brominated compounds are synthesized near the surface of the algae. An interesting observation was also made by Sauvageau (1926), who claimed that the red seaweeds *Anthithamnionella sarni-*

ensis and *Anthithamnion plumula* were able to form free Br_2 which converted fluorescein to its brominated derivative, eosin. Thus, it is conceivable that some of these seaweeds also release hypobromous acid and bromine in sea water. Bromine oxidants will rapidly react (Jaworske and Helz, 1985) with dissolved organic matter, which decays to form bromoform. Thus, part of the brominated compounds may originate from a direct reaction between HOBr generated by algae and organic matter (Helz and Hsu, 1970) in seawater.

High levels of brominated organic species, of which bromoform is the main component (Berg et al., 1984), are found in the Arctic atmosphere. Measurements of atmospheric bromoform at Point Barrow, Alaska, show that the presence of bromoform is a function of the season, bromoform concentrations being maximal in winter and minimal in summer (Cicerone et al., 1988). Similarly, the bromine content of Arctic aerosols at Point Barrow, Alert (Canadian Arctic), and Spitsbergen, Norway, shows a sharp maximum between February and May with concentrations which are the highest anywhere in the world (Berg et al., 1983; Sturges and Barrie, 1988). This maximum is found every year just after Arctic dawn. There is substantial evidence that the source of the bromine-containing particles and bromoform is biogenic and that they originate from the oceans in the northern part of the northern hemisphere. Dyrssen and Fogelqvist (1981) showed that bromoform is present at all depths in the Arctic Ocean near Spitsbergen. The profile of depth versus concentration shows clearly that algal belts produce bromoform. Data for fjords in south Sweden show the same (Fogelqvist et al., 1982; Fogelqvist and Kryssel, 1988). However, algae are probably not the only source of bromoform; phytoplankton may also contribute (Fogelqvist, 1985). Depth profiles of bromoform point to a source near the surface, and a correlation was observed between high concentrations of bromoform and high concentrations of chlorophyll *a* (Dyrssen and Fogelqvist, 1981).

There is considerable interest in the formation of these brominated compounds, in particular because Barrie et al. (1988) showed that, at polar sunrise, ozone destruction occurs in the lower Arctic atmosphere at Alert and that there was a strong correlation with the presence of filterable bromine in aerosols. Organobromides such as bromoform are precursors of filterable bromine, and bromoform is photochemically active (Barrie et al., 1988). Sturges and Barrie (1988) favor the theory that the inorganic Arctic Br originates from the photochemical conversion of marine organic bromine. It is very likely that the vanadium bromoperoxidases present in seaweeds found in the Arctic Ocean are responsible for the production of bromoform and the presence of bromine in aerosols. Thus I propose, in line with my earlier studies (Wever, 1988), that the biological activity of these seaweeds, which may be triggered by light in this part of the world, is linked to the observed ozone destruction at ground level in the Arctic. This requires, however, that these vanadium enzymes be found in most algae present in the Arctic Ocean. The vegetation in this ocean is dominated (Kjellman, 1883) by two orders of brown seaweeds (*Laminariales* and *Fucales*). Indeed, the brown seaweeds *Alaria esculenta*, *Ascophyllum nodosum*, and *Fucus distichus*, which were collected along the shores of Iceland, do contain such enzymes (Wever et al., 1988). However, data concerning the presence of bromoperoxidases in other seaweed species are still lacking. What are presently clearly needed are more direct measurements of bromoform in the lower troposphere and in seawater at locations such as rocky coastal areas where defined seaweeds are present. These data, collected as a function of the season, will indicate whether seaweeds in the Arctic Ocean are responsible for the production of organic gaseous bromine (Sturges and Barrie, 1988) and indirectly for the decrease of ozone concentrations in the lower troposphere (Barrie et al., 1988; Wever, 1988).

Acknowledgments. I thank M. van de Kaaden for her help in the literature research and M. L. Dutrieux for the preparation of the manuscript. This work is supported in part by the Netherlands Foundation for Chemical Research (S.O.N.) with financial aid from the Netherlands Organisation for Scientific Research (N.W.O.).

REFERENCES

Barrie, L. A., J. W. Bottenheim, R. C. Schnell, P. J. Crutzen, and R. A. Rasmussen. 1988. Ozone destruction and photochemical reactions at polar sunrise in the lower Arctic atmosphere. *Nature* (London) **334:**138–141.

Berg, W. W., L. E. Heidt, W. Pollock, P. D. Sperry, and R. J. Cicerone. 1984. Brominated organic species in the Arctic atmosphere. *Geophys. Res. Lett.* **11:**429–432.

Berg, W. W., P. D. Sperry, K. A. Rahn, and E. S. Gladney. 1983. Atmospheric bromine in the Arctic. *J. Geophys. Res.* **88:**6719–6736.

Burreson, J. A., R. E. Moore, and P. P. Rohler. 1976. Volatile halogen compounds in the alga *Asparagopsis taxiformis* (rhodophyta). *J. Agric. Food Chem.* **24:**856–861.

Cicerone, R. J. 1981. Halogens in the atmosphere. *Rev. Geophys. Space Phys.* **19:**123–139.

Cicerone, R. J., L. E. Heidt, and W. H. Pollock. 1988. Measurements of atmospheric methyl bromide and bromoform. *J. Geophys. Res.* **93:**3745–3749.

Class, T., R. Kohnle, and K. Ballschmiter. 1986. Chemistry of organic traces in air. VII. Bromo- and bromochloromethanes in air over the Atlantic Ocean. *Chemosphere* **4:**429–436.

De Boer, E., M. G. M. Tromp, H. Plat, G. E. Krenn, and R. Wever. 1986a. Vanadium(V) as an essential element for haloperoxidase activity in marine brown algae: purification and characterization of a vanadium(V) containing bromoperoxidase from *Laminaria saccharina*. *Biochim. Biophys. Acta* **872:**104–115.

De Boer, E., Y. Van Kooyk, M. G. M. Tromp, H. Plat, and R. Wever. 1986b. Bromoperoxidase from *Ascophyllum nodosum*: a novel class of enzymes containing vanadium as a prosthetic group. *Biochim. Biophys. Acta* **869:**48–53.

De Boer, E., and R. Wever. 1987. Some structural and kinetic aspects of vanadium bromoperoxidases from the marine brown alga *Ascophyllum nodosum*. *Receuil. Trav. Chim. Pays-Bas* **106:**409.

De Boer, E., and R. Wever. 1988. The reaction mechanism of the novel vanadium bromoperoxidase: a steady-state kinetic analysis. *J. Biol. Chem.* **263:**12326–12332.

Dyrssen, D., and E. Fogelqvist. 1981. Bromoform concentrations of the Arctic Ocean in the Svalbard area. *Oceanol. Acta* **4:**313–317.

Faulkner, W. J. 1984. Marine natural products: metabolites of marine algae and herbivorious marine molluscs. *Nat. Products Rep.* **1:**251–280.

Fenical, W. 1979. Molecular aspects of halogen-based biosynthesis of marine natural products. *Recent Adv. Phytochem.* **13:**219–239.

Fenical, W. 1982. Natural products chemistry in the marine environment. *Science* **215:**923–928.

Fogelqvist, E. 1985. Carbon tetrachloride, tetrachloroethane, 1,1,1,-trichloroethane and bromoform in Arctic sea water. *J. Geophys. Res.* **90:**9181–9193.

Fogelqvist, E., B. Josefsson, and C. Roos. 1982. Halocarbons as tracer substances in studies of the distribution pattern of chlorinated waters in coastal areas. *Environ. Sci. Technol.* **16:**479–482.

Fogelqvist, E., and M. Kryssel. 1988. The anthropogenic and biogenic origin of low molecular weight halocarbons in a polluted fjord, the Idefjorden. *Mar. Pollut. Bull.* **17:**378–382.

Gschwend, P. M., J. K. MacFarlane, and K. A. Newman. 1985. Volatile halogenated organic compounds released to sea water from temperate marine macroalgae. *Science* **227:**1033–1036.

Harper, D. B. 1985. Halomethane from halide ion—a highly efficient fungal conversion of environmental significance. *Nature* (London) **315:**55–57.

Harper, D. B. 1986. Effect of growth conditions of halomethane production by *Phellinus* species: biological and environmental implications. *J. Gen. Microbiol.* **132:**1231–1246.

Helz, G. R., and R. Y. Hsu. 1970. Volatile chloro- and bromocarbons in coastal waters. *Limnol. Oceanogr.* **23:**859–869.

Jaworske, D. A., and G. R. Helz. 1985. Rapid consumption of bromine oxidants in river and estuarine waters. *Environ. Sci. Technol.* **19:**1188–1191.

Kjellman, F. R. 1883. The algae of the Arctic Sea. *Kongl. Svenska Vetensk. Akad. Handl.* **20:**1–61. (Reprinted by Otto Koeltz Antiquariat, Koenigstein-Taunus, Federal Republic of Germany, 1971.)

Korzh, V. D. 1984. Ocean as a source of atmospheric iodine. *Atmos. Environ.* **12:**2707–2710.

Krenn, B. E., Y. Izumi, H. Yamada, and R. Wever. 1989a. A comparison of different (vanadium) bromoperoxidases: the bromoperoxidase from *Corallina pilulifera* is also a vanadium enzyme. *Biochim. Biophys. Acta* **998:**63–68.

Krenn, B. E., H. Plat, and R. Wever. 1987. The bromoperoxidase from the red alga *Ceramium rubrum* also contains vanadium as a prosthetic group. *Biochim. Biophys. Acta* **912:**287–291.

Krenn, B. E., M. G. M. Tromp, and R. Wever. 1989b. The brown alga *Ascophyllum nodosum* contains two different vanadium bromoperoxidases. *J. Biol. Chem.* **264:**19287–19292.

Kylin, H. 1929. Über das Vorkommen vom Jodiden, Bromiden und Jodidoxydasen bei den Meeresalgen. *Hoppe-Seyler's Z. Physiol. Chem.* **186:**50–84.

Kylin, H. 1930. Über die Jodidspaltende Fähigkeit der Phäophyceen. *Hoppe-Seyler's Z. Physiol. Chem.* **191:**200–210.

Liss, P. S., and P. G. Slater. 1974. Flux of gases across the air-sea interface. *Nature* (London) **247:**181–184.

Lovelock, J. E. 1975. Natural halocarbons in the air and in the sea. *Nature* (London) **256:**193–194.

Lovelock, J. E., R. J. Maggs, and R. J. Wade. 1973. Halogenated hydrocarbons in and over the Atlantic. *Nature* (London) **241:**194–196.

Moore, R. E. 1977. Volatile compounds from marine algae. *Acc. Chem. Res.* **10:**40–47.

Neidleman, S. L., and J. Geigert. 1986. *Biohalogenation: Principles, Basic Roles and Applications.* Ellis Horwood Ltd., Chichester, U.K.

Penkett, S. A., B. M. R. Jones, M. J. Rycrofft, and D. A. Simons. 1985. An interhemispheric comparison of the concentrations of bromine compounds in the atmosphere. *Nature* (London) **318:**550–553.

Ramanathan, V., R. J. Cicerone, H. B. Singh, and J. T. Kiehl. 1985. Trace gas trends and their potential role in climate changes. *J. Geophys. Res.* **90:**5547–5566.

Rasmussen, R. A., M. A. K. Khalil, R. Gunawardena, and S. D. Hoyt. 1982. Atmospheric methyl iodide (CH_3I). *J. Geophys. Res.* **87:**3086–3090.

Rasmussen, R. A., L. E. Rasmussen, M. A. K. Khalil, and R. W. Dalluge. 1980. Concentration distribution of methyl chloride in the atmosphere. *J. Geophys. Res.* **85:**7350–7356.

Renganathan, V., K. Miki, and M. H. Gold. 1987. Haloperoxidase reactions catalyzed by lignine peroxidase, an extracellular enzyme from the basidiomycete *Phanerochaete chrysosporium. Biochemistry* **26:**5127–5132.

Sauvageau, C. 1926. Sur quelques alques floridées renformant du brome a l'état libre. *Bull. Stat. Arc.* **23:**5–23.

Singh, H. B., L. J. Salas, H. Shigeishi, and E. Scribner. 1979. Atmospheric halocarbons, hydrocarbons and SF_6: global distributions, sources and sinks. *Science* **203:**899–903.

Singh, H. B., L. J. Salas, and R. E. Stiles. 1982. Distribution of selected gaseous organic mutagens and suspect carcinogens in ambient air. *Environ. Sci. Technol.* **16:**872–880.

Singh, H. B., L. J. Salas, and R. E. Stiles. 1983. Methyl halides in and over the eastern Pacific. *J. Geophys. Res.* **88:**3684–3690.

Sturges, W. T., and L. A. Barrie. 1988. Chlorine, bromine and iodine in Arctic aerosols. *Atmos. Environ.* **22:**1179–1194.

Theiler, R., J. C. Cook, L. P. Hager, and J. F. Siuda. 1978. Halohydrocarbon synthesis by bromoperoxidase. *Science* **202:**1094–1096.

Turner, E. M., M. Wright, T. Ward, D. J. Osborne, and R. Self. 1975. Production of ethylene and other volatiles and changes in cellulase and lactase activities during the life cycle of the cultivated mushroom. *J. Gen. Microbiol.* **91:**167–176.

Vilter, H., K.-W. Glombitza, and A. Grawe. 1983. Peroxidases from Phaephyceae. I. Extraction and detection of the peroxidases. *Bot. Mar.* **26:**331–340.

Waaland, R. J. 1981. Commercial utilization, p. 726–741. *In* C. S. Lobban and M. J. Wynne (ed.), *The Biology of Seaweeds*. University of California Press, Berkeley.

Wever, R. 1988. Ozone destruction by algae in the Arctic atmosphere. *Nature* (London) **335:**501.

Wever, R., G. Olafsson, B. E. Krenn, and M. G. M. Tromp. 1988. Ozone destruction and bromoform production in the Arctic: pieces of a puzzle. Abstr. no. 210. 32nd IUPAC Congress, Stockholm, Sweden.

White, R. H. 1982a. Analysis of dimethyl sulfonium compounds in marine algae. *J. Mar. Res.* **40:**529–536.

White, R. N. 1982b. Biosynthesis of methyl chloride in the fungus *Phellinus pomaceus*. *Arch. Microbiol.* **132:**100–102.

Whitehead, D. C. 1984. The distribution and transformations of iodine in the environment. *Environ. Inter.* **10:**321–339.

Wuosmaa, A. M., and L. P. Hager. 1990. Methyl chloride transferase: a carbocation route for biosynthesis of halometabolites. *Science* **249:**160–162.

Zafiriou, O. 1974. Photochemistry of halogens in the marine atmosphere. *J. Geophys. Res.* **79:**2730–2732.

Zafiriou, O. C. 1975. Reactions of methyl halides with sea water and marine aerosols. *J. Mar. Res.* **33:**75–80.

Chapter 16

Research Needs in the Microbial Production and Consumption of Radiatively Important Trace Gases

William B. Whitman and John E. Rogers

The chapters of this volume addressed the state of our knowledge concerning the emissions of the microbially produced radiatively important trace (RIT) gases and the biological processes involved. From this discussion, a number of themes emerged which may be able to guide future research in this area. At this point it is useful to summarize these observations, many of which were made repeatedly by workers in quite different areas.

What are reasonable research goals? Much of the current work is directed to determining the current sources and sinks for the RIT gases. In addition, there is a need to understand the processes which control these sources and sinks. While great progress has been made on both these topics, our current knowledge is far from complete. A more remote goal is to learn how changes in global climate will affect the microbial processes which drive RIT gas production and consumption. Of special interest are the possibilities for positive and negative feedbacks. Thus, if increases in the atmospheric concentrations of the RIT gases force climate change, will the climate change stimulate further increases in these gases or stimulate processes which return the gases to their normal levels?

Developing global budgets for the trace gases is a major component of understanding the current fluxes or the sources and sinks. Budgets identify the processes and ecosystems which are most significant, and they are an important step in setting priorities. Budgets also identify which human activities are likely to be contributing to changes in the RIT gas composition of the atmosphere. From this information, intervention strategies may be proposed. Budgets also provide a test of the data base. If the sources and sinks for a gas are well understood, the budget should be self-consistent.

These points are illustrated by examination of the current source budget for CH_4 (Tyler, this volume). Global estimates for the production of methane have an uncertainty of about twofold, 355 to 870 Tg of CH_4 per year. Typically, the estimates

William B. Whitman • Department of Microbiology, University of Georgia, Athens, Georgia 30602. *John E. Rogers* • Environmental Research Laboratory, Athens, U.S. Environmental Protection Agency, College Station Road, Athens, Georgia 30613-7799.

of the strengths of individual sources vary from twofold to an order of magnitude. Even given these large uncertainties, some methane sources are clearly more important than others. Four sources, natural wetlands, rice paddies, livestock, and possibly termites, account for 70 to 80% of the emissions to the atmosphere. Two of these, rice paddies and livestock, are directly related to human activities, and they are likely candidates for intervention strategies. Moreover, if climate changes affect any of these four sources, they are likely to have a large influence on the global emissions of methane. Therefore, understanding the microbial processes involved in methane emissions from these sources should have a high priority. Equally important, a number of anthropogenic sources including coal mining, gas venting and flaring, industrial losses, pipeline losses, and automobiles, are quantitatively less significant. Changes in public policy to reduce these emissions are unlikely to have a major impact on global emissions.

Similarly, the budgets for the nitrogen oxides establish some facts clearly. For N_2O, major anthropogenic sources such as combustion of biomass and coal or oil and fertilization of agricultural land are unlikely to represent major sources of this gas (Davidson, this volume). A much larger source is likely to be tropical and subtropical forests, although the magnitude of this source is not known precisely. In contrast, fossil fuel combustion and biomass burning dominate the source budget for NO (Jacob and Bakwin, this volume), and these sources are amenable to regulation.

Do we have accurate estimates of the current sources and sinks for the RIT gases? If the uncertainties in the methane source budget are used to evaluate the completeness of our understanding of current fluxes, large gaps in our knowledge must exist. This situation is certainly worse for the other RIT gases, where the budgets are even more speculative. The budgets for N_2O and NO are based on very few measurements from only a limited number of ecosystems (Davidson, this volume). Likewise, large uncertainties exist in the budgets for the haloalkanes (Wever, this volume). Improving the budgets for all the RIT gases must have a high priority. Because there is an inherent risk in basing public policy on incomplete information, accurate budgets should be constructed in a timely fashion.

Why are the current budgets inaccurate or incomplete? One problem is that the RIT gases are minor components of the major geochemical cycles. Therefore, identifying their sources (and sinks) is analogous to finding pinholes in very large pipes. The large pipes represent the large fluxes of carbon, nitrogen, and sulfur in the geochemical cycles. The pinholes are fluxes of methane, N_2O, and other trace gases, and it is difficult to predict likely sources and sinks for the RIT gases by casual observation. For instance, methane emissions to the atmosphere represent less than 1% of the global carbon cycle, and these emissions are not well correlated with measures of biological activity such as primary production or biomass. Thus, the oceans, which contain a major component of the carbon cycle, make a very small contribution to methane emissions. In contrast, livestock, which represent a minor component of the carbon cycle, are a major source of atmospheric methane.

In the absence of clear predictors of RIT gas production or consumption, the alternative is to sample representative sites to determine typical fluxes. This sampling strategy is guided by our prior knowledge of the processes involved. For instance, methane-producing bacteria are a major source of atmospheric methane, and these bacteria are strict anaerobes. Therefore, anaerobic environments are examined as potential sources of methane. This strategy is limited by our understanding of these processes. In the case of methane, there is reason to believe that methane may also be produced in some aerobic habitats (Sieburth, this volume). If methane emissions are substantial from aerobic environments, these sites may not be sampled, and the budget would be incomplete.

During the sampling process, problems are also encountered while making the field measurements. Upon examining a particular site, it is not always obvious how many and what kind of measurements are needed to gain an adequate representation of its flux characteristics. Variability is usually large, and it includes differences in location within the site as well as diurnal and seasonal differences. Large differences are also found between comparable sites. With few exceptions, sampling is seldom guided by statistical theory (Crill et al., this volume). Moreover, if a theoretical basis for the sampling problem was utilized, there is no a priori reason to assume that this solution for a particular gas or site will generalize for all gases or sites. In practice, this issue is avoided by making a large number of measurements, which can be very expensive in remote locations.

In addition, field measurements have to consider a large number of complexities associated with the gas or site. For instance, flux measurements at the soil surface may not always be sufficient. In certain tropical forests, only about 25% of the NO_x emitted from the soil is expected to leave the forest canopy (Jacob and Bakwin, this volume). Thus, field measurements need to consider a large number of special circumstances, some of which may be unique to the site or gas. These problems are discussed in greater detail by Crill et al. and Davidson in this volume.

After flux measurements are made at representative sites, these measurements are "scaled up" by multiplying the fluxes by the total area of the sites to obtain a global estimate for that class of sites. The scaling up process is not straightforward. To go from measurements on the order of a square meter to planetary estimates involves large assumptions. Sometimes these assumptions can be tested by increasing the number of field measurements or measuring regional fluxes by meteorological techniques. For most of the current budgets, these assumptions are largely untested. For instance, estimates of methane production by cattle are based largely on measurements of herds in the industrial nations. These measurements are then extrapolated globally to all cattle, even though management practices and diets vary greatly throughout the world. Likewise, methane production by termites is based upon small data sets, and large uncertainties exist in estimates of the total number and number of different types of termites. For these reasons, estimates of the global methane emissions from termites vary by 2 orders of magnitude.

During scaling, it is important to identify ecological or environmental factors which are strong predictors of RIT gas flux (Crill et al. and Groffman, this volume). Vegetation, soil type, soil moisture, and temperature are some of the parameters which have been examined. Knowledge of the controlling factors for gas flux is necessary to determine how representative a site is. If gas fluxes correlate well with the plant community or soil type, large-scale measurements of these features provide good estimates of regional fluxes. If the controlling factors are not understood, flux data for one site may be incorrectly assumed to predict fluxes for sites which are similar, but different in some key factors. For example, freshwater lakes are sites of very active methane production. However, in deep lakes much of this methane is oxidized and never released to the atmosphere (Kiene, and Topp and Hanson, this volume). Therefore, lake depth is an important factor in controlling methane flux, and an average flux value for one lake is a poor predictor of fluxes from all lakes.

After scaling, RIT gas sources and sinks are summed to create the global budget. The budget may be used in general circulation models to compare it with the atmospheric composition of the gas. If the model is sensitive to small changes in the budget and predicts the RIT gas composition well, it is likely that the budget is accurate. If the budget is inaccurate, sensitivity tests of a general circulation model could determine possible sources of error and establish priorities for future research.

The validity of the budget can also be tested by a number of other criteria. First, if the concentration of the gas is relatively constant, the sources should be equal to the sinks. The isotopic composition of the gas should be consistent with the isotopic composition of the sources and any fractionations which occur in the sinks (Tyler, this volume). The budget should also be able to explain any regional or hemispheric differences in the gas composition.

Current models of trace gas fluxes are largely empirical and not mechanistically or process based. This situation is unavoidable considering the incomplete knowledge of the current RIT gas fluxes and the microbial processes which control them. Process-based models have a number of clear advantages, and they are an important research goal. One, they provide a clearer understanding of what is happening and an increased opportunity for doing something about it. Two, they have great predictive power, especially if climate change creates novel circumstances which affect trace gas fluxes. Three, they aid in the process of data collection and scaling. This last factor is of critical importance for constructing accurate budgets. Because many of the RIT gas sources are incidental to the major geochemical cycles, process models are necessary to locate potential sources.

Our current understanding of the microbial processes which affect RIT gas production and consumption is extremely incomplete. In some cases, the important organisms have not yet been identified. Although many methane-oxidizing bacteria have been described, it is not clear that these organisms are catalyzing this reaction in soil (Topp and Hanson, this volume). Likewise, the soil is a major sink for CO, and the ecologically significant CO-oxidizing bacteria have not been established. Anaerobic methane oxidation has also been proposed to be a major sink in some environments, and the microbes catalyzing this reaction have not been described (Kiene, this volume). Likewise, the mechanisms of methyl halide production are poorly understood (Wever, this volume), and the ecological importance of microbes which dehalogenate hydrocarbons is unknown.

Even when the important microbial agents are known, the development of process models is hindered by the absence of detailed knowledge of their physiology and ecology. The recent discoveries of aerobic denitrification and heterotrophic nitrification (Robertson and Kuenen, this volume) are clear examples of physiological data which have not been integrated into models of N_2O and NO production. Likewise, the relative contribution of methane-oxidizing and ammonium-oxidizing bacteria to methane oxidation is not known. Even for well-studied organisms like the methane-producing bacteria, new substrates are continually being discovered. Two recent examples whose ecological significance has yet to be evaluated include dimethyl sulfide and secondary alcohols (Jones, this volume). Likewise, while many of the general properties of the microbial ecology of methanogens are understood, a great deal is not well explained, such as the importance of formate or H_2 in interspecies electron transfer and the competition between acetogens and methanogens (Boone, Miller, this volume).

In terrestrial systems, there is a great need for more detailed information about plant-microbe interactions. In addition to serving as a major conduit of trace gases to and from the soils, plants are frequently the major carbon and energy source through exudates or sloughed tissue. If plants are a controlling factor of RIT gas emissions, they could serve as a major predictor of such emissions.

Current efforts concentrate on environments or ecosystems which are significant sources or sinks for RIT gases. In many aquatic and marine systems, RIT gases like methane may be cycled (Kiene, Sieburth, this volume). Thus, production and consumption occur at nearly equal rates so that the net fluxes to the atmosphere are small. Similarly, in marine systems much of the nitrogen is cycled, while only a small amount of the N_2O that is produced is released to the atmosphere (Capone, this volume). Because of the large capacity of these systems, they have the potential to become major sources or sinks, and the effects of climate change have not been explored.

In conclusion, a great deal remains to be learned about the microbial production and consumption of RIT gases. Two immediate goals are clear. Accurate budgets for the gases need to be developed and tested to design intervention strategies and set research priorities. Greater knowledge of the microbial processes which drive RIT gas production and consumption is also needed to facilitate budget construction and to develop process-based models for predictions of the effect of climate change on RIT gas emission. These endeavors would be greatly aided by a strong background in bacterial physiology, microbial ecology, process modeling, and data base development.

Index